Halide Glasses for Infrared Fiberoptics

NATO ASI Series

Advanced Science Institutes Series

A Series presenting the results of activities sponsored by the NATO Science Committee, which aims at the dissemination of advanced scientific and technological knowledge, with a view to strengthening links between scientific communities.

The Series is published by an international board of publishers in conjunction with the NATO Scientific Affairs Division

A	Life Sciences	Plenum Publishing Corporation
B	Physics	London and New York
C	Mathematical and Physical Sciences	D. Reidel Publishing Company Dordrecht, Boston, Lancaster and Tokyo
D	Behavioural and Social Sciences	Martinus Nijhoff Publishers Boston, Dordrecht and Lancaster
E	Applied Sciences	
F	Computer and Systems Sciences	Springer-Verlag Berlin, Heidelberg, New York
G	Ecological Sciences	London, Paris, Tokyo
H	Cell Biology	

Series E: Applied Sciences – No. 123

Halide Glasses for Infrared Fiberoptics

edited by:

Rui M. Almeida

Centro de Fisica Molecular
Instituto Superior Tecnico
1000 Lisboa
Portugal

1987 **Springer-Science+Business Media, B.V.**

Proceedings of the NATO Advanced Research Workshop on "Halide Glasses for
Infrared Fiberoptics", Vilamoura (Algarve), Portugal, March 31–April 4, 1986

Library of Congress Cataloging in Publication Data

NATO Advanced Research Workshop on "Halide Glasses
    for Infrared Fiberoptics" (1986 : Vila Moura,
    Portugal)
    Halide glasses for infrared fiberoptics.

    (NATO ASI series. Series E, Applied sciences ; no.
123)
    "Proceedings of the NATO Advanced Research Workshop
on "Halide Glasses for Infrared Fiberoptics", Vilamoura
(Algrave), Portugal, March 31–April 4, 1986"--CIP
t.p. verso.
    "Published in cooperation with NATO Scientific
Affairs Division."
    Includes index.
    1. Fiber optics--Materials--Congresses. 2. Metallic
glasses--Congresses. 3. Infrared technology--Congresses.
I. Almeida, Rui M.   II. North Atlantic Treaty Organiza-
tion. Scientific Affairs Division.   III. Title.
IV. Series.
TA1800.N367 1986   621.36'92   87–1508
ISBN 978-94-010-8093-4

ISBN 978-94-010-8093-4   ISBN 978-94-009-3561-7 (eBook)
DOI 10.1007/978-94-009-3561-7

To *Professor J.D. Mackenzie ...*
for his teachings, his encouragement and his help,
during the past ten years.

PREFACE

The field of heavy metal halide glasses (namely fluorides) is only ten years old now, but it has developed rapidly since the discovery of fluorozirconate glasses by the group at the University of Rennes (France). The main reason for this was the early demonstration of the enormous potential of such glasses for use as long-haul ultra-low loss middle infrared waveguide materials, aided in part by the scientific interest held by their unusual short range structures. As a result, significant research efforts were initiated in the academic, government and industrial sectors in Europe, the United States and Japan. However, the search for a finished product has perhaps led to a partial overlooking of some of the more fundamental aspects by the scientific community.

After the initial excitement, the workers in this field are perhaps at a crossroads where attenuations lower than 1 dB/Km need to be obtained for long lengths of fiber of good chemical and thermal stability, in order to guarantee continual R&D sup ports. Therefore, there is a strong need for a critical assessment of the potential of halide glasses for infrared fiberoptics and the formulation of recommendations for future research in this area and other related fields.

The NATO Advanced Research Workshop on Halide Glasses for Infrared Fiberoptics, held in Vilamoura (Portugal) from March 31-April 4, 1986, assembled a group of 35 experts from fourteen different countries. The main results were:

a) A critical review by leading experts regarding the state of science, state of development and future needs for these new fibers as of April, 1986.

b) An authoritative summary of the future potentials of these fibers and a definition of needs in research and needs in development for the next five years.

c) An identification of R&D centers in different NATO coun tries where cooperative research may be done and in which specific areas of work can more efficiently be covered.

d) The editing of the present volume in the NATO ASI series, whose publication will hopefully contribute to stimulate interest in this field in other countries.

The program of the meeting was organized around seven topics: (1) Raw Materials and Purity, (2) Structure, (3) Thermal and Mechanical Behavior, (4) Optical Properties of Glasses and Fibers, (5) Fiber Fabrication, (6) Fiber Systems and (7) Other Vitreous Halides. It included twenty seven oral presentations and a few poster presentations as well. The organization of the book followed a similar scheme.

I would like to thank the members of the organizing committee - Dr. D.C. Tran, Prof. J. Lucas, Prof. J.D. Mackenzie and
Dr. D.R. Ulrich - for all their invaluable help during the
early stages of the preparation of this meeting. In particular,
I am very grateful to Dr. D.C. Tran, the co-Director of the
Workshop, for helping me to get the meeting off the ground and
for his major contributions with the organizational and finan
cial aspects of the Workshop. This would not have been possible
without him.

I also thank all the session chairmen, speakers and participants for the wonderful job they did. In addition, I wish
to express my sincere appreciation to all those speakers who
delivered their manuscripts on time, thus helping to render
the editing task slightly more bearable.

For those who should have been invited to this workshop and,
for one reason or another, where not, I take a major share of
the responsability and, therefore, I sincerely apologize.

I also thank the generous support of this meeting by the
following organizations:

- NATO Scientific Affairs Division, who provided∿ 60% of
  the funds through its excellent Advanced Research Workshop program.

- The following U.S. companies, who provided small grants,
  corresponding to the remaining support: GTE Laboratories,
  Inc., Corning Glass Works, Owens Corning Fiberglas,
  SpecTran and Hughes Research Laboratories.

Finally, I would like to thank the devoted secretarial help
of Manuela P. Nunes, the assistance of Fernanda Serrenho
with partial typing of this manuscript and the dedicated organizational contributions of José L. Grilo and Clara H.
Gonçalves during the meeting.

Rui M. Almeida

Lisboa

June 1986

CONTENTS

THE FIRST TEN YEARS

J. LUCAS

Laboratoire de Chimie Minérale, Unité Associée au C.N.R.S. n° 254, Université de Rennes, 35042 Rennes Cédex (France)

ABSTRACT
The state of the art before and after the discovery of heavy metal fluoride glasses is discussed. The key points on the fundamental knowledge of the material and on the technological developments which resulted in the fabrication of optical fibers are presented. This presentation will deal with the crossed fertilization process which is the main feature of this new scientific community originating from diverse domains of physics, chemistry and materials science.

1. INTRODUCTION
It is clear that glass science has been and is still dominated by the oxide-based glasses. Among them, the silicates, phosphates, borates and their combinations represent the most important glassy materials used by the industry either for large scale applications like package, isolation, ..., or for more sophisticated devices like optical fibers for telecommunications. All these glasses contain light elements like B, Si, P, with strong chemical M-O bonds and an elementary module which is a MO_4 tetrahedron.

The direct consequence of this situation is that the optical properties and specially the infrared transmission is limited to the near I.R. region with an I.R. cut-off usually located in the 2-3 μm region.

The need for optical glasses operating at longer wavelengths has encouraged research on new glasses. The chalcogenide and halide combinations are potential candidates for this objective because of their relatively weak bond strength and the possibility of associating heavy atoms. Except for some very well-known glass former materials like As_2S_3, GeS_2, BeF_2, $ZnCl_2$, the number of vitreous families in this chemistry is extremely poor.

Most of the halide materials, because of the electronegativity of the halogens X = F, Cl, Br, I tend to give ionic materials governed by coulombic forces like NaCl, BaF_2... Also, because of the monovalence of X, they tend to form isolated molecules like $SiCl_4$ or UF_6.

Consequently, the glass-crystal competition leading to glassy materials is not very favorable for stabilizing aperiodic giant 3D framework which is characteristic of the vitreous state. This explains the almost quasi desertic situation of this part of solid state chemistry.

2. THE HALIDE GLASS CHEMISTRY BEFORE 1974
This presentation deals exclusively with halide glasses and the glasses derived from chalcogens will not be discussed.

Only three halides have been claimed to be glass-formers before 1974 (1) : the beryllium fluoride BeF_2, known to be isostructural of SiO_2, gives a viscous melt above the liquidus and ,when cooling, a glassy state is easily obtained. This behaviour has been known for more than fifty years, but these glasses are highly toxic and hygroscopic and, also because their I.R. transmission is very limited, applications have not been found. The zinc chloride $ZnCl_2$ is also known to melt giving a viscous liquid and is easily obtained as a glass on cooling. Despite their interesting transmitting properties, the I.R. cutoff being located near 12 μm, the $ZnCl_2$ glasses are extremely hygroscopic and their optical properties are immediately destroyed in normal atmospheric conditions. Sun was the first to demonstrate that in combinating MF_3 fluorides, here AlF_3, with other fluorides like PbF_2, it was possible by quenching to synthesize glasses. But because of their high tendency to devitrify, these glasses were not suitable for optical applications.

3. THE PECULIAR CRYSTALCHEMISTRY OF FLUOROZIRCONATES

During the period 1960-1975, solid state chemistry had been a very fast growing field, in the domain of new compounds synthesis, and crystal structure determination. The crystalchemistry of ZrF_4 and fluorozirconates had been extensively investigated and it had been demonstrated that this specific part of chemistry was extremely rich in new combinations and was characterized by a huge diversity in crystal structure.

This is due to the extreme ability of Zr^{4+} to modify its own environment when surrounded by F^- ions. Figure 1 shows the different polyhedra ZrF_n with n = 6, 7, 8 and the multiple geometries which have been proved to exist in many fluorozirconates (2).

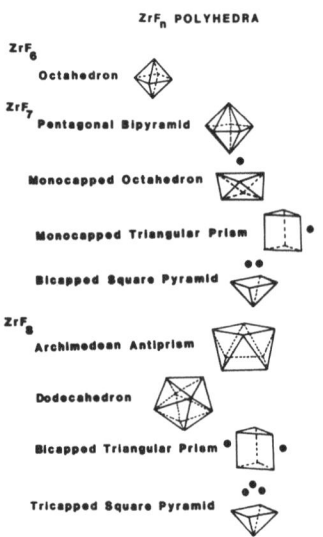

ZrF_n POLYHEDRA

ZrF_6
  Octahedron

ZrF_7
  Pentagonal Bipyramid

  Monocapped Octahedron

  Monocapped Triangular Prism

  Bicapped Square Pyramid

ZrF_8
  Archimedean Antiprism

  Dodecahedron

  Bicapped Triangular Prism

  Tricapped Square Pyramid

FIGURE 1. The ZrF_6, ZrF_7 and ZrF_8 polyhedra usually found in fluorozirconate crystalchemistry.

When n = 6, ZrF_6 is a regular octahedron ; when n = 7, the ZrF_7 polyhedra are either pentagonal bipyramids or monocapped octahedra, or monocapped triangular prisms or bicapped square pyramids. When n = 8, the geometry of the ZrF_8 polyhedron is a square antiprism, a dodecahedron, a bicapped triangular prism or a tricapped square pyramid.

At this time, these very diversified observations were creating some "confusion" in the mind of the solid state chemists working in the field. This ambiguous situation is certainly the main feature of the Zr^{4+} ions when in a fluoride melt, they are surrounded by F^- ions and must choose, in such a diversity, a specific stereochemistry and coordination number.

4. THE FLUORIDE GLASS CHEMISTRY AFTER 1974

Consequently, the first application of the so called "confusion principle" in fluoride chemistry was probably demonstrated when Michel Poulain, at the University of Rennes, in 1974, opened a nickel sealed tube containing a complex mixture of fluorides mainly based on ZrF_4. The crystalline compounds which were expected from the mixture ZrF_4-NaF-BaF_2-NdF_3 turned out to be largely amorphous. The author's coworkers and among them Marcel Poulain, Maydom Chanthanasinh, decided from this observation to explore systematically the glass formation in ZrF_4-based multicomponent systems.

This lead to the determination of the glass area in some ternary systems. Among them, we will select the diagram ZrF_4-BaF_2-LaF_3 shown on fig. 2 with the corresponding glass called ZBL. It was clear at this time, in referring to the traditional terminology that ZrF_4 was the glass-former and BaF_2 played the role of modifier. The fluorides LaF_3 as well as ThF_4 used in a few percentages played an important role in decreasing the devitrification rate and were called stabilizers.

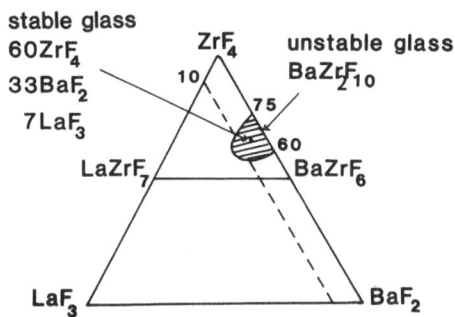

FIGURE 2. One of the first diagrams where fluorozirconate glasses have been isolated The binary glasses BaF_2/ZrF_4 are of interest for structural investigations.

These ternary glasses were still very unstable and one had to to refer to the systematic work of Lecoq (3) in his PhD for the development of a good composition having a low tendency to devitrify and called ZBLA : 57 ZrF_4, 34 BaF_2, 5 LaF_3, 4 AlF_3. Some other compositions suitable for fiber drawing have also been proposed by different authors, for example : 51 ZrF_4, 16 BaF_2, 5 LaF_3, 3 AlF_3, 20 LiF, 5 PbF_2 by Tran (4) and 60 ZrF_4, 32 BaF_2, 4 GdF_3, 4 AlF_3 by Mitachi (5).

As discussed in another chapter called Zirconium-Free Fluoride Glasses, the author's group developed around 1980 several new classes of heavy metal fluoride glasses (HMFG) not containing zirconium, but based on the combination of other fluorides. Two compositions leading to glasses having a weak tendency towards devitrification are as follows :

a) the glass BZYbT developed by Fonteneau and Slim (6) has the composition : 15 BaF_2 – 28 ZnF_2 – 28 YbF_3 – 28 ThF_4, and can be obtained in samples of 1 cm thickness.

b) the glass BIZYT, with 30 BaF_2 – 30 InF_3 – 20 ZnF_2 – 10 YF_3–10 ThF_4 has been very recently discovered by Bouaggad in the author's laboratory and samples of 2 cm thickness have been prepared.

At the end of the seventies, another French group located in Le Mans and headed by Jacoboni, discovered that the transition metal fluorides MF_3 (7), when combined with other fluorides like MF_2 or PbF_2, were able to give glasses which have been

4

proved structurally speaking to belong to a new kind of vitreous material ; a typical composition is PbMnFeF$_7$.

5. THE NEED OF CUMPLICITY FOR DEVELOPPING NEW MATERIALS

During the few years after their discovery, fluoride glasses had been of purely academic interest and the only potential property which made them original was their better transmission in the mid-I.R. region than the oxide glasses.

Attention had been first paid to these materials by the community of glass laser for nuclear fusion. M. Weber at Lawrence Livermore Laboratory was looking for Nd^{3+} glass matrix having the lowest linear and non linear refractive indices. The fluoride glasses were good candidates with a special figure of merit for beryllium fluoride glasses. But, because of the need for high optical quality and large size samples, the choice had been made in favour of fluorophosphate glasses.

The decisive step which put the fluoride glasses on the launch pad was the potential of their ultratransparency in the mid IR region. Different authors have speculated on the intrinsic ultra-low loss in different materials in combining the two main loss factors : scattering and multiphonon absorption (8) (9) (10). The now famous V shape curve resulting from the combination of these two loss mechanisms had been first published by the NTT group (11) for fluoride glasses. Figure 3 shows this valley shaped curve for SiO$_2$ glasses and fluoride glasses. It is obvious that the ultimate transparency for SiO$_2$ glass is 0.2 dB/km at 1.50 μm and 10^{-2}–10^{-3} dB/km at about 3 μm for fluoride glasses.

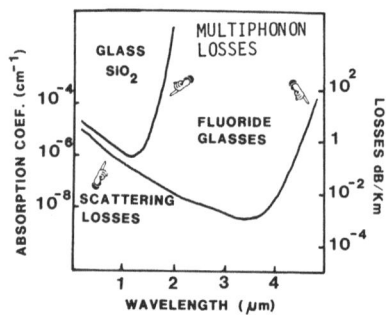

FIGURE 3. Evolution of the intrinsic losses for SiO$_2$ glass and fluoride glasses. Two absorption mechanisms are considered : the Rayleigh scattering and the multi-phonon loss giving a typical V shaped curve.

This target was a strong motivation for the entrance in the field of the optical fiber community like Nippon Telegraph Telephone, Naval Research Laboratory, British Telecom Research Labs, Centre National d'Etude des Télécommunications, Hughes Research Laboratory, Standard Telecom Laboratories, Furukawa Industries, Corning Glass, Rome Air Development Center, Laboratoires de Marcoussis, A T & T Bell Labs, G.T.E. Laboratories, etc...

This potential superiority of fluoride glasses on silicates was also a good argument for us to negociate contracts with different research agencies.

Another domain of optical science was also pushing for the development and research on these new glasses which were considered as new matrices for laser applications. Because of the original ligand field acting on transition metal or rare-earth ions and also because of the large transmission range giving a low multiphonon emission rate, and a broad spectral domain for excitation and emission, the doped fluoride glasses have received a lot of attention for optical applications.

6. A CONVERGENCE OF PLURIDISCIPLINARY TALENTS

The most exciting feature in the evolution of this domain of materials science has been unquestionably the formation of a new scientific community having different origins and concentrating their efforts and talents on making this fast-growing field progress.

Among the various contributions to the field, we will select those which have been the most significant in the sense that they have increased our knowledge on these new glasses either on a technical or academic point of view.

The long distances repeaterless communications have of course motivated the fabrication of low loss optical fibers. From the leading groups such as NRL,NTT, CNET, STL, BTL..., we learn that the key problems are :

1) How to prepare a good preform and how to draw fibers. The development of the so called rotational casting process by Tran (4) and the teflon jacketing by Mitachi (12) for the protection of the fibers might be considered as significant technical contributions.

2) How to avoid crystallization during the preform fabrication and the fibering process. This key factor which is responsible for the increase in the scattering loss has been investigated from a fundamental point of view by different groups (13) (14).

3) How to avoid the extrinsic absorption losses due to impurities such as transition metals, rare-earth and also the critical parasitic complex anion OH^- absorbing in the ultratransparency region. The need of ultrapurified materials has been demonstrated by these studies.

4) What was the relation between the composition of the glass and the multiphonon edge, which affects severely the shape of the valley curve (15).

In addition to these investigations mainly related to optical fiber fabrication, several fundamental contributions have marked the knowledge of these new materials.

Investigation on the chemical durability was important because of the well known sensitivity of halide materials to moisture and liquid water and their dramatic consequences on optical properties. Although the intimate mechanisms of water corrosion are still not well understood, some significant contributions from Simmons and others (16) have helped to understand such phenomena.

Despite the difficulties in investigating the structure of amorphous materials the last five years have been extremely rich in new results : the convergence of investigations by vibrational spectroscopy (17), molecular dynamics, X-ray scattering (18) and also local probe spectroscopies lead to models which give a good representation of the structure of this new family of glasses. Figure 4 represents the structure of the simple binary glass $BaZr_2F_{10}$.

This fundamental knowledge of the short range and medium range order in the structure as well as the kinetic phenomena and especially the diffusion mechanisms are essential in understanding the nucleation and crystal growth phenomena in these types of glasses. The competition glass versus crystal is very critical here and has direct consequences on the scattering losses in fibers.

In this way, one has to consider as being very important the information obtained in studying the OH diffusion profile during the corrosion of the glass surface even in an atmosphere with a low moisture content (19).

For the same reasons, Ravaine's investigation (20) on the electrical conductivity of fluoride glasses, giving information on F ions diffusion mechanisms and the nature of conductivity, are fundamental steps in understanding the intimate

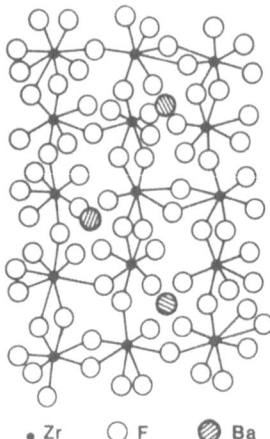

FIGURE 4. A structural model for the simple fluorozirconate glass $BaZr_2F_{10}$. The ZrF_7 and ZrF_8 polyhedra sharing corners or edges form a 3D network in which Ba^{2+} is inserted.

• Zr   ○ F   ⊘ Ba

electrochemical and ionic dynamic phenomena and the potential applications of these new materials as solid electrolytes or F sensitive electrodes.

Many other studies have been conducted to assess the specific physical properties of fluoride glasses such as :

- investigations of magnetic properties (21) at low temperatures and the discovery of spin glass behaviour in glasses containing a high concentration of paramagnetic cations,

- determination of the most important mechanical properties (22) such as hardness, fracture toughness, fracture strength...,

- spectroscopical investigations (23) using transition metals or rare earths as local probes to measure the ligand field parameters.

7. CONCLUSION

There are many ways to evaluate the good health of a scientific domain. Among them, the number of published papers or presentations in International Conferences could be used as reference material. Figure 5 shows the evolution of the number of papers and participants to the first, second and third international Symposiums on halide glasses.

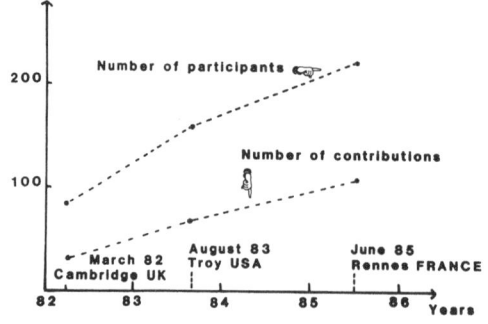

FIGURE 5. Evolution of the number of participants and contributions to the International Halide Glasses Symposia.

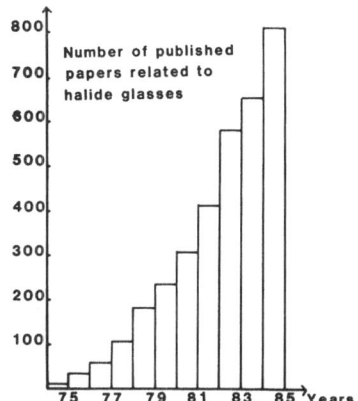

FIGURE 6. Evolution of the number of published contributions including proceedings, papers, reports..., related to halide glasses during the last ten years.

Figure 6 indicates the number of published papers relative to fluoride glasses from 1974 up to now. There is no doubt that we are faced with a very fast growing field where the interactions of interdisciplinary scientific domains appear to be specially fruitful. Another fact which does not appear in these two figures is the high scientific level of the contributions and their diversity. But probably, the most fascinating discovery during those last ten years is the quality of the human relationship in this international community represented at the Rennes symposium by 22 nations. If, by misfortune, the fluoride glasses have not in the future all the expected exciting applications, they will have had, at least, the merit of being at the origin of many strong and sincere friendships. This is often a first and necessary step before solving scientific problems.

REFERENCES

1. Baldwin, C.M., Almeida, R.M. and Mackenzie, J.C., J. Non-Cryst. Solids 43, 309 (1981).
2. Laval, J.P., Lucas, J. and Frit, B., Materials Science Forum, Trans. Tech. Publ., 6, 457 (1985).
3. Lecoq, A., Thèse de 3ème cycle, Université de Rennes, 1980.
4. Tran, D.C., Fisher, C.F. and Sigel, G.H., Electron. Lett., 18, 657 (1982).
5. Mitachi, S., Ohishi, Y. and Miyashita, T., Proc. 4th IOOC 83, Tokyo, 102 (1983).
6. Fonteneau, G., Slim, H., Lahaie, F. and Lucas, J., Mat. Res. Bull., 15 (10) 1425 (1980).
7. Miranday, J., Jacoboni, C. and De Pape, R., Rev. Chim. Min., 16, 277 (1979).
8. Van Uitert, L.G. and Wemple, S.H., Appl. Phys. Lett., 33, 57 (1978).
9. Goodman, C.H.L., Solid State and Elect. Dev., 2, n° 5 (1978).
10. Gannon, J.R., J. Non-Cryst. Solids, 42, 239 (1980).
11. Shibata, S., Horiguchi, M., Jinguji, K., Mitachi, S., Kanamori, T. and Manabe, T., Electron Lett., 17, 21, 775 (1981).
12. Mitachi, S., Shibata, S. and Manabe, T., Electron. Lett., 17, 128 (1981).
13. Lucas, J. and Moynihan, C.T., Materials Science Forum, Trans. Tech. Publ., 6, 193 (1985).
14. Drexhage, M.G., Heavy metal fluoride glasses, Treatise on Materials Science and Technology : ed. Tomozawa M, Doremus R, Academic Press, 1985.

8

15. Bendow, B., Banerjee, P.K., Drexhage, M.G., El Bayoumi, O., Mitra, S.S., Moynihan, C.T., Gavin, D., Fonteneau, G. and Lucas, J., J. Amer. Ceram. Soc., 66, n° 4, C64 (1983).
16. Simmons, C.J., Azali, S.A. and Simmons, J.H., Proceedings of the 2nd Int. Symposium Halide Glasses, Troy, 1983, paper 47.
17. Almeida, R., Materials Science Forum, Trans. Tech. Publ., 6, 427 (1985).
18. See Lucas, J., Angell, C.A. and Louer, D., pp. 449, and Kawamoto, Y., pp. 417, Materials Science Forum, Trans. Tech. Publ., 6 (1985).
19. Tregoat, D., Fonteneau, G., Moynihan, C.T. and Lucas, J., J. Am. Ceram. Soc., 68, n° 7, C171 (1985).
20. Ravaine, D., Materials Science Forum, Trans. Tech. Publ., 6, 761 (1985).
21. Dupas, C., Materials Science Forum, Trans. Tech. Publ., 6, 731 (1985).
22. Mecholsky, J.J., Pantano, C.G. and Gonzalez, A.C., Materials Science Forum, Trans. Tech. Publ., 6, 699 (1985).
23. Sibley, W., Materials Science Forum, Trans. Tech. Publ., 6, 611 (1985).

DISCUSSION

Q: A.C. Wright. Why not mix heavy metal fluoride glasses with beryllium fluoride?

A: J. Lucas. The mixture would probably be phase-separated.

Q: P.W. France. How did you obtain SmF_2?

A: J. Lucas. SmF_3 was reduced by using radiation and it was stable.

Q: R.M. Almeida. Would you care to comment on other halide glasses?

A: J. Lucas. They have limited potential as fiber materials. Many of these halides have a high devitrification tendency. Also I am not optimistic about them because of their hygros copic character.

HIGH PURITY COMPONENTS FOR FLUOROZIRCONATE GLASS OPTICAL FIBERS [1]

M. ROBINSON

HUGHES RESEARCH LABORATORIES
3011 MALIBU CANYON ROAD
MALIBU, CA 90265 USA

ABSTRACT

Some of the pertinent research and development dedicated to the preparation of ultra-high purity starting materials necessary for the fabrication of low-loss heavy metal fluoride glass (HMFG) fiber is described. The specific areas of cation and anion purification techniques addressed are sublimation, [1] crystallization, electrolytic processing, [2] chemical vapor purification (CVP), [3] and solvent extraction. Additionally, reactive atmosphere processing (RAP) [5,6,7] is discussed in relation to removal of IR active anion impurities, the most troublesome of which is OH^- due to the ubiquitous nature of its source, water. Moreover, RAP data is given relating to the elimination of black optical scattering centers emanating from fluorine deficiencies in the glass.

INTRODUCTION

It is well known that the primary contributors to IR loss for HMFG aside from scattering arise from contamination by water related impurities, primarily OH^-, along with several transition metal ion and rare earth ion impurities. Fe^{+2}, Co^{+2} and Ni^{+2} are representative of the former and Nd^{3+} the most bothersome of the latter in terms of its strong IR activity. In recent years, however, great strides have been made in achieving highly IR transparent fiber in the 2.5 μm region of the spectrum and much of this accomplishment stems from a concentrated materials effort directed to complete removal of the above mentioned impurities. Consequently, IR fiber losses at 2.55 μm have been steadily reduced from the several hundreds dB/km loss values seen only a few years ago to current values of only a few dB/km. [8]

The purpose of this presentation is to give results on the elimination of impurities in the raw materials and glass which add to IR loss in the 2-4 μm IR region and to direct attention to new and novel methods of achieving the ultra low-loss fiber objective---the intrinsic loss value of 0.01 dB/km.

[1]

This work is supported in part by Naval Research Contract N00014-85C-2524.

PURITY RELATED PROBLEMS

The heart of the purity problem is that all fluorides by
their nature are quite dirty when compared to pure metals,
semiconductors, and oxides. The cause stems from typical
wet preparation techniques involving aqueous
hydrofluorination of the oxide or precipitation of a
precursor from water solution. Such methods discourage
establishment of high purity (HP) with respect to water-
associated impurities such as OH^-, OF^{-3}, O^{-2}, and metal ion
impurity---specifically, transition element (TE) ions. In
many cases the anhydrous fluoride is not attained due to the
formation of very stable hydrates, and TE impurity is high
due to reagent contamination and reagent-container
interaction.

If more stringent methods are utilized such as reaction of
HP oxide with gaseous HF, SF_6, NF_3, [9] or excess NH_4HF_2,
the purity and dryness problems are easier to overcome.
Ultra HP oxides are readily available commercially and the
fluorinating agents may be vapor transported to enhance
purity-dryness to yield suitable fluoride-glass forming
materials without any additional purification steps.
However, high purity crucibles and dry furnace environments
are of utmost importance to ensure production of the desired
product. Additionally, once the component is satisfactorily
prepared the concern is with impurity pick-up through
handling and interaction with the environment. Thus,
airborne particulates, hydrolysis from environmental
moisture, and pick-up from container etching (physical and
chemical) must all be minimized.

From the preceding discussion it is obvious that development
of ultra HP fluoride components and the glasses derived
therefrom is an extremely difficult but not insurmountable
task requiring emphasis on new and novel methods of
materials preparation and handling. In addition to the
above, it must be pointed out that no single preferred
analytical method exists for accurately determining sub ppm
impurity concentration in either the pure component or
glass. Perhaps the most satisfactory method at this
juncture, although roundabout and time consuming, is optical
absorption spectroscopy on the drawn fiber itself.
Evaluation involves drawing many meters of fiber for the
required sensitivity. However, this method does offer the
potential for consistent reproducible results.

COMPONENT PREPARATION

The discussion in this section is limited to preparation of
component fluorides having purities adequate for preparation
of high quality bulk glass but perhaps unsuitable for

production of ultra low-loss fiber without further
purification steps. The materials will be carbon free,
exhibit little or no oxygen containing impurity, and show
small but detectable TE contamination. For this purpose,
each component should be prepared by a non-aqueous method
utilizing the highest purity reagents. The ideal approach
makes use of HP oxide or metal in conjunction with a
fluorinating agent such as HF [5], SF_6, or CF_4. For
example, ZrO_2 is readily available in 6-9's purity at a
price about 10 times lower than the highest-available-purity
ZrF_4 (3-9's purity). In the case of ZBLAN glass, ZrO_2 or
the carbonates of Ba, La and Na may be easily converted to
their respective fluorides at a moderate temperature of
about 300°C. A system for conversion of oxide to fluoride
utilizing anhydrous HF is shown in Fig. 1. The circular end

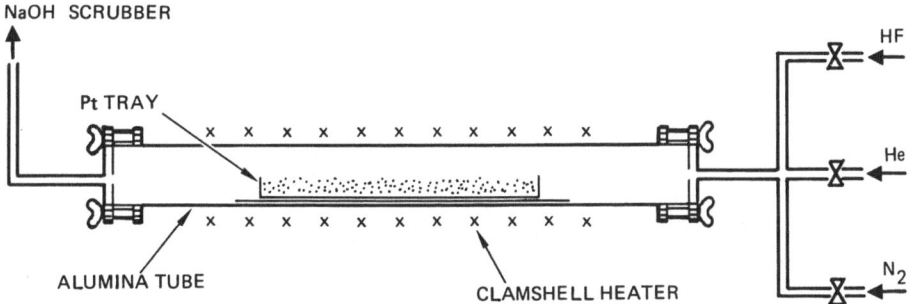

Figure 1.   Oxide to Fluoride Conversion Apparatus.

plates are constructed of stainless steel and have been gold
plated to minimize TE ion contamination. Also, all gas
lines and fittings are made of Teflon. When operational,
the $H_2O(g)$ is rapidly removed from the reaction site by the
flowing HF gas. The chemical reaction is

$$O^{-2}_{(s)} + 2HF_{(g)} \rightleftharpoons 2F^-_{(s)} + H_2O_{(g)}. \tag{1}$$

In the above equation s = solid and g = gas.

In all cases the reaction-free energy ΔF_R is favorable for
fluoride formation at moderate temperatures. This is shown
in Fig. 2 for ZrF_4 and AlF_3. However, in the AlF_3 case one
should note that although the thermodynamics are favorable
the kinetics are slow enough to permit utilization of an
Al_2O_3 tube as the reaction chamber. Therefore, to produce
AlF_3 in a reasonable time period other schemes such as
reaction of Al metal with HF or NF_3 are considered. Figure
3 is a photo of LaF_3 and NaF prepared by interaction of HF
with La_2O_3 and Na_2CO_3, respectively, followed by melting and
slow crystallization at 1500°C and 1000°C. The high
temperatures involved provide for further purification by
vapor transport of volatile TE fluoride contaminants. Owing
to the absence of nucleating sites in such high purity

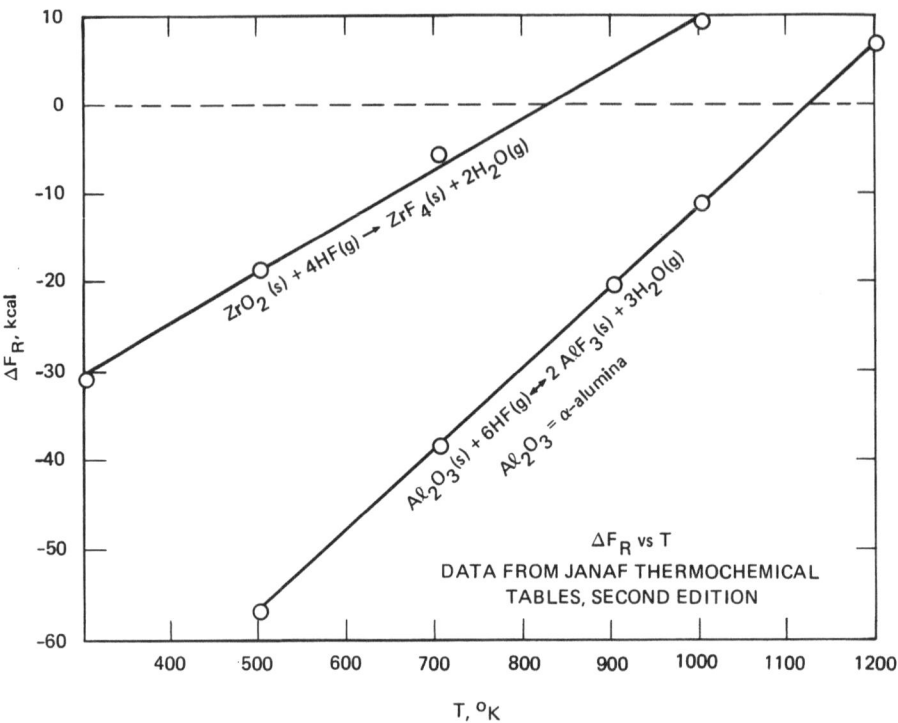

Figure 2. Free Energy of Reaction (ΔF_R) Versus Temperature, ($^\circ$K).

melts, large crystalline, high purity fluoride ingots are routinely cast simply by slow cooling through the melting temperature.

PURIFICATION SCHEMES

From the recent work of France, Carter, Moore, and Williams [10] it can be shown that only fractional ppb impurity levels of Fe^{+2}, Ni^{+2}, and Co^{+2} along with a few ppb OH^- and Cu^{+2} can be tolerated to realize transmission loss of <0.1 dB/km in fluorozirconate fiber. To approach such impurity levels in ZrF_4, the component typically comprising greater than 50 mole percent of the glass composition, commercially available starting material is inadequate and novel purification techniques must be considered. Table 1 shows several promising methods related to production of ultra high purity ZrF_4, free from TE impurities as well as detrimental anion impurities such as OH^-.

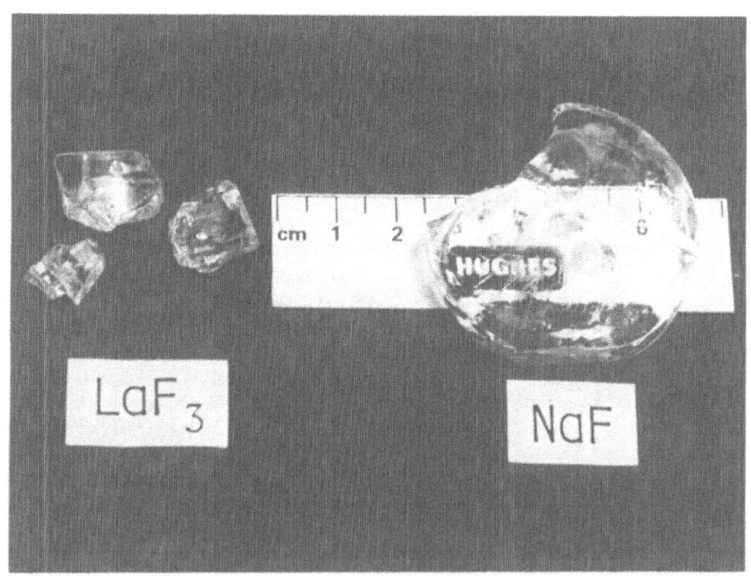

Figure 3.   LaF$_3$ and NaF Crystallized from their
Melts at 1500°C and 1000°C in HF.

METHOD	PROCEDURE	TE ESTIMATED IMPURITY	COMMENT
CHEMICAL VAPOR PURIFICATION (CVP)	PRODUCTION OF ZrCl$_4$ BY CHLORINATION OF Zr; THEN FLUORINATION	SUB PPB PREDICTED	Zr RICH CONDITION IN CHLORINATION SUP-PRESSES TE TRANSPORT
ELECTROMOTIVE SERIES DISPLACEMENT (ESD)	PLATE OUT TE IMPURITIES FROM Zr RICH MELT	PPB	CAPABILITY OF TECHNIQUE UNEXPLORED
SUBLIMATION/ DISTILLATION	VAPOR TRANSPORT AT 1100°K	PPM LEVEL	HIGH TEMP LEACHING OF TE FROM CRUCIBLE A PROBLEM
SOLVENT EXTRACTION (SE)	EXTRACTION OF ZrOCl$_2$ (AQ) WITH ORGANIC SOLVENTS	PPB LEVEL	CONVERSION OF ZrOCl$_2$ TO ZrF$_4$ POSES H$_2$O RELATED IMPURITY PROBLEMS
REACTIVE ATMOSPHERE PROCESSING (RAP)	ATMOSPHERE REACTS WITH AND REMOVES OXYGEN CONTAINING IMPURITIES IN THE MELT AND ELIMINATES ATMOSPHERIC MOISTURE		NECESSARY FOR OH$^-$ REMOVAL AND VALENCE CONTROL

Table I.   Synthesis of High Purity ZrF$_4$.

CHEMICAL VAPOR PURIFICATION (CVP)

The CVP of ZrF_4 currently being developed by Folweiler and Guenther [3] makes use of two halogenation reactions utilized in sequence. The first,

$$Zr + 2Cl_2 \rightleftharpoons ZrCl_4 \uparrow \tag{2}$$

is followed by fluorination with agents such as SF_6 or HF. When the chlorination is carried out under Zr rich conditions the vapor transport of TE metal contaminants such as Fe, Co, and Cu is highly unfavorable. Thus the ultimate product is anticipated to exhibit TE levels near undetectable limits. The process is suitable for production of volatile components such as $ZrCl_4$ (sublimation point, 604°K) and $AlCl_3$ (boiling point, 720°K). The implication drawn from table II is that for the examples cited ($ZrCl_4$ and $AlCl_3$), when formed near equilibrium and metal-rich-conditions, the occurrence of involatile TE metals is

REACTION	$-\Delta F_R$, kcal	OXIDATION POTENTIAL, V
$Al + 3/2\ Cl_2 \rightarrow AlCl_3$	144	2.1
$Zr + 2\ Cl_2 \rightarrow ZrCl_4$	192	2.1
$Fe + 3/2\ Cl_2 \rightarrow FeCl_3$	72	1.0
$Fe + Cl_2 \rightarrow FeCl_2$	66	1.4
$Co + Cl_2 \rightarrow CoCl_2$	60	1.3
$Cr + 3/2\ Cl_2 \rightarrow CrCl_3$	103	1.5
$Cu + Cl_2 \rightarrow CuCl_2$	36	0.8
$Ni + Cl_2 \rightarrow NiCl_2$	54	1.2

Table II. Redox Values of Some Chlorides at 500°K.

strongly favored over their respective chlorides. The oxidation potentials given for Al and Zr are both 2.1 V, substantially high enough to plate out TE from their chlorides precluding transport by volatilization.

ELECTROMOTIVE SERIES DISPLACEMENT (ESD) [2]

Electrolytic methods for purification of ZrF_4 include ESD, a technique in which ZrF_4 is distilled from a ZrF_4-BaF_2 melt containing disbursed Zr metal. Both processes, CVP and ESD capitalize on the electroplating nature of Zr when in contact with TE halides. For example, in the case of ESD the standard redox potentials, ϵ_T (T=1000°K) for the fluorides of Fe as well as Co, Ni and Cu are all considerably lower than 4 V. This in contrast to ϵ_T for

ZrF_4 and BaF_2 which are 4.1 V and 5.3 V respectively, provides efficient reduction and plating out of only TE ions contained in the melt. Separation and collection of the purified ZrF_4 is accomplished by distillation at a moderate temperature of 1100°K, where its vapor pressure is 137 torr [11]. Preliminary experiments show a factor of 600 reduction in Fe content for ZrF_4 prepared by ESD.

Table III gives some results on iron analysis of ZrF_4 treated by various purification methods. A single pass sublimation reduces Fe by a factor of 10 while a distillation from a BaF_2–ZrF_4 melt gives a reduction of a factor of 80, 320 ppm to 4 ppm. For the case of ESD where Zr metal was added to the distillation source, the Fe level in the distillate is only 0.5 ppm. This result perhaps will be further improved when totally iron free distillation crucibles are utilized.

● 99.5% ZrF_4 (CERAC CO.)			
MATERIAL	Fe, ppm	METHOD	YIELD, %
ZrF_4 (AS RECEIVED)	320	ESA	—
FIRST HF SUBLIMATION	<100 (~30)	ESA (PEA)	70
RESIDUE	11000	ESA	
ZrF_4 – BaF_2 (HF DISTILLATION)	4	ZAA	50
ZrF_4 – BaF_2 – Zr (He DISTILLATION)	0.5	ZAA	50
VITREOUS CARBON CRUCIBLE	10	ZAA	

ZAA = ZEEMAN ABSORPTION ANALYSIS $\begin{Bmatrix} \text{REL ERROR} - 50\% \\ \text{ABS ERROR} - 200\% \end{Bmatrix}$

ESA = EMISSION SPECTROGRAPHIC ANALYSIS $\begin{Bmatrix} \text{REL ERROR } 20\% \\ \text{ABS ERROR } 100\% \end{Bmatrix}$

PEA = PLASMA EMISSION ANALYSIS

Table III. Iron Analysis of ZrF_4.

Figure 4 is a photograph of ZrF_4 crystals prepared with ESD then subsequently recrystallized at 700°C. Many large clear and transparent crystals are formed and the IR spectrum of this ZrF_4 is given in Fig 5. At the recrystallization temperature a fine grained specimen is transformed in time to a clear and transparent ingot consisting of cm size crystals. The large sample size is achieved when one grain grows at the expense of its fine grained neighbors. Recrystallization if set-up as a zone process may be used to further enhance ZrF_4 purity.

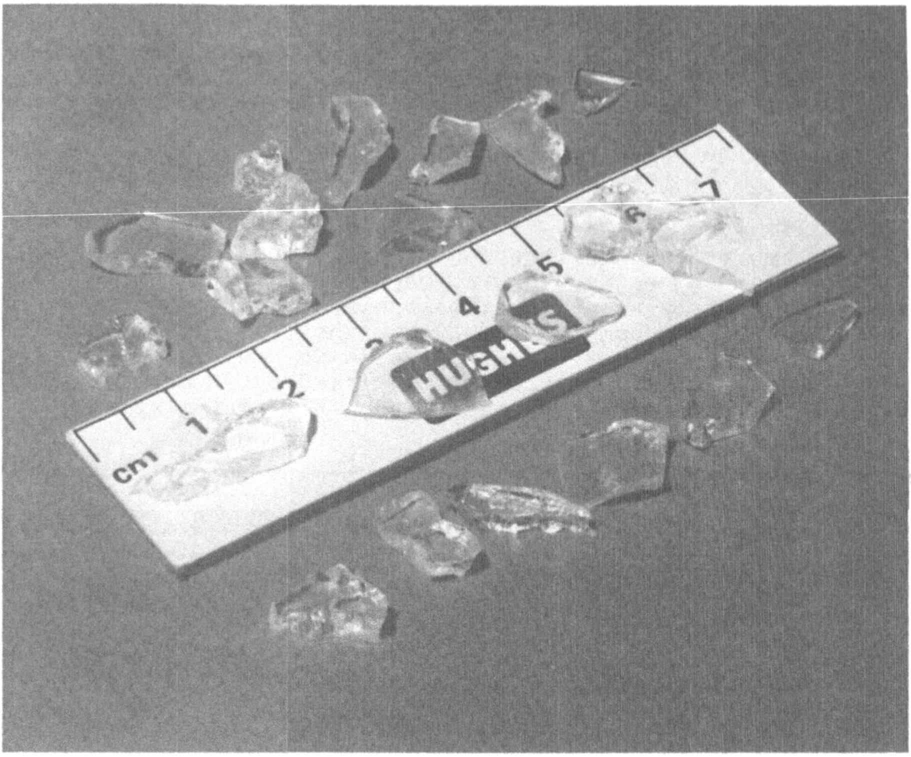

18

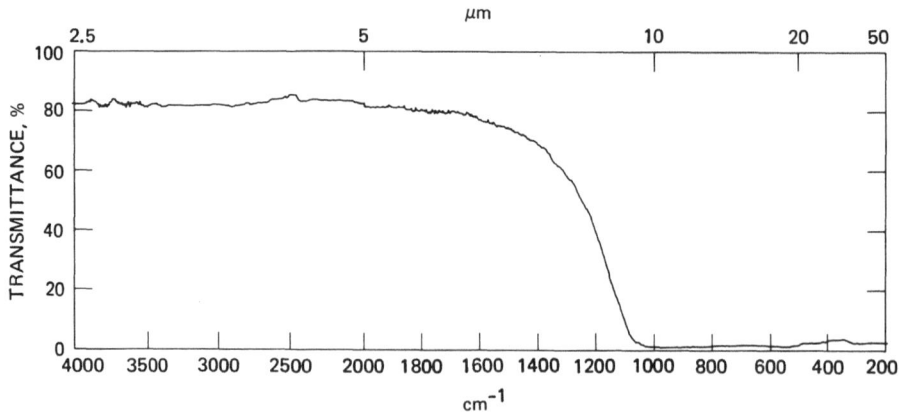

Figure 4.   Recrystallized ZrF$_4$.

Figure 5.   IR Spectrum of ZrF$_4$ Crystal (taken with a
Pye Unicam 3-300 Spectrophotometer.   Sample
Thickness 2.7 mm).

SUBLIMATION-DISTILLATION

The most convenient and rapid methods for ZrF_4 purification involve either sublimation of as received material, or distillation from a melt containing about 30% BaF_2 at 1100°K. Either technique provides expeditious separation of typical impurities such as oxides, carbon, and TE fluorides. However, distillation is more efficient in that the vapor pressure of TE fluorides are lowered as a consequence of Raoult's Law. Since the vapor pressures of TE fluorides are not immeasurably low (see table IV) total elimination even after several attempts may not be achievable [2,12]. A typical result is as follows: high purity ZrF_4 (BDH brand) shows by ZAA analysis to have an Fe content of 4 ppm. One sublimation brings down the Fe content to 1-2 ppm and subsequent attempts show that approximately 1 ppm is the limiting value even when the highest-available purity vitreous carbon crucibles are utilized. Figure 6 is a thermogram of HP crystalline ZrF_4. It points to a prime difficulty in obtaining oxide free ultra HP materials. In this experiment dry helium contains enough H_2O to permit some hydrolysis of the ZrF_4, especially at elevated

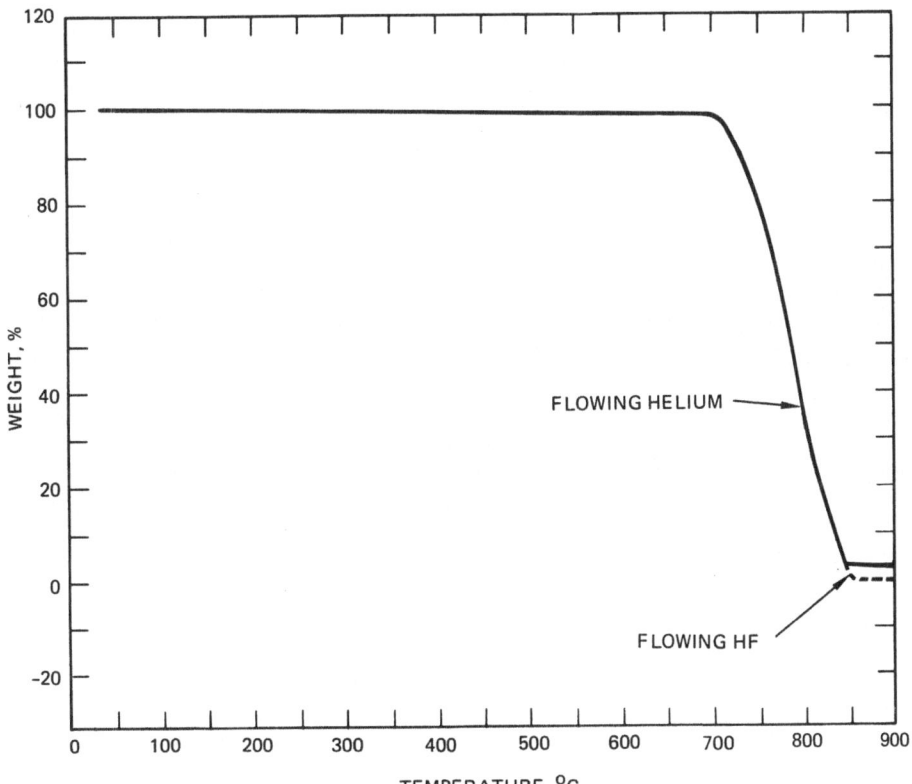

Figure 6. TGA of ZrF_4.

COMPOUND	VAPOR PRESSURE (Torr)	
ZrF_4	137	
FeF_3	2.5	APPROXIMATED FROM ΔH_v GIVEN IN ARGONNE NATIONAL LABORATORY REPORT NO. 5750 BY A. GLASSNER
FeF_2	0.014	
CuF_2	0.026	
CuF	2.9	
NiF_2	0.004[1]	

[1]MASS SPECTROMETRIC STUDIES AT HIGH TEMPERATURES. VI. THE SUBLIMATION PRESSURE OF NICKEL (II) FLUORIDE, T.O. EHLERT ET AL., JACS 86, 5093 (1964).

Table IV.   Vapor Pressures at the ZrF_4 Sublimation Temperature, $1100^\circ K$.

temperatures. The residue (~4%) analyzed by x ray to be ZrO_2 is totally absent when sublimation is carried out in a dry HF atmosphere. Thus, the implication is that some form of reactive atmosphere is required to prevent hydrolysis and formation of unwanted oxygen containing impurities in the components and certainly the glasses as well.

SOLVENT EXTRACTION

Purification by solvent extraction (SE) [13,14] applied to TE clean-up of aqueous zirconium salt solutions is currently under development by Fisher, Tran, Hart, and Sigel [4]. With careful attention to clean handling procedures, TE impurity levels in the ppb range are realizable, and perhaps ppt levels for iron are achievable in the aqueous solutions. The method utilizes chelating agents along with appropriate high purity organic solvent for extraction. However, isolation of anhydrous ZrF_4 poses problems in that stable hydrates, $ZrF_4 \cdot XH_2O$, are known which are not easily dehydrated by normal drying techniques. Also, conversion of purified $ZrOCl_2$ (Aq.), the operational solution, to ZrO_2 followed by synthesis of ZrF_4 opens up many avenues of impurity pick-up perhaps not readily obviated by ultra HP reagents and specialized handling methods.

REACTIVE ATMOSPHERE PROCESSING (RAP)

RAP is a purification technique specifically tailored to favor high fluoride purity by eliminating oxygen containing impurities in the molten & solid state. In RAP a reactive gaseous environment is maintained during all phases of materials synthesis and processing ranging from sublimation or melting of raw materials to various stages of glass preparation. Gases such as BF_3, SF_6, F_2, CF_4 [15] and NF_3 [9] at elevated temperatures 300°K to 1200°K eliminate IR absorbing impurities such as OH^-, O^{-2}, and OF^{-3} as well as H_2O, by chemical reaction. Using CF_4 as an example,

$$1/2CF_4(g)\ OH^-(s) \rightleftharpoons F^-(s) + HF(g) + 1/2\ CO_2(g) \qquad (3)$$

$$1/2CF_4(g) + H_2O(g) \rightleftharpoons HF(g) + 1/2\ CO_2(g). \qquad (4)$$

Thus CF_4 at ~1200°K reacts and eliminates OH^- in the fluoride in addition to eliminating the OH^- source, H_2O--and is considerably more efficient than HF alone. In any fluoride, OH^- is the most difficult anionic impurity to eliminate due to the ubiquity of H_2O. Figure 7 shows the IR transmission spectrum of 5-9's BaF_2 grown by the Stockbarger method in an flowing atmosphere of CF_4 and HF diluted with helium at 1500°C. Note that no IR absorptions can be observed from 2 μm to the IR cut-off. However, if the RAP CF_4 is eliminated, several unspecified absorption bands would be readily observed in the 3 μm to 9 μm region.

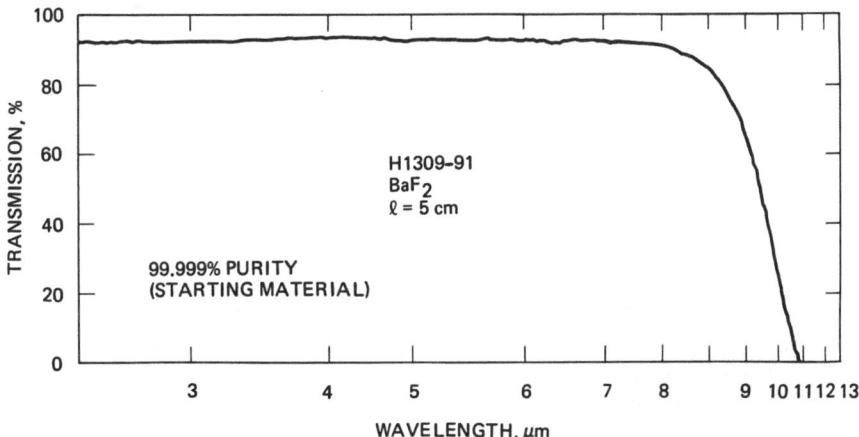

Figure 7. Transmission Spectrum of a BaF_2 Crystal, $(HF+CF_4)$/He Atmosphere. 99.999% Purity Starting Material.

Fluorozirconate glass (ZBLAN) prepared without the benefit of RAP always exhibits a black precipitate permeating the entire bulk of the sample. Such specimens are unusable for fiber drawing owing to the high degree of optical scatter. Electron microprobe analysis indicates that the black substance contains only one-half the fluorine of the transparent phase, and other analytical techniques confirm that it is not carbon. Formation of this black material proceeds regardless of whether a platinum or vitreous carbon crucible is used, and copious amounts are rapidly formed in glass melts where either inert gas or HF gas is used as the blanketing atmosphere. The black material is believed to be a reduced species of ZrF_4,

$$\text{ZrF}_4(\text{White}) \rightleftharpoons \text{ZrF}_{(4-x)}(\text{Black}) + X/2\ F_2 \quad (X>0). \tag{5}$$

This result is substantiated by residual gas analysis (RGA) of crystalline ZrF_4 which in fact shows evolution of fluorine at $T \geq 773^\circ K$ [7]. The black $\text{ZrF}_{(4-x)}$ can be totally eliminated by the reaction of molten glass with an oxidizing agent such as chlorine or oxygen according to the following:

$$\text{ZrF}_{(4-x)} + X/2\ Cl_2 \rightleftharpoons \text{ZrF}_{(4-x)}\ Cl_x \tag{6a}$$

$$\text{ZrF}_{(4-x)} + X/4\ O_2 \rightleftharpoons \text{ZrF}_{(4-x)}O_{0.5x} \tag{6b}$$

In the above, the chemically reduced black species is replaced by $\text{ZrF}_{(4-x)}Cl_x$ or $\text{ZrF}_{(4-x)}O_{0.5x}$ (both white in color) and this gives rise to clear and transparent glass totally devoid of any colored particulate. Figure 8 shows the result of reprocessing highly colored ZBLAN glass. Reprocessing was accomplished by remelting in an atmosphere of O_2 diluted with CF_4 for one hour at 850°C in a platinum crucible. The resultant specimen on the right is a clear cylindrical disk about 30 mm in diameter and 4 mm thick. The few bubbles observed are only detected at or near the melt/crucible interface and may be removed easily by

 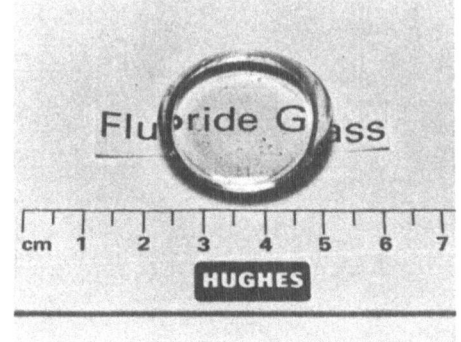

Figure 8. ZBLAN Glass (RH Specimen Remelted in O_2 + CF_4 ATM).

polishing. ZBLALI glass prepared from scratch using high
purity fluoride components and processed with O_2 and CF_4
shows no evidence of OH^- contamination when measured
spectrophotometrically. Figure 9 is an IR transmission
spectrum of ZBLALI glass prepared by this process. CF_4 at
the melt process temperature (850°C) reacts with trace
oxide, OH^-, and moisture to yield high purity fluoride.
Moreover, O_2 prevents reduction of ZrF_4, yielding remarkably
clear and transparent glass free of colored particulate.

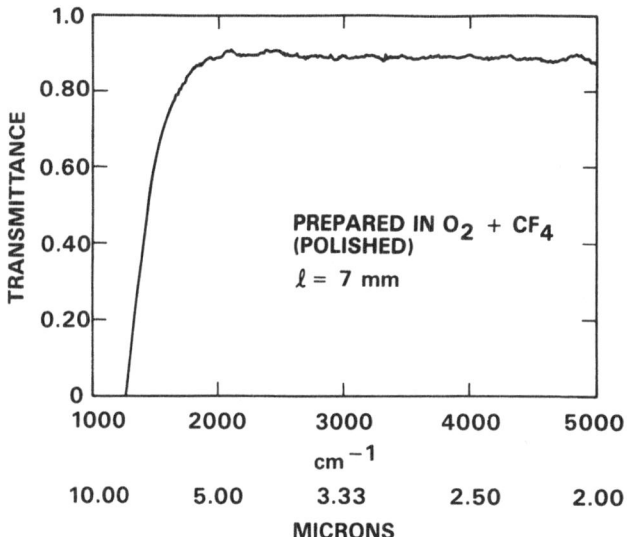

Figure 9. IR Spectrum of ZBLALI Glass.

SUMMARY AND CONCLUSIONS

Numerous IR-active impurities are germane to fluoride-glass
raw materials. TE ions, rare earth (RE) ions, and oxygen
containing anions are some of the most detrimental to IR
transparency. In the glass, anionic impurities give rise to
degradation of optical and mechanical properties. Trace
quantities of some cationic impurities lead to large losses
through optical absorptions, and although total removal in
an absolute sense is an impossibility, the levels are being
brought under control suitable for fabrication of fiber
having loss values rapidly approaching 1 dB/km at 2.5 μm.
CVP, ESD, SE, and RAP are promising methods being currently
developed for near total elimination of the most harmful
impurities. These methods along with specialized handling
methods to minimize impurity pick-up during glass forming
and fiber drawing hopefully will lead to unprecendented low
losses approaching intrinsic values.

REFERENCES

1. M. Robinson, Extended Abstract, 3rd International Symposium on Halide Glasses, Rennselaer Polytechnic Institute, U.S.A., Aug 1983.

2. M. Robinson, Mat. Sci, Forum 5, 19 (1985).

3. R.C. Folweiler and D.E. Guenther, Mat. Sci. Forum 5, 43 (1985).

4. C.F. Fisher, D.C. Tran, P. Hart, and G.H. Sigel, Mat. Sci. Forum 5, 51 (1985).

5. M. Robinson and D.M. Cripe, J. Appl. Phys. 5, 2072 (1966).

6. R.C. Pastor and A.C. Pastor, Mat. Res. Bull. 10, 117 (1975).

7. M. Robinson, R.C. Pastor, R.R. Turk, D.P. Devor, M. Braunstein and R. Braunstein, Mat. Res. Bull. 15, 735 (1980).

8. D.C. Tran, M.J. Burk, K.H. Levin, C.F. Fisher, P. Hart, L. Busse, G. Lu, and G.H. Sigel, Jr., Mat. Sci. Forum, 5, 339 (1985).

9. T. Nakai, Y. Minura, H. Tokiwa, and O. Shinbori, J. Lightwave Tech. LT4, No. 1 (1986).

10. P.W. France, S.F. Carter, M.W. Moore and R. Williams, SPIE Proc. vol. 618 (1986) to be published.

11. K.A. Sense, M.J. Snyder and R.B. Filbert, J. Phys. Chem. 58, 995 (1954).

12. H. Poignant, J. LeMellot, Y. Bossis, A. Rupert, M. Minier, M. Gauneau, Mat. Sci. Forum 5, 63 (1985).

13. J. Stary, The Extraction of Metal Chelates, Pergamon Press, London, 1964.

14. D.R. Gabbe, Mat. Sci. Forum 5, 85 (1985).

15. M. Robinson, R.C. Pastor, R.R. Turk, D.P. Devor, and M. Braunstein, SPIE Proc. 266, 78 (1981).

DISCUSSION

Q: C. Moynihan. What was the effect of the Pt itself? What was the composition of the O_2/CF_4 atmosphere? Were they previously mixed?

A: M. Robinson. We did not worry about the platinum, because ZrF_4 was going to be sublimed afterwards. No loss of platinum was detected. The atmosphere was a mixture of 50 cm^3/min of O_2 and 350 cm^3/min of CF_4.

Q: R. Almeida. Concerning the distillation from the ZrF_4-BaF_2 melt, what is the state of the contaminant species? Also, how do you dry the ammonium bifluoride?

A: M. Robinson. I expect the contaminant species to be in solution in the melt, although some compound may be formed. The ammonium bifluoride may be purified by distillation.

Q: H. Poignant. Does the fact that you obtain large ZrF_4 crystals pose a problem, since they have to be ground and probably contaminated in the process?

A: M. Robinson. We do not grind the large pieces of ZrF_4 and the losses are small. We compensate for these by adding a little extra ZrF_4.

Q: J. Lucas. What is the melting point of ZrF_4?

A: M. Robinson. In a sealed ampoule, it melts at 940°C under 1.5 atm of pressure.

Q: C. Moynihan. How is the dissolution process for the larger chunks of each component?

A: M. Robinson. When melting is done under a CCl_4 atmosphere, LaF_3 is the last to melt. The other components melt readily.

CHEMICAL ANALYSIS OF TRACE IMPURITIES

PHILIPP H. KLEIN
U. S. Naval Research Laboratory
Washington, D.C.   20375-5000

ABSTRACT
   Ideally, analytical methods for materials used in halide glasses must detect and differentiate impurity levels down to the parts-per-billion or parts-per-trillion level. These sensitivities are required for transition-metal and rare-earth ions, as well as for ammonium, hydroxide, and some other nonmetallic ions. Dissolved oxygen, carbon dioxide, and even argon are also of concern, as is dispersed elemental carbon.
   The materials-purification community has met this broad range of analytical demands by preparing glass samples and determinimg their absorption spectra. While this method may be necessary when total attenuation falls below the 1.0-dB/km range, it is cumbersome, and often nondifferentiating. It would be preferable use sensitive and specific analytical methods to guide purification of raw materials. Five such methods are compared.
   Spark-source mass spectrography and neutron activation analysis give evidence of being useful at the ultimate-purity level (<1ng/g), but they are relatively expensive and have long turnaround times. Atomic absorption and plasma emission spectroscopy are inexpensive and convenient, but lack sufficient sensitivity for use at the highest purity level. Ion-exchange chromatography may combine sensitivity with convenience at the 1-ng/g level, but it has not been widely used in this field. We conclude that analysis is most useful at the start of purification work. Glassmelting, fiber-drawing, and absorption spectrometry will probably continue to guide the attainment of the highest purity levels.

1. INTRODUCTION

   Transparency in fluoride glasses is adversely affected by extremely small amounts of cationic, anionic, particulate, or gaseous contamination. It has been shown that contamination by 3d transition elements [1] and rare-earth ions [2] in the nmol/mol range will prevent attainment of the theoretical limits [3] of transmission in these glasses. Hydroxide, and other oxygen-containing anions are similarly detrimental. Carbon and other particulates decrease effective transparency by introduction of scattering centers.

   Preparation of a glass and measurement of its transparency has served to guide much materials-purification work. The obvious drawbacks - enormous increases in labor, uncertainty as to sources of specific contaminants or their identities -

make it essential   that means be perfected for analysis of
each ingredient as it is processed.

This paper reviews the requirements for any useful
analytical method and compares them with the characteristics
of several sensitive techniques. Examples of analyses of
fluoride-glass components performed by their use are given.
The work concludes with a ranking of the utility of the
methods discussed.

2. ANALYTICAL SENSITIVITY REQUIRED AND AVAILABLE

Table 1 illustrates the severity of the analytical problem
for cations. Even though an absorption peak does not coincide
with the wavelength of interest, the skirts of a large peak

TABLE 1. Concentrations of metals
in ng/(g glass) yielding
0.001dB/km absorption [1, 2, 4].

| Metal | Wavelength, micrometers: | | | | |
	2.0	2.5	3.0	3.5	4.0
Fe	0.1	0.4	5.0	11.	---
Co	0.1	0.3	2.5	13.	---
Ni	0.1	0.6	20.	---	---
Cu	3.3	---	---	---	---
Ce	---	---	1.3	0.2	0.2
Pr	---	5.7	---	2.2	0.5
Nd	---	0.5	16.	---	15.
Sm	2.3	3.9	1.8	8.5	1.4
Eu	0.8	7.4	3.2	0.7	5.4
Tb	0.4	---	0.7	5.8	---
Dy	---	15.	---	9.9	---

"---" = >20 ng/g

may overlap and increase absorption there. By taking into
account this "skirt" absorption, the data in Table 2 were
obtained [5]. While their application to fluoride glasses
may be only qualitative, it is evident that nonmetallic ions

TABLE 2. Nonmetallic ions yielding 0.01 dB/km absorption
in KCl at 2.7 micrometers [5]

Ion	Concentration Yielding 0.01 dB/km, ppb
Hydroxide	0.007
Ammonium	0.02
Cyanate	0.2
Borate	0.7
Hydrosulfide	1.1
Bicarbonate	1.1
Chlorate	2.0

must be measured at extremely low concentrations, if glass transparency is to be assured. Therefore, it is well to seek analytical methods which can yield results in the 0.01-ng/g range.

The implication of this sensitivity requirement is that a truly useful analytical method must be capable of detecting picogram amounts of the ion sought. The extra senstivity is needed to allow for dilution in sample preparation. We have incorporated this requirement in Table 3, along with some of

---

TABLE 3. Desirable features of analytical techniques for materials purification.

1. Capable of determining all ions of concern with 0.1-1-g sample.
2. Simultaneous determination of all impurities in 0.001-1000-ng range.
3. Free of interelement interferences (no matrix effects).
4. Minimal sample preparation or manipulation.
5. Fast enough to guide next step in process perfection.
6. Appropriate cost for equipment, manpower, supplies, etc.

---

the other features desired in an analytical method. Characteristics of analytical methods may be compared with the ideal properties shown in the table.

3. ANALYTICAL METHODS

In this section we discuss the methods listed in Table 4. No more than 0.1-1.0 gram of sample is required for any listed technique. Where listed methods have been successfully applied to materials-purification work, we present illustrative examples.

3.1. Spark-Source Mass Spectrography (SSMS).

From its entry in Table 4, SSMS would appear to be a method of general utility, with few serious problems. One of its principal virtues is the fact that a photographic plate is made during analysis. The darkness of the traces recorded on the plate represent the abundances of every element detected in the sample. Presence of unsuspected contaminants becomes evident by reading these plates.

One example of the completeness with which SSMS reveals impurity spectra is a case in which the gold used to prepare the sample for sparking was found to contain 7 ppm of Fe [6]. Furthermore, samples may be checked at later times for elements whose presence had not been of concern at the time of analysis. An example would be Nd, whose absorption at 2.55 micrometers has assumed great importance as interest in that wavelength has increased.

Matrix and memory effects are sometimes serious in SSMS. If one is analysing a zirconium sample as the oxide, the nonvolatility of the matrix may tend to trap impurities. Nonvolatility of impurity oxides may tend to lead to a low estimate of their abundances. In contrast, analysis of the

same zirconium sample, but after conversion to the fluoride,
would tend to give a different result. Higher volatility of

TABLE 4. Analytical techniques for glasses and
their constituents.

	SSMS	AAS	PES	NAA	IEC
Minimum detectable mass, ng	0.0001	0.01	0.1-10	0.001	0.01-1
Elemental limitations	None	Metals	None	None	None
Matrix complications	Few	Many	Some	Some	Few
Background complications	Few	Few	Many	Some	Few
Simplicity of sample preparation	Fair	Fair	Poor	Good	Fair
Sample in-out time, days	1-2	0.5	0.5	>2	0.5
Cost per analytical sample, $	300-500	150*	150*	200-400	150*

SSMS = Spark-source mass spectrography
AAS = Atomic absorption spectroscopy
PES = Plasma emission spectroscopy
NAA = Neutron activation analysis
IEC = Ion-exchange chromatography

*Assumes ownership of the equipment, and thus assumes an
initial investment of about $35,000-60,000 for AAS; of
$40,000-65,000 for PES; and about $12,000-25,000 for IEC.

both matrix and impurities could lead to a different impurity
profile than was obtained with the oxide.

Matrix effects are less serious when one is concerned
primarily with removal of impurities. Consistent use of the
same matrix for preparation of analytical samples can yield
self-consistent relative data.

Memory effects in mass spectrography are carried-over
indications of elements abundant in previous runs. They are
the result of contamination of beam-forming parts of the
spectrograph by previous samples. For example, if one runs
samples of an oxide after analysing a fluoride, a spurious
indication of fluorine in the oxide is not uncommon. Memory
effects can virtually be eliminated by replacement of slits
and other beam-forming parts of the spectrograph.

Isotope dilution is a method of analysis which makes use of
the mass spectrograph. It has the virtue of not requiring
quantitative isolation of the element to be determined. It
should be immune to matrix effects. The element to be
determined must exist in at least two isotopes. One adds to
the sample a measured "spike" of the element in which the
isotope ratio is not the same as in the sample. The mass
spectrometer is used to determine the isotope ratios in the
sample, spike, and spiked sample. From the three isotope
ratios, one can calculate the amount of the element present in
the unspiked sample.

The general principle of isotope dilution was used by

Fisher et al. [7] to determine the effectiveness of solvent
extraction in removal of Fe from zirconium salts. The authors
added a radioactive iron "spike" to their sample, and
determined Fe concentration with a nuclear counter. By this
sensitive detection method, they were able to demonstrate that
solvent extraction can permit preparation of zirconium
oxychloride containing only 4 pg/g of iron.

Acquisition of a mass spectrograph solely for use in a
glass-research program is uneconomical. It is sometimes
possible to obtain complete analyses of samples within one or
two days of their receipt by the analyst. Costs for this
service are under $300 for reporting all elements (in a list
of eighty) which are present in parts-per-million concentrations. For sensitivity in the parts-per-billion (ng/g)
range, the cost is about $500 for the eighty-element list.

3.2. Atomic Absorption Spectroscopy (AAS).

Virtually every metal can be detected by AAS, at levels near
those desired for fluoride-glass work. Backgrounds and interferences once posed serious problems. However, modifications
of the basic technique – such as Zeeman correction for
spectral-line interferences, or use of a pyrolytic-graphite
furnace for increasing reproducibility of sample
volatilization – have greatly diminished these difficulties.

AAS is available in many laboratories involved in
fluoride-glass research. Where that is the case, low costs
per sample (about $150, neglecting cost of equipment) and
half-day turnaround times are attractive. In addition, the
individual spectral sources required for each element are
likely to be on hand.

The principal disqualifying characteristic of AAS is that
detection limits are above the desired parts-per-billion
level. Purity of aluminum fluoride was measured [8] in a
study of alkyl intermediates. Synthesis of these proved to be
extremely effective as a means for purification – as far as
could be seen by AAS. The cited [8] detection limits in an
aluminum fluoride matrix are, in parts per billion (ng/g):
Fe, 149; Cr, 17; Cu, 35; Mn, 22; and Ni, 47. AAS has also
been used for assessing contamination in borohydride-purified
zirconium fluoride [9].

Comparison with Table 1 shows clearly that AAS is inadequate
for determining ultimate purity in fluoride-glass ingredients.
Nevertheless, its cost and convenience may make it attractive
for guiding initial purification studies.

3.3 Plasma Emission Spectroscopy (PES).

This method represents the most refined form of a convenient
and reasonably sensitive analytical method. As shown in Table
4, it has some serious drawbacks for analysis of final purity.
Nevertheless, because it is available in many laboratories, it
is listed. PES is not sufficiently sensitive to be used for
analyses of fluoride-glass components of ultimate purity.
Still, a form of emission spectroscopy - possibly involving
volatilization by plasma – was used [10] in a study of
purification of zirconium and hafnium fluorides by
sublimation.

In our own laboratory, we have used PES to follow purification of a variety of cations by ion exchange [11]. It has required utmost care to avoid problems with shifting spectral backgrounds. As the detection limit is approached, background intensity often exceeds the increase of intensity at the spectral line of interest. We have found that the background-suppression problem can be as time-consuming as the impurity-removal problem under study. Nevertheless, we have found PES to be quite useful in the initial stages of purification.

3.4 Neutron Activation Analysis (NAA).

Matrix and background problems can seriously negate the sensitivity of NAA. For example, it is virtually impossible to determine directly the 3d transition elements in Zr because the host becomes excessively radioactive. In Hf, the high neutron cross-section of the host complicates the irradiation. However, if one is willing to remove complicating species, the method can be sensitive indeed.

It is unlikely that NAA can be used for routine guidance of a materials-purification program. Matrix problems are one reason. Another is the long time required for shipment, irradiation, and counting of a sample, even when the expected impurity profile is known. The cycle time listed in Table 4 is virtually an ideal; one or two weeks is far more usual. Its sensitivity and accuracy can nevertheless make NAA useful for verifying the effectiveness of a purification scheme.

3.5 Ion-Exchange Chromatography (IEC).

This method has not been applied, to our knowledge, to the materials-purification problems of fluoride glasses. However, as Table 4 illustrates, IEC has several characteristics which may represent improvements over current techniques. It should be at least as sensitive as AAS and PES, have fewer matrix and background complications, and have the short in-out times and low costs which come from its location in or near the purification laboratory.

The method is relatively new [12]. In its most frequent form, IEC involves attachment of host ions and impurity ions to an ion-exchange resin. Care is taken - by resin and eluant selection - to elute only the impurity ions. They are detected by the change in electrical conductance they induce as they pass out of the ion-exchange column.

When properly operated, IEC eliminates matrix problems: the host is separated during the first step. Concentration of impurities on the column increases effective sensitivity. However, optimum separation of impurity ions may be impossible for every resin-eluant combination. Consequently, a complete determination of impurities in a given sample may require use of two or more sequences of resin and eluant. Evolution of these sequences may require considerable effort, possibly modification of a resin or perfection of an eluant mixture.

Once it becomes routine, IEC can function as an integral part of a purification laboratory. Its relatively rapid turnaround time and low acquisition and operation costs can favor such integration. It will require direct experimental investigation to determine the limits of this method.

4. SUMMARY AND CONCLUSIONS

We have discussed five analytical methods which meet some of the requirements for use in fluoride-purification research. Two - spark source mass spectrography and neutron activation analysis - appear to be useful for analysis of materials of the ultimate purity. They are relatively expensive, not often available under the same roof and management as the purification laboratory. Two other techniques - atomic absorption spectrscopy and plasma emission spectroscopy - are only sufficiently sensitive to guide the initial stages of a purification program. They compensate partially for this defect by their general availability and relatively short cycle time and low cost.

The fifth method - ion-exchange chromatography - has yet to be fully investigated for its fluoride-glass applicability. Its known characteristics strongly suggest that it will combine the benefits of relatively rapid and low-cost analyses with a sensitivity and simplicity of technique intermediate between the two groups above.

One hoped-for benefit of an optimum analytical technique is that it would eliminate the need for glassmelting, fiber-drawing, and inrared spectroscopy in order to determine the purity of glass ingredients. It must be concluded, at this writing, that none of the techniques described here can supplant this complex procedure. They can, however, simplify all but the last steps in the iterative purification-verification cycle.

REFERENCES

1. Ohishi, Y., Mitachi, S., and Kanamori, T., Japan J. Appl. Phys. 20, L787 (1981).
2. Ohishi, Y., et al., ibid., L191.
3. Shiibata, S., Electronics Lett. 17, 775 (1981).
4. Ohishi, Y., Mitachi, S., Kanamori, T., and Manabe, T., Phys. Chem. Glasses 24, 135 (1983).
5. Duthler, C. J., J. Appl. Phys. 45, 2668 (1974).
6. Poignant, H., Le Mellot, J., Bossis, Y., Rupert, A., Minier, M., Gauneau, M., Mat. Sci. Forum 5, 63 (1985).
7. Fisher, C. F., Tran, D. C., Hart, P., and Sigel, G. H., Jr., Mat. Sci. Forum 5, 116 (1985).
8. Thompson, D., Mat. Sci. Forum 5, 69 (1985).
9. Bridenne, M., Folcher, G., and Marquet-Ellis, H., Mat. Sci. Forum 5, 59 (1985).
10. Churbanov, M. F., Rudnevsky, N. K., Tumanova, A. N., Zvereva, V. I., and Maslov, Yu. V., Mat. Sci. Forum 5, 73 (1985).
11. Klein, P. H., Nordquist, P. E. R., Jr., and Singer, A. H., Opt. Eng. 24, 516 (1985).
12. Small, H., Stevens, T. S., and Baumann, W. D., Anal. Chem. 17, 1801 (1975).

DISCUSSION

Q: C. Pantano. Can laser-induced mass spectrometry be used?

A: P. Klein. It works well for semiconductors because of the fixed matrix. But this is not true in the present case, where we have a variable matrix. One would also need a laser that is highly absorbed by the fluoride glass.

Q: C. Moynihan. The OH extinction coefficient values are currently being questioned and there is on-going research in this area. Was the use of ion exchange chromatography suggested for OH analysis?

A: P. Klein. I did not suggest that, but, in principle, it can be done.

C: M. Robinson. There are no reproducible ways of measuring ppb levels of OH because of lack of controls.

ROLE OF IMPURITIES IN HALIDE GLASSES

H. POIGNANT
CNET LANNION B - ROC/FOG, Route de Trégastel, 22301 LANNION, France

ABSTRACT

Many impurities are usually encountered in fluoride glasses, coming from both the melting and casting methods used for their fabrication and the starting materials used. Among these elements which contaminate the optical glass quality, the most frequently met are the transition metals, the rare earths, the oxides, the carbon traces and some ions (NH_4^+, Cl^-...) resulting from NH_4HF_2 or CCl_4 treatment. The effects of such impurities on the glass properties (optical, physical...) are discussed. Most of them cause absorption peaks responsible for excess loss in fluoride glass optical fibers, in the 1-4 μm wavelength range (transition metals, rare earths, CO, CO_2...) while extrinsic scattering may appear when using too much ammonium bifluoride in the glass melting process.

1. INTRODUCTION

The fluoride glasses based on ZrF_4 are expected to be quite suitable for the achievement of optical fiber communications operating in the mid-infrared (2-4 μm) since, in such wavelength range, transmission losses much lower than for silica fibers have been predicted. However, though the optical quality of these new glasses has considerably increased these last years, due to improvements in the glassmaking, their ultra low loss properties have not yet been confirmed. Besides the excess scattering losses which are mainly due to glass manufacturing, the other concern is the presence of impurities in the glass. This contamination may cause drastic absorption loss at different wavelengths, depending on the nature of the impurity. This paper deals exclusively with this impurity problem. The results of absorption measurements performed on transition metal (Fe, Co, Cr, Ni) or rare earth (Nd, Ce, Tb, Sm, Er, Eu) doped fluoride glasses are given, as well as the effects of oxides or carbon traces which occur in the melt and will affect the glass properties.

2. NATURE AND ORIGIN OF IMPURITIES

2.1 Transition metals

In this group of impurities, we mainly find the following species: Fe, Co, Ni, Cr, Cu, Mn, V, which essentially originate from insufficient purity starting materials (metallic fluorides such as ZrF_4, BaF_2, AlF_3, LaF_3... or corresponding oxides ZrO_2, Al_2O_3, La_2O_3...).

Typical analysis of some chemical components currently used in the fluoride glass melting are listed in Table I [1].

Though they may be purified, the total impurity amount will likely exceed the ppm level. As a comparison, the 3d transition metal absorptivity measurements carried out by [2] in silica fibers showed that the presence

TABLE 1: Starting materials used and their typical analysis.

Sample	Manufacturer	Impurities (from reported analysis in ppm)						
		Cr	Fe	Co	Ni	Cu	V	Mn
NH_4HF_2 Optipur Ref. 1184	Merck	10^{-3}	10^{-2}	10^{-3}	10^{-3}	10^{-3}	5.10^{-3}	10^{-3}
NH_4HF_2 RP Normapur	Prolabo	0.1	0.5	0.1	0.1	0.1		
ZrO_2 UPH	Criceram	2	5		8	2.5	3	1
ZrO_2 Optipur Ref. 8906	Merck	7	1	0.1	0.1	0.1		5
La_2O_3 5N	O.S.I.							
Al_2O_3 Optipur Ref. 15103	Merck	20	10		1	10		
ZrF_4 ZLA 756	Merck		7	0.15	0.6	0.1		
ZrF_4 OFG 4N Purity	RMC – Total metallic impurity less than 1.6 ppm	1	1		1	1		
AlF_3 OFG 4N Purity	RMC		2		1	0.1		1
LaF_3 OFG 4N Purity	RMC		1					

of such impurities at the ppb level could give db/km absorption loss in the 0.4-1.6 μm wavelength range (see Table 2).

The technical equipment involved in the fluoride glass fabrication constitutes a second source of transition metal impurities. Although platinum, gold, or vitreous carbon are currently used for melting the glass, which, of course, will minimize the potential glass contamination, the glass casting often occurs in metallic molds (brass, e.g.). So, even if they are gold lined (such protective coating having a tendency to disappear after a few pouring operations and to diffuse in the fluoride glass), some metallic particles can appear in the glass which originate from such mold.

TABLE 2: Absorption loss due to 1 ppb of transition metals at
0.85 µm in silica [2].

Ion	Absorption peak wavelength (µm)	Loss at 0.85 µm db/km/ppb
Cr^{3+}	0.625	1.6
Co^{2+}	0.685	0.1
Cu^{+}	< 0.30	0.01
Cu^{2+}	0.85	1.1
Fe^{2+}	1.10	0.68
Fe^{3+}	< 0.40	0.15
Ni^{2+}	0.65	0.1
Mn^{3+}	0.46	0.2
V^{4+}	0.725	2.7

2.2 Rare earth elements

The fluoride glasses most investigated for optical applications (fibers, IR windows) may be divided in two main groups: the ZrF_4 (or HfF_4) and ZnF_2/ThF_4 based glasses. In both cases, the most stable compositions always involve a rare earth fluoride at a level which is not negligible. For example, the ZrF_4 glasses typically contain 3-6 mole % LaF_3 or GdF_3 [3][4] while in the zinc/thorium fluoride ones, YbF_3 or LuF_3 are present at a level up to 25-27 mole % [5]. Thus, the rare earth impurities (Sm, Tb, Ce, Pr, Nd, Dy, Eu...) will likely originate from this specific component. If LaF_3 and GdF_3 do not affect the fluoride glass transmission spectrum over the 0.4-7 µm range, it is quite different with most of the other rare earths which are well known to exhibit strong absorption peaks in the 0.8-6 µm wavelength region. Table 3 lists the absorption peak wavelengths for some troublesome rare earths.

2.3 Oxides

Fluoride glasses are prepared using melting techniques involving starting materials which can be on the fluoride (ZrF_4, BaF_2, AlF_3, NaF, LaF_3...) or oxide (ZrO_2, La_2O_3, Al_2O_3...) form or both, the conversion of oxides into fluorides being ensured by NH_4HF_2.

2.4 Carbon species

Carbon traces are often found in the fluoride glasses, even when the melt is submitted to an oxidizing treatment (O_2, e.g.). Some explanations may be advanced to justify the existence of such particles in the glass:

- use of melting furnace involving graphite or vitreous carbon systems, or graphite susceptor in the case of R.F. heating [6].

- use of a halogenated carbon gas in the reactive atmosphere processing to decrease the OH absorption peak located at 2.9 μm (CF_4, CCl_4, $CClF_3$...).

However, even when the glass fabrication method avoids these two contamination possibilities (use of platinium crucibles exclusively and no RAP), carbon impurities are found too, which indicates that the starting materials are also responsible.

TABLE 3: Rare earth absorption peak wavelengths.

ION	Ce^{3+}	Pr^{3+}	Nd^{3+}	Sm^{3+}	Eu^{3+}	Tb^{3+}
λp(μm)	4.59	4.69	5.07	4.13	5.15	4.65
		1.96	2.51	2.64	3.63	2.94
		1.56	1.72	1.99	3.36	2.29
		1.46		1.51	2.85	1.99
				1.39	2.63	1.88
				1.24	2.21	
				1.09	2.09	

ION	Dy^{3+}	Ho^{3+}	Er^{3+}	Tm^{3+}	Yb^{3+}
λp(μm)	2.83	1.96	1.54	1.68	0.98
	1.71	1.16	0.98	1.22	
	1.29				
	1.10				

2.5 OH⁻/Water

Typical IR transmission spectra of fluoride glasses show an absorption band around 2.9 μm, which is attributed to the fundamental OH stretching vibration. In silica glass fibers, the presence of such hydroxil species is well known to cause strong absorption loss, as indicated in Table 4.

In fluoride glasses, this water contamination may be attributed to different sources:

- starting materials: The fluoride components are not completely dehydrated or have absorbed water when handling...

- glass processing: a) The atmosphere around the glass is not dry enough.

b) NH_4HF_2 treatment for oxides → fluorides conversion generates water which may be incorporated in the glass melt. In fact, for ZrO_2(e.g.), the fluorination reaction is expressed as:

$$2ZrO_2 + 7NH_4HF_2 \rightarrow 2(NH_4)_3ZrF_7 + NH_3 + H_2O$$
$$\hookrightarrow ZrF_4 + 3NH_4F$$

(1)

TABLE 4: Absorption loss resulting from 1 ppm(OH^-) in silica [7]
- F_3 = Fundamental OH vibration
- F_1 = SiO_4 tetrahedron vibration
- nF_i = Harmonic n (F_i)

Wavelength (µm)	Frequency	Absorption loss (db/km)
0.72	$4F_3$	0.07
0.82	$2F_1 + 3F_3$	0.04
0.88	$F_1 + 3F_3$	0.09
0.945	$3F_3$	1
1.13	$2F_1 + 2F_3$	0.11
1.24	$F_1 + 2F_3$	2.8
1.38	$2F_3$	65
2.73	F_3	10^4

2.6 Ionic species

This group includes chlorine, ammonium, sulfate and phosphate ions.

3. EFFECT OF IMPURITIES

3.1 Effect of transition metals [8]

The influence of transition metals on the transmission loss in fluoride glasses was determined only for Co, Ni, Fe, Cr. Difficulties were encountered to achieve good quality Cu-doped glasses. The experimental conditions are given in Table 5.

TABLE 5: Transition metal doped fluoride glasses

Doping transition metal	Concentration c(wt %)	sample thickness(mm)	
		t_1	t_2
Fe	0.67	21	9.5
Co	0.11	21	9.4
Ni	0.28	39	21
Cr	0.02	49.5	39.5

The absorption spectra where recorded using a CARY 14 spectrophotometer, working in the 0.4-2.2 µm range. The absorption loss, ν, due to one ppm Fe, Co, Ni, Cr was deduced from the expression:

$$\nu = 10^3 \; \epsilon \; \frac{d}{Mi} \; (db/km)$$

where:

d is the glass density,

Mi is the impurity molar weight,

ε is the extinction coefficient deduced from the following equation:

$$\epsilon = \frac{1}{c} \frac{OD_1 - OD_2}{t_1 - t_2} \qquad \text{in mol}^{-1} \text{ cm}^{-1} \text{ liter,}$$

OD_i and Li being respectively the optical density and length of sample i at a given wavelength. Figure 1 shows the absorption loss spectra for Fe, Co, Ni, Cr in a ZBLA fluoride glass matrix, while Table 6 summarizes the results compared to those published by [9] and [10].

TABLE 6: Absorption loss due to transition metals (db/km/ppm) at different wavelengths (+ : NF_3 atmosphere).

WAVELENGTH /μm

ELEMENT	2.0			2.5			3.0			3.5		
(Ref.)	(9)	(10)	(8)	(9)	(10)	(8)	(9)	(10)	(8)	(9)	(10)	(8)
Fe	90		75	28			2			0.9		
Co	130		44	31			4			0.75		
Ni	90		50	6	2.6 / 2.4+		0.5			0.05	0.09 / 0.07+	
Cu			3	0.14	4.1 / 2.7+		0.01			1.2 / 7.1		
Cr					6.3 / 10^{-3+}					2.7 / 10^{-3+}		

The 2.0 μm absorption losses due to Fe, Ni and Co appear to be lower than those reported by Y. Ohishi [11]. This may result from slightly different experimental melting conditions.

3.2 Effects of rare earth elements

Up to now, the only work devoted to the study of the absorption loss involved by rare earth ions in a fluoride glass was performed by [9]. We decided to carry out the same study at CNET with the following rare earths: Ce, Nd, Sm, Tb, Er, Eu, Pr. All the glasses were fabricated under the same experimental conditions (N_2/O_2 melting atmosphere; melting time: 20 mn; doped glass charge: 25 g). The glass, which was R.F. heated, was allowed to cool- in the crucible. After annealing, all the samples were polished

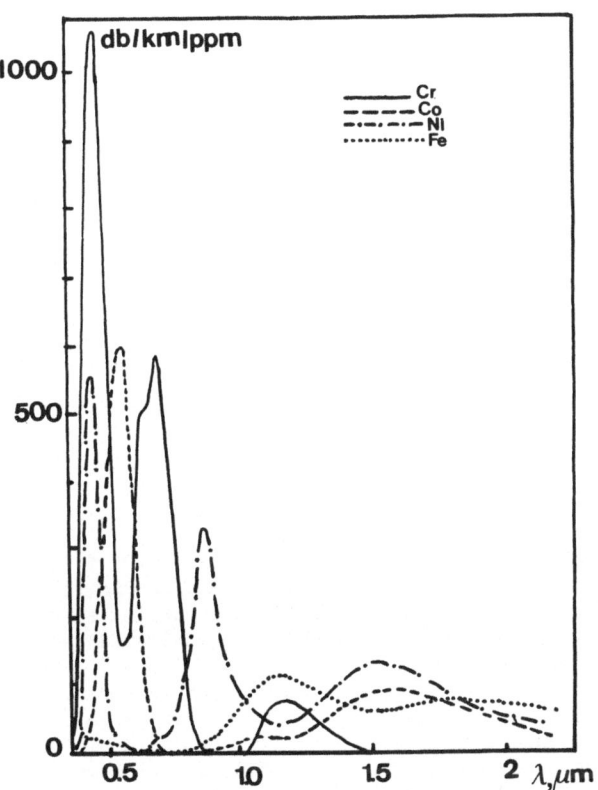

Figure 1: Absorption loss spectra for
Fe, Ni, Co, Cr in ZrF_4 based
glass [8].

and thinned to a 7 mm thickness (diameter : 25 mm). The Table 7 presents the different rare earth concentrations studied.

The absorbance spectra were recorded in the 7000-1000 cm^{-1} frequency range (1.40-10.0 µm) using a Fourier transform spectrophotometer (Bruker IFS 113). The figures 2.a, b., c., d., e., f., g., show the typical spectra for the 4 % rare earth doped glasses.

On Figure 3 we have plotted the optical density (cm^{-1}) of doped samples versus the rare earth addition (in weight %) for Eu, Tb and Er. The straight lines obtained (similar plots were obtained with Ce, Nd, Sm and Pr) indicate that the absorption losses due to rare earth in the fluoride glasses vary linearly with the dopant concentration.

TABLE 7: Rare earth doped fluoride glasses (ZBLAN matrix).

Element	Rare earth content (weight %)				
Ce	0.5	1	1.5	4	
Pr	0.1	1	2	4	
Nd	0.1	0.2	1	2	
Sm	0.22	1	2	4	
Eu	0.1	0.2	1	2	4
Tb	0.24	1	2	4	
Er	0.2	1	2	4	

From a general point of view, our values are 20-25 % lower. It must be recalled that the dopant concentrations are related to the starting batch and maybe some rare earth fluoride would have sublimated when preparing the glass. However, we can estimate that the indicated values are in agreement with the real rare earth glass content within 8-10 %.

Table 9 lists the absorption loss due to one ppmw rare earth at different wavelengths while Table 10 presents the rare earth concentration (in ppbw) required to give 10^{-2} db/km absorption loss in the 2-4 µm range. As can be seen, Nd^{3+} appears the most troublesome since only 0.48 ppb are tolerated at 2.5 µm, which is the wavelength where minimum loss occurs for the floride glass fibers.

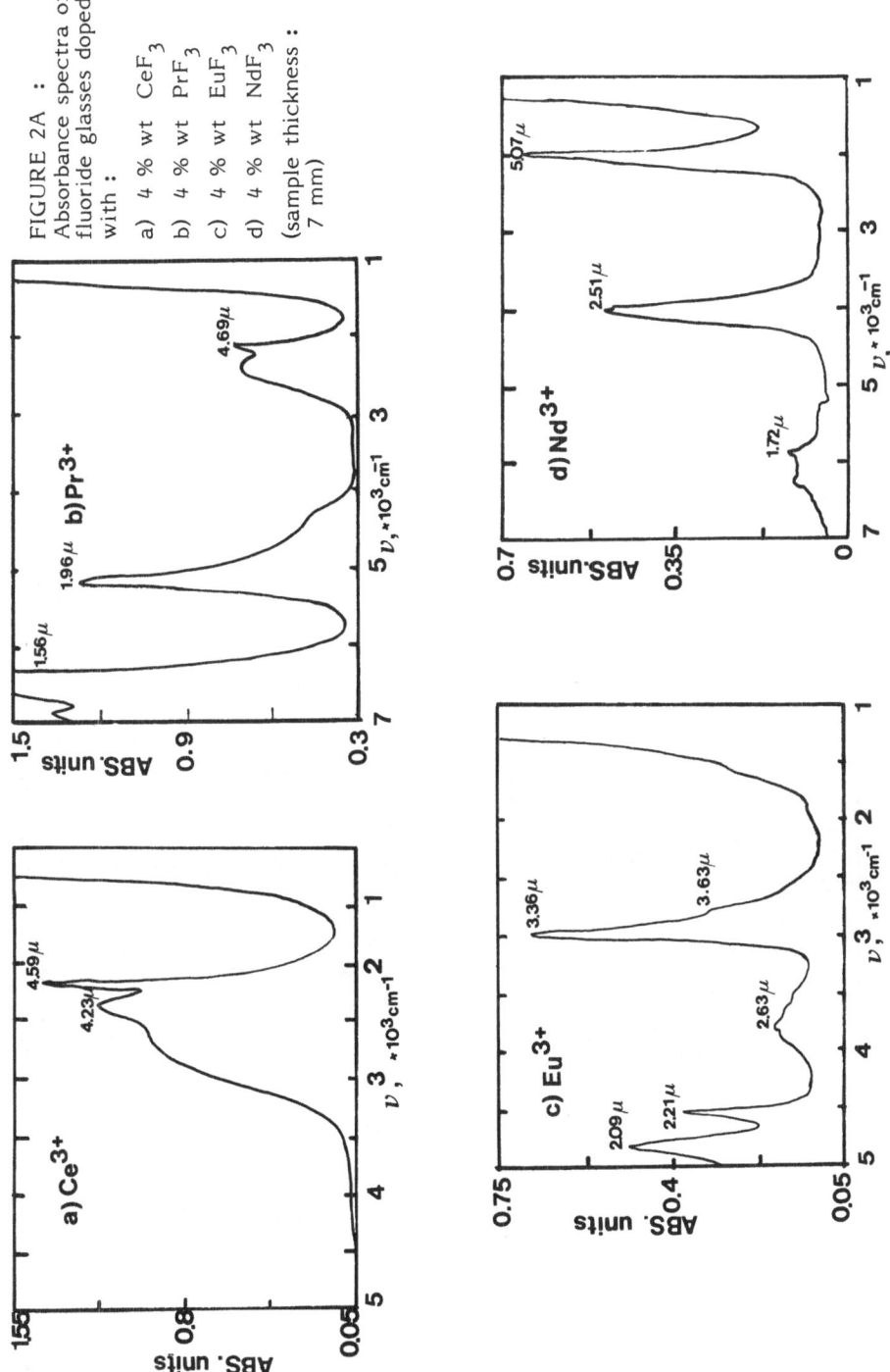

FIGURE 2A :
Absorbance spectra of fluoride glasses doped with :

a) 4 % wt CeF$_3$
b) 4 % wt PrF$_3$
c) 4 % wt EuF$_3$
d) 4 % wt NdF$_3$

(sample thickness : 7 mm)

44

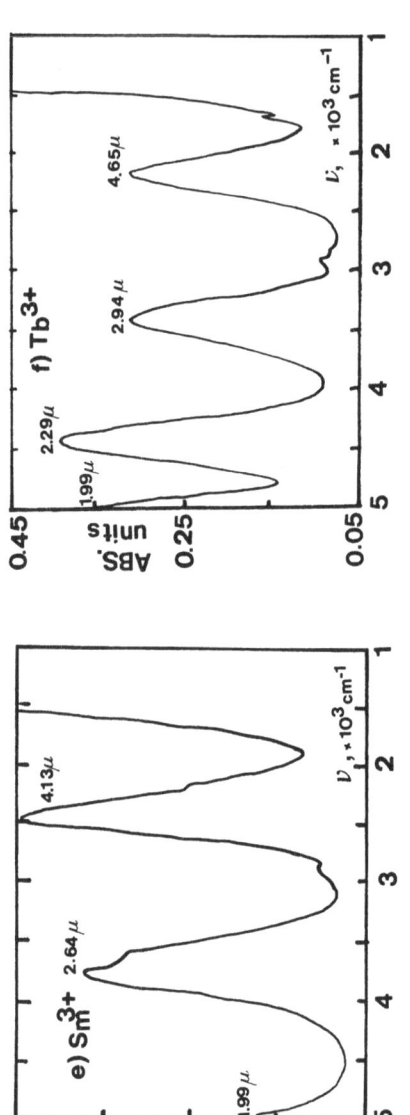

FIGURE 2B :

Absorbance spectra of fluoride glasses doped with

e) 4 % wt SmF$_3$

f) 4 % wt TbF$_3$

g) 4 % wt ErF$_3$

(sample thickness : 7 mm)

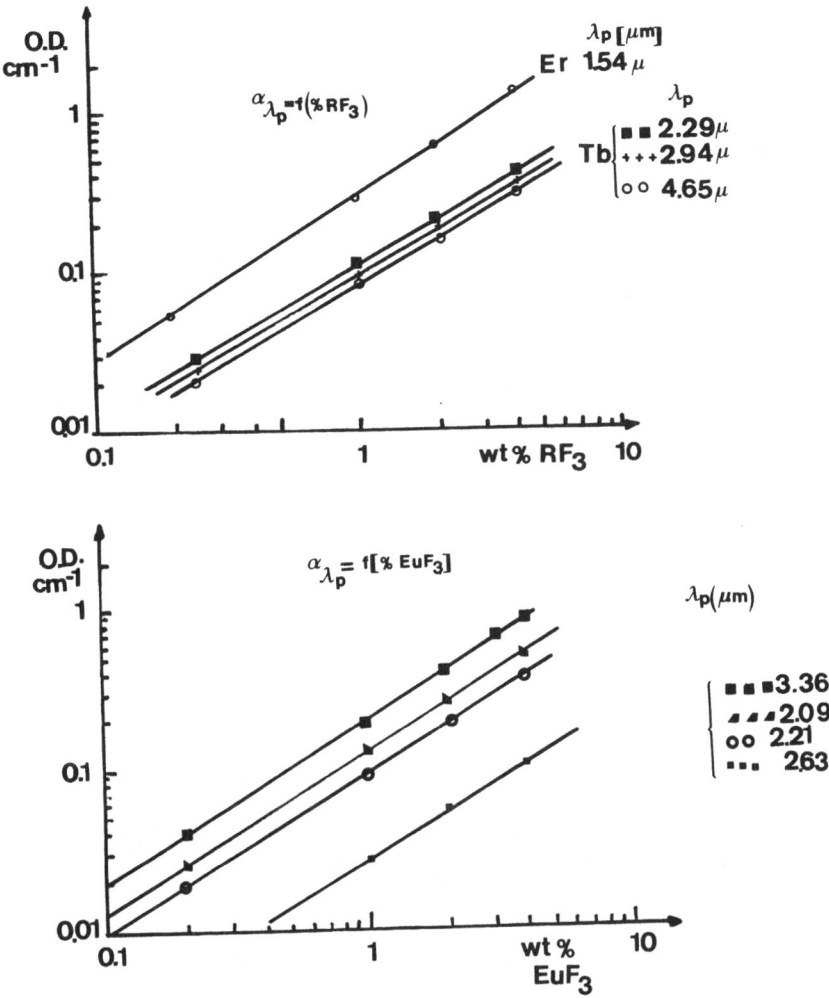

FIGURE 3 : Plot of optical density (cm^{-1}) versus rare earth addition for ErF_3, TbF_3 and NdF_3 at their absorption peak wavelengths.

TABLE 8: Absorption loss due to one ppmw rare earth at their
peak wavelengths.

Ion	$\lambda_p(\mu m)$	db/km/ppm	
		Y. OHISHI	This work
Ce^{3+}	4.59	60.8	46.3
Pr^{3+}	4.69	21.5	15
	1.96	5.3	3.5
Nd^{3+}	5.07	31.7	20
	2.51	21.1	16
	1.72	4.02	1
Sm^{3+}	4.13	8.4	6
	2.64	6.23	5
Eu^{3+}	3.63	9.13	7.8
	3.36	25.2	20
	2.63	3.43	2.8
	2.21	15.3	10
	2.09	20.8	14
Tb^{3+}	4.65	8.01	8
	2.94	14	9.3
	2.29	13.1	11
Er^{3+}	1.54	38.6	32

TABLE 9: Absorption loss (db/km) due to one ppmw rare earth at different
wavelengths (n.m= not measured)

Wavelength (µm) Ion	2.0		2.5		3.0		3.5		4.0	
	Ref. 9	This Work	Ref. 9	This Work	Ref. 9	This Work	Ref. 9	This Work	Ref. 9	This Work
Ce^{3+}	-	-	-	-	7.7	4	50	28	50	37.8
Pr^{3+}	27	21	1.8	1.2	0.4	0.3	4.5	3.5	20	13
Nd^{3+}	-	0.3	20	21	0.6	0.4	-	-	0.7	-
Sm^{3+}	4.4	2.5	2.6	2	5.6	3.5	1.2	2	7.2	6
Eu^{3+}	12	13	1.4	1.5	3.1	2	14.3	11.8	1.8	1.5
Tb^{3+}	25	20	-	-	14.3	12.7	1.1	0.8	-	1.1
Dy^{3+}	0.06	n.m	0.7	n.m	8.3	n.m	1	n.m	0.2	n.m

TABLE 10: Rare earth concentration (ppbw) which would give
10^{-2} db/km loss at 2-2.5-3-3.5-4 μm (this work)

Wavelengths (μm) / Ion	2	2.5	3	3.5	4
Ce^{3+}	-	-	2.5	0.36	0.26
Pr^{3+}	0.5	8	33	3	0.76
Nd^{3+}	33	0.48	25	-	-
Sm^{3+}	4	5	2.9	5	1.7
Eu^{3+}	0.76	6.7	5	0.85	6.7
Tb^{3+}	0.5	-	0.79	12.5	9.1

3.3 EFFECT OF OH HYDROXYLES AND WATER

3.3.1 2.9 μm absorption band (fundamental OH stretching vibration)

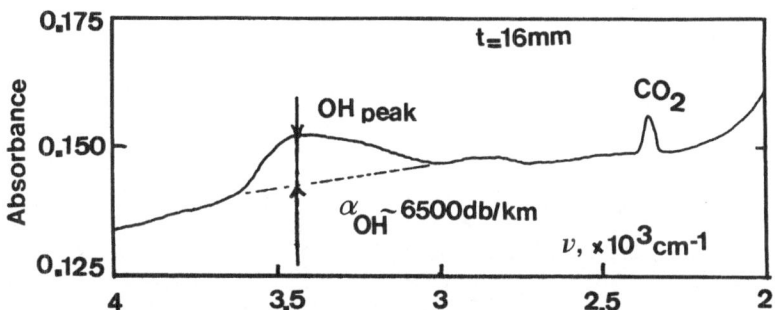

Figure 4: Typical fluoride glass spectrum in the 2000-4000 cm^{-1}
frequency range.

The figure 4 shows a typical absorbance fluoride glass spectrum, with
the well known OH peak centered at about 2.9 μm. The importance of such
absorption peak is strongly depending on the glass processing conditions:
fluorination temperature, atmosphere, melting time and temperature, melting furnace system... Figure 5   has plotted the 2.9 μm absorption
peak intensity for different experimental conditions (fluorination tempera
ture, inert gas flow), the glasses being prepared using a resistance heating system with gas flowing directly above the melt. As can be seen, the
lowest loss ($\sim$ 1800 db/km) was obtained for moderate fluorination temperature ($\sim 300°C$) and gas flow (1-2 ℓ/min N_2). The platinum crucible assembly
capacity was about 250 cm^3.
   The effect of melting time, melting temperature, and batch size upon
the 2.9 μm attenuation was reported by [13].

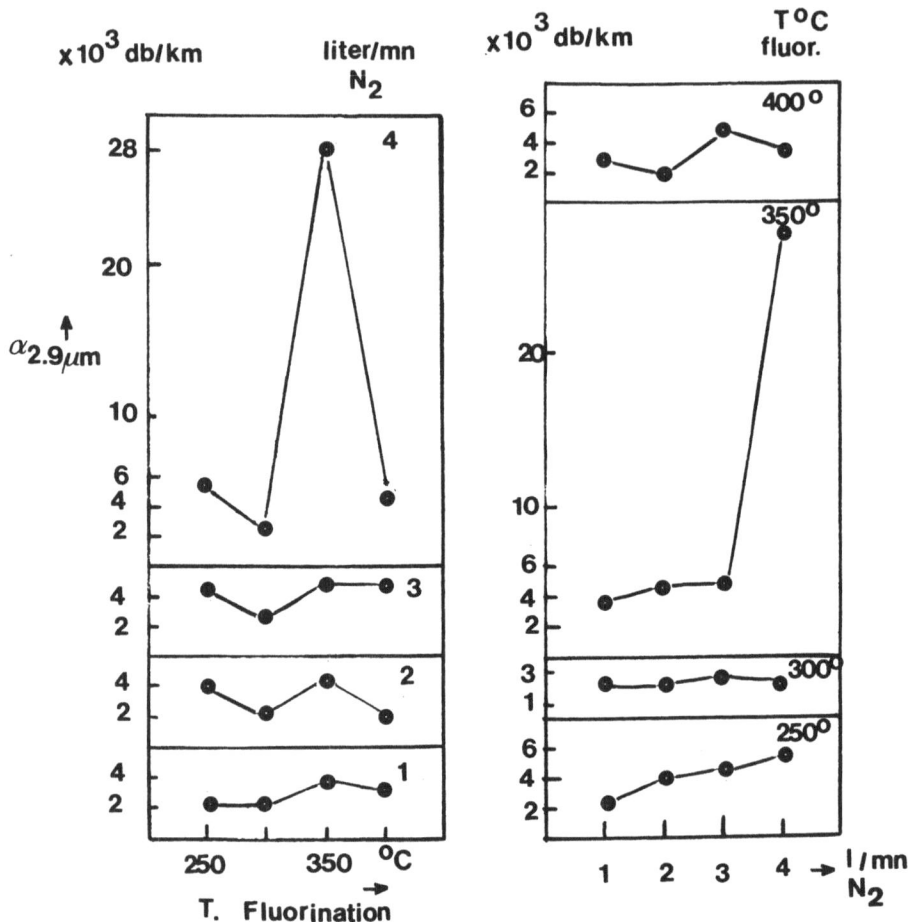

FIGURE 5: 2.9 μm absorption peak intensity versus fluorination
temperature and gas flow [2].

When using a melting furnace system involving the platinum crucible
maintained in a dry atmosphere vessel (the gas flows around the glass and
not directly in the melt), lower absorption peaks were achieved
($\alpha_{2.9\mu m} \sim 500$ db/km; sometimes, the peak is not detected, see Figure 6).
The expression between the 2.9 μm absorption and the OH content is not
yet clearly defined, though some authors have tried to estimate the loss
due to one ppm OH⁻ at this wavelength. Their results are shown in Table
11.

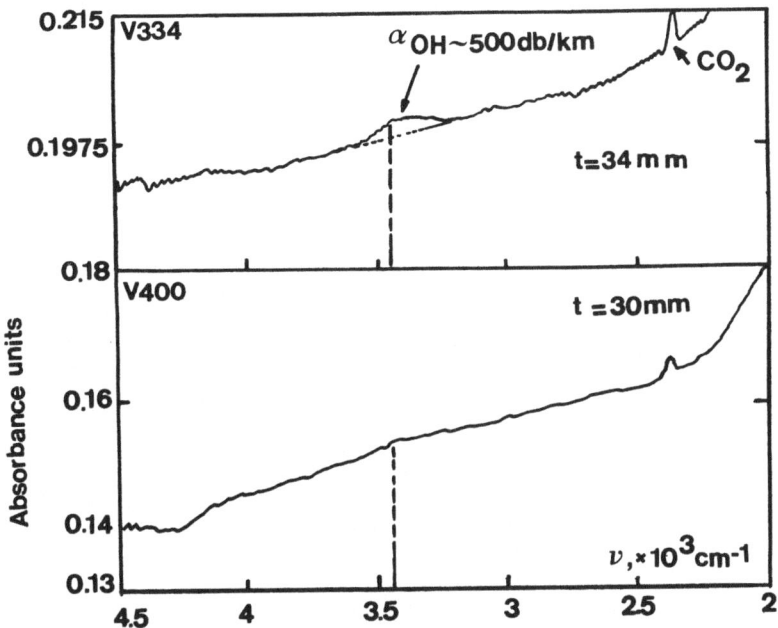

FIGURE 6: Absorbance spectra of two low-OH content fluoride
glass samples.

TABLE 11: Absorption loss due to one ppm OH^- at 2.9 μm in fluoride
glasses.

Glass	$\alpha_{OH}(2.9\ \mu m)$ db/km/ppm	Reference
ZBL	135	14
ZBLA	5200	15
BTYbZ	11690	15
ZBGdA	4370	15 (from NTT communication)

3.3.2 Reactions with water

Besides the 2.9 µm absorption, water in fluoride glasses may cause the formation of alien species in the melt, according to the following chemical reactions between the fluoride components (MF_n, where $MF_n = ZrF_4$, BaF_2, AlF_3, LaF_3...) and H_2O [13]:

$$MF_n + H_2O \rightarrow MF_{n-1} OH + HF \tag{2}$$

$$MF_n + H_2O \rightarrow MF_{n-2}O + 2HF \tag{3}$$

with an intermediate stage:

$$MF_n OH \rightarrow MF_{n-2}O + HF \tag{4}$$

Thus, it appears clearly that oxyfluorides ($MF_{n-2}O$) or hydroxyfluorides ($MF_n OH$) can be generated in the glass melt depending on the temperature and, consequently, will likely create excess scattering and absorption loss, refractive index inhomogeneities...

This is illustrated in figure 7, where both the scattering and absorption loss measured on bulk samples ($35 \times 5 \times 5$ mm^3 size) are plotted versus the NH_4HF_2 amount used for fluorinating the starting oxides (ZrO_2, Al_2O_3, La_2O_3). As reported in equation (1), (section 2.5) this fluorination generates water which may react with the fluorides as shown by equations (2)(3)(4).

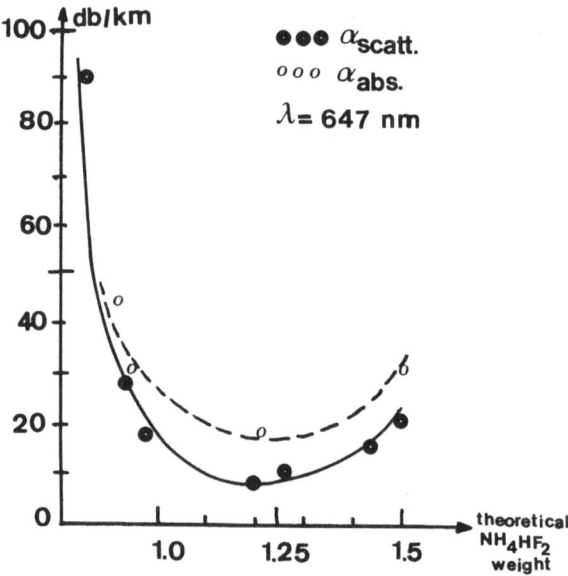

Figure 7: Effect of NH_4HF_2 (and subsequent water) on fluoride glass absorption and scattering loss [16].

As can be observed, there is a minimum loss in the curve associated with the most convenient NH_4HF_2 quantity which may correspond to the lowest hydroxyfluoride or oxyfluoride formation.

3.4 Effect of carbon traces

The occurrence of carbon in a fluoride glass melt will yield two chemical reactions with both CO and CO_2 formation as indicated below:

a) Reaction with oxide impurities:

$$C + O_{melt} \rightleftarrows CO_{gas}$$

$$CO_{gas} + O_{melt} \rightleftarrows CO_{2gas}$$

b) Reaction of CCl_4 RAP gas with the melt water and hydroxyles such as:

$$CCl_{4gas} + 2OH^-_{mel} \rightleftarrows CO_{2(g)} + 2HCl_{gas} + 2Cl^-_{melt} \qquad (5)$$

$$CCl_{4gas} + 2H_2O_{gas} \rightleftarrows CO_{2(g)} + 4HCl_{gas} \qquad (6)$$

Undoubtedly, absorption peaks owing to CO or CO_2 gas will appear in the transmission spectra at characteristic frequencies (2352 cm^{-1} and 2220 cm^{-1} for CO_2 and CO/CO_2 intermediate, respectively [17]). Though the latter one is less observed, the former appears for most of the fluoride glasses, even if carbon or graphite equipment are avoided and whatever may be the melting atmosphere (neutral or oxidizing). The figure 8 shows the typical CO_2 peaks observed in the transmission spectra of fluoride glass samples with different thicknesses.

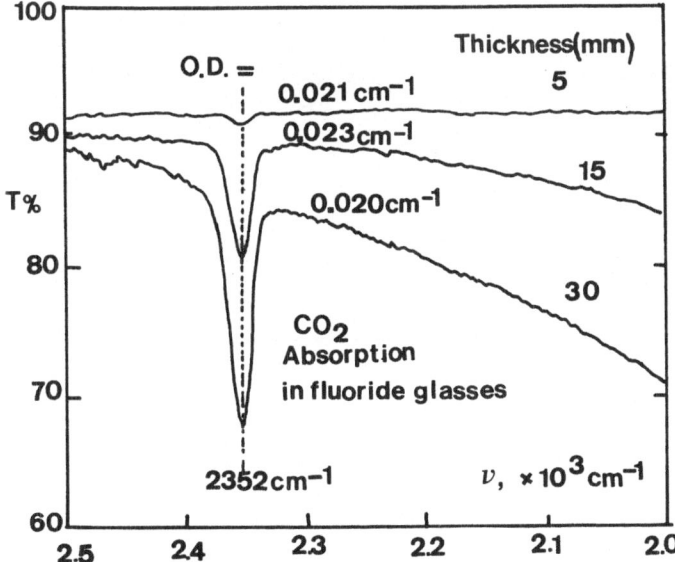

FIGURE 8: Effect of gaseous CO_2 dissolved in fluoride glasses of different lengths (5, 15, 30 mm).

The absorbance intensity was found to be proportional to the thickness, indicating that the CO_2 was dissolved inside the bulk glass; the corresponding absorption loss was estimated at about 21000 db/km at the peak wavelength. Given the experimental glass preparation conditions, such CO_2 likely results from chemical reactions between carbon traces from starting materials and oxides (or oxygen) in the melt.

3.5 Effect of oxides (and oxygen)

In fluoride glasses, oxygen may occur under numerous species: anions (OH^-, SO_4^{2-}, PO_3^-), oxides (ZrO_2, Al_2O_3, La_2O_3...), oxyfluorides or dissolved gas...

The effect of oxides in a fluoride glass is clearly seen in Figure 9 which shows the absorbance spectra for two samples LA_1 and LA_2, respectively, doped with 0.25 and 0.50 wt % La_2O_3. Examination of this figure allows the following assessment:

- excess absorption loss occurs in the 2.8-4 µm wavelength range, which may be evaluated to approximately 0.5-1 db/km/ppm oxide.

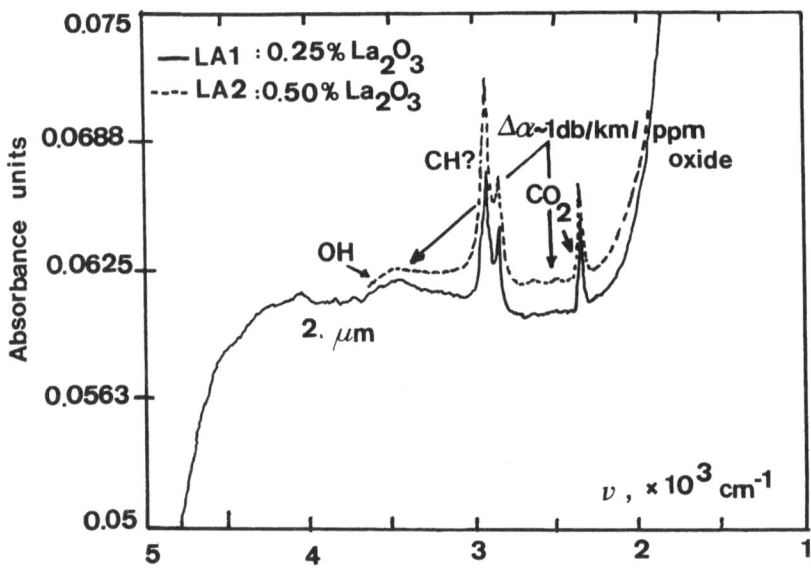

FIGURE 9: Effect of La_2O_3 doping in the absorption fluoride glass spectra.

Oxygen and oxides will induce excess scattering loss too, as can be seen in Figure 7 and as was reported by Mitachi [18].

3.6 Effects of some ionic species

3.6.1 Ammonium ions (NH_4^+)

Since NH_4HF_2 is always used in the fluoride glass preparation in large or small amounts for ensuring the local oxide fluorination, it is not surprising that the glass melt may be contaminated by NH_4^+ ions, due to the involved chemical reactions (see equation (1) for example, section 2.5). The effect of such ions in fluoride glasses was studied by France [19], which pointed out two absorption peaks located at 2.96 and 3.04 µm due to N-H vibration modes.

3.6.2 Chlorine ions (Cl^-)

The occurence of chlorine in fluoride glass is mainly associated with the use of chlorinated gas (such as CCl_4, $CFCl_3$...) for two purposes:

a) Decreasing the OH^- absorption (see equations section 3.4)

b) Oxidizing some reduced species in the glass melt, e.g. for ZrF_4:

$$ZrF_{4-x} + x/2 \ Cl_2 \rightarrow ZrF_{4-x} \ Cl_x$$

The effect of this chlorine incorporation in the glass is not yet well defined and not enough studies have dealt with this problem. The work performed by Neilson [20] only indicates that chlorine can affect the crystallization products when the glass is heated at $350^\circ C$. In fact, both α - and β - $BaZrF_6$ phases appear in heated glasses while only β phase is observed in Cl free glass. Up to now, nothing may be concluded regarding the chlorine effects upon the fluoride glass stability and crystallization behaviour.

3.6.3 Complex anions (SO_4^{2-}, PO_3^-)

Besides the well known impurities which are the most studied and are the focus of great attention in order to decrease them (transition metals, rare earths, carbon, oxides, OH...), others are also currently encountered in the fluoride glass starting materials: sulfates, phosphates, at the trace level. The effect of such complex anions was investigated by Poulain [14] who deduced that quite important absorption loss may occur in the 4-5 µm wavelength range due to such species. His results are summarized in Table 12.

TABLE 12: Absorption loss due to complex anions (db/km/ppm) [14]

Ions	λ in µm		
	4.0	4.2	4.7
Sulfates (S)	45	230	700
Phosphates (P)	6	45	900

CONCLUSION

Most of the impurities which may be found in the fluoride glasses have been reviewed and their effects on the glass properties investigated. If low loss fluoride glass fibers(say $\sim$ 0.01 db/km at 2.5-2.6 µm) are to be achieved, this will necessitate drastic efforts in the fluoride starting materials purification and in the glass processing. The impurity concentration required to produce such a low loss at these wavelengths should be reduced in fact at very low a level for most of them as indicated in Table 13.

TABLE 13: Impurity concentration (ppbw) which would give 10^{-2} db/km absorption loss at 2.5 µm.

Impurity	Fe	Co	Ni	Cu	Nd	Pr	Sm	Eu
Concentration (ppbw)	0.35	0.3	1.6	70	0.5	8	2.9	6.7
Reference		[9]				This work		

The most troublesome, Fe, Co, Nd should not exceed the few tenths of ppb level, while Sm and Ni will be tolerated at a 1-2 ppb concentration. This clearly shows the difficulties to solve, without forgetting the scattering loss which will have to be decreased down to the intrinsic value ($\sim 10^{-2}$ db/km at 2.5 µm [21]).

REFERENCES

1. Poignant H, el al. : Analysis of fluoride glass starting materials. Proc. 3rd Int. Simp. Halide glasses, Rennes (France/1985).
2. Schultz P.C.: J. Am. Ceram. Soc., 57,7 (1974).
3. Lecop A., et al. : Lanthanum fluorozirconate glasses. J. Non Cryst. Sol., 34 (1979) p. 101.
4. Miyashita T., et al. : Infrared optical fibers. IEEE Trans, Microwave Theory Techn., MTT 30, 10 (1982).
5. Drexhage M.G., et al. : J.Am. Ceram. Soc., 65, C168 (1982) p. 1420.
6. Drexhage M.G., et al. : CO_2 absorption in HMF glasses. Proc. 3rd Int. Symp. Halide Glasses, Rennes (France/1985).
7. Kaiser P., et al. : J. Opt. Soc. Am., 63.9 (1973) p. 1141.
8. Maze G., et al. : Contract report LVF/CNET nº 8235232.
9. Ohishi Y., et al. : Optical absorption of 3d transition metals and rare earth elements in ZrF_4 glasses. Phys. Chem. Glasses, 24, 5 (1983) p. 135.
10. Nakai T., et al. : Redox states and absorption losses of 3d transition metals in fluoride glasses. J.J.A.P., 24, 9, (1985) p. L714.
11. Ohishi Y. et al. : Impurity absorption loss in the IR region due to 3d transition elments in fluoride glass. JJAP, 20, 3 (1981).
12. Falcou C. : Internal CNET study, not published.

13. Maze G. et al. : Reduction of OH absorption in fluoride glasses. J. Lightwave Techn., Lt2, 5 (1984) p. 596.
14. Poulain M. et al. : Absorption loss due to complex anions in fluoro-zirconate glasses. J. Lightwave Technol., Lt2, 5 (1985) p. 599.
15. Fonteneau G. et al. : Determination du coefficient d'extinction molaire de OH^- dans les verres fluorés à base de métaux lourds. Mat. Res. Bull. 20, (1985) p. 1047.
16. Poignant H.: Internal CNET study, not published.
17. Moore M.W., et al : Gaseous absorption in ZrF_4 based glasses. Proc. 3rd Int. Symp. Halide glasses, Rennes (France 1985).
18. Mitachi S., et al. : Relationship between scattering loss and oxygen content in fluoride glasses. J.J.A.P., 24, 10 (1985) p. L827.
19. France P.W., et al. : NH_4^+ absorption in fluoride glass infrared fibers. Communications of the American Ceramic Society., Nov., (1984), p. C244.
20. Neilson G.F., et al. : Effect of chlorine incorporation on the crystallization of ZBLA fluoride glasses. J. Am. Ceram. Soc., 68, 11 (1985) p. 629.
21. Poignant H., : Dispersive and scattering properties of a fluoride glass. Electron. Lett., 17, 25/26, (1981), p. 978.

DISCUSSION

Q: G. Frischat. Is there any corrosion of platinum or gold crucibles by heavy metal fluoride glasses?

A: H. Poignant. Apparently, there is no such problem.

VIBRATIONAL SPECTROSCOPY STUDIES OF HALIDE GLASS STRUCTURE

RUI M. ALMEIDA

Centro de Física Molecular, Complexo I, Instituto Superior Técnico, Av. Rovisco Pais, 1000 Lisboa, PORTUGAL

ABSTRACT

Vibrational spectroscopy is one of the most powerful techniques for the study of glass structure. This article presents infrared absorption and polarized Raman scattering spectroscopy results for a series of heavy metal halide glasses, including binary glasses based on HfF_4, ThF_4, $ZnCl_2$ and $ZnBr_2$.

The short range structures which best agree with the vibrational spectra are discussed, based on a model which includes two types of stretching force constants and one bending force constant. Tentative conclusions are derived concerning the main structural features which appear to favor glass formation in halide systems.

1. INTRODUCTION

Halide glasses comprise several different glass-forming systems, including the more ionic fluorides and chlorides or the more covalent bromides and iodides, as well as several mixed halide systems. With the exception of vitreous BeF_2 ($v-BeF_2$), which is a *weakened version* of the structure of $v-SiO_2$[1], all the other pure halides are either marginal glass formers or they do not form glass even under moderately severe quenching rates. There may be several reasons for such behavior, according to each case: (1) simple lack of glass-forming ability, requiring extremely high quenching rates and drastically limiting the sample size, as in the case of ThF_4 or $ZnBr_2$, (2) sublimation under 1 atm of external pressure, as for ZrF_4 and HfF_4 or (3) the requirement of extremely anhydrous conditions, as in the case of $ZnCl_2$. Therefore, the simplest heavy metal halide systems which can easily be prepared and studied are binary, including a *network-forming* and a *network-modifying* halide. The present paper discusses the short range structures of such binary systems, considering a typical example in each case and taking only the most stable and stoichiometrically simple compositions.

One of the objectives of this paper will be the identification of the main structural features (if any) which appear to be common to the most stable halide glass-forming compositions. Coordination of the netwok-forming cation is hardly a major factor, since it can vary from approximately fourfold in zinc halide-based glasses to six or sevenfold in fluorohafnates and fluorozirconates, or eightfold in thorium fluoride-based glasses. On the other hand, although the simultaneous presence of *bridging* and *non-bridging* anionic species appears to be important, i.e. *modified* glasses are always more stable than the pure network-forming compounds, the particular value of the average *bridging* angle θ (M-$\bar{X}$-M bridges, with M=Zr, Hf, Th, Zn and X=F, Cl, Br, I) may not be important. This varies widely, from close to $180°$ for fluorohafnate[2] or ThF$_4$-based glasses[3], to close to $110°$ in the anion-random-close-packed v-ZnCl$_2$[4] (and presumably also in other ZnCl$_2$ or ZnBr$_2$ - based glasses), passing through intermediate values for v-BeF$_2$ at $\sim 156°$ [4]. Therefore, one should look for common structural features in the main types of structural units which stable modified glasses tend to form.

2. EXPERIMENTAL

The fluoride glass samples were melted ($\sim$10g batches) in a dry-box filled with high purity nitrogen, at $\sim$900–1000°C and they were cast between stainless steel plates without annealing. The raw materials were HfF$_4$ of optical grade (Cerac Pure, 99.9%) further purified by sublimation, optical grade ThF$_4$ (Cerac Pure, 99.99%) and Ultrapure PbF$_2$ (Alfa Products). The zinc chloride and zinc bromide-based glasses were melted at $\sim$400°C in sealed pyrex ampoules (8 mm ID) after vacuum dehydration and they were cooled in air. Their Raman spectra did not reveal the presence of water. The different glass compositions are given in Table 1.

TABLE 1. Compositions of halide glass samples.

Sample	Composition (mol %)						
	HfF$_4$	ThF$_4$	PbF$_2$	ZnCl$_2$	ZnBr$_2$	KBr	KI
A	70		30				
B	43	57					
C				50			50
D				50		50	
E					50		50

Infrared absorption spectra were recorded for the fluoride glasses as thin films or KBr pellets in a Perkin-Elmer 683, grating, double-beam spectrophotometer above 300 cm^{-1} and in a Perkin-Elmer 180, double-beam spectrophotometer for polyethylene pellets, below 300 cm^{-1}. Raman spectra were taken for small fluoride glass plates or encapsulated zinc halide-based glasses at $90°$, in a Cary 82 Raman spectrometer, with the 488.0 nm or the 514.5 nm lines of an Argon ion laser, in the HH and HV configurations[3]. All spectra were recorded at room temperature.

3. RESULTS AND DISCUSSION
3.1. Fluoride glasses

Fig.1 shows the IR absorption and polarized Raman spectra of
lead *dihafnate* glass A, which was representative of the most
stable modified fluorohafnate and fluorozirconate glass compositions. The two spectra were mutually exclusive, except for

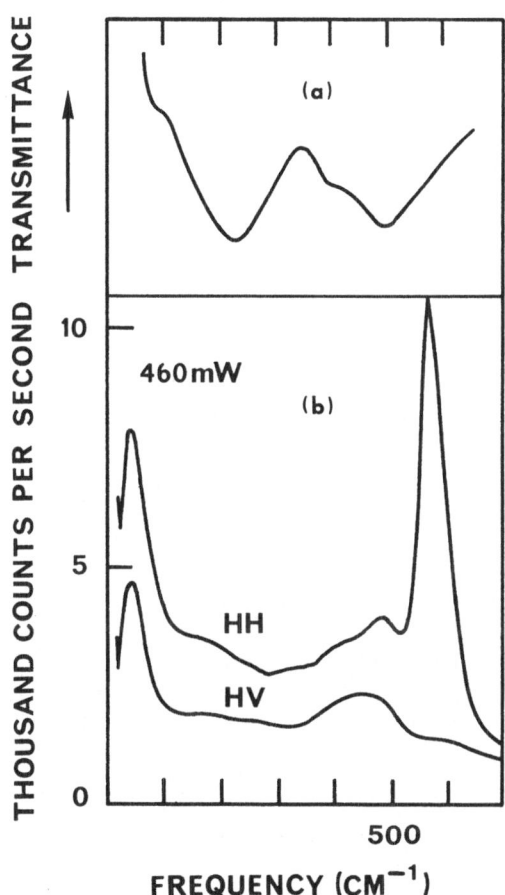

Fig. 1. a) IR transmission and b) polarized
Raman spectra of 2 HfF$_4$ · PbF$_2$ glass.

the strong 490 cm^{-1} IR band which appeared to be weakly Raman
active. A comparison with the spectra of lead *dizirconate*
glass[5] showed that the dominant Raman band at 575 cm^{-1} did not
involve Hf(Zr) cation motions. A previous X-ray diffraction

study of the structure of *isomorphic* barium *dihafnate* glass[2] yielded a Hf coordination number of ~7.7. However, the occurren ce of a single, sharp (FWHM= 63 cm^{-1}), completely polarized (depolarization ratio ~0.1) Raman line involving no Hf cation motion, near 580 cm^{-1} - the same frequency of the dominant Raman bands of six-coordinated Li_2ZrF_6 and Cs_2ZrF_6 crystals[6] - is in best agreement with highly symmetrical F atom environments about Hf(Zr), predominantly sixfold coordinated. Fig. 2 shows a possible structure compatible with the 2:1 stoichiometry and all the available vibrational data for fluorohafnate and fluorozirconate glasses[5]. The possibility that, for each

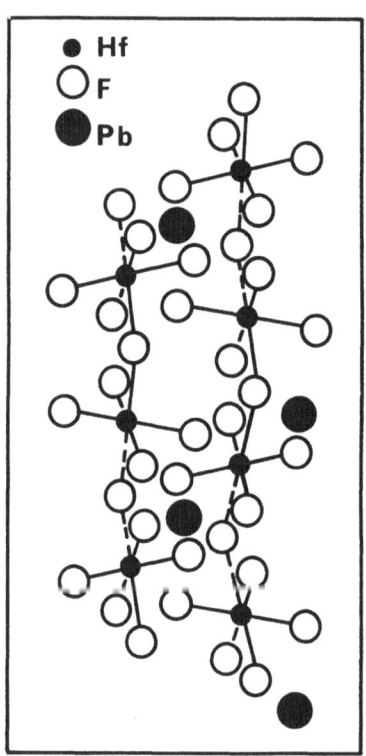

Fig. 2. Chain-like structure of 2 HfF$_4$·PbF$_2$ glass.

Hf atom, there were one or two additional F atoms from neighboring chains which randomly overlaped at slightly larger distances (~ + 0.01 nm), might explain the higher Hf coordination number obtained by X-ray diffraction[7]. From the diffraction results, one also calculated a bridging angle θ at ~ 172° for barium *dihafnate* glass, which appeared to be *isomorphic* with lead *dihafnate* glass. In conjunction with a neutron diffraction analysis of the barium *dizirconate* composition[8], this shows that the anion packing is not compact in these glasses, as one would expect if the Hf coordination number were larger than six.

The vibrational spectra of Fig. 1 may be interpreted via a generalization of the central force network model of Sen and Thorpe[9,10], which includes two types of stretching force constants and a bending force constant[10]. According to the original model[9], which was developed for AX_2 network glasses (e.g. SiO_2 or BeF_2) where the non-central forces were neglected, there is a *critical bridging angle* $\theta_c = \cos^{-1}(-2m/3M)$, m and M being the masses of the bridging ànion and the network-forming cattion, respectively, below which *molecular* effects predominate and above which *solid-state* effects predominante. Taking this critical angle as a guideline in the present case (despite the structural differences of the glass networks), one has $\theta_c = 94^\circ$ for the lead *dihafnate* glass (or any other modified fluorohafnate glass) and one would have $\theta_c = 98^\circ$ for a fluorozirconate glass. Therefore, solid-state effects with extended vibrational modes are expected to predominate in the glass of Fig. 1, for which the average angle θ is likely to be close to 180° ($\gg 94^\circ$). The 575 cm^{-1} dominant Raman band, however, involves fairly localized motions of *non-bridging* F atoms (F_{nb}) about fixed Hf atoms, which do not depend upon θ. This mode corresponds to the symmetric stretch of F_{nb} atoms[7] - SS(F_{nb}) - with an angular frequency given by[10]:

$$\omega^2_{SS(F_{nb})} = \frac{k_1}{m_F} \tag{1}$$

where k_1 is the *non-bridging stretching* force constant and m_F is the mass of the fluorine atom. Thus, for $\bar{\nu}_{SS(F_{nb})} = \omega_{SS(F_{nb})}/2\pi c = 575$ cm^{-1}, one calculates $k_1 = 370$ Nm^{-1}. The highest frequency IR mode is an antisymmetric stretch of *bridging* F atoms (F_b) perpendicular to the bissector of θ, with simultaneous Hf cation motion in the opposite direction - AS(C) - whose angular frequency is given by[10]:

$$\omega^2_{AS(C)} = \frac{k_2}{\mu}(1 - \cos\theta) \tag{2}$$

where k_2 is the *bridging stretching* force constant and μ is the reduced mass of the F and HfF$_4$ *particles*. Here, for $\bar{\nu}_{AS(C)} = 490$ cm^{-1} and $\theta = 172^\circ$, one finds $k_2 = 126$ Nm^{-1}, although the exact value will be slightly dependent upon the actual average value of θ; when θ varies between 170° and 180°, k_2 will vary only between 126 Nm^{-1} and 125 Nm^{-1}. Finally, the lower frequency strongly IR-active mode at 235 cm^{-1} is a symmetric stretch of F_b atoms parallel to the bissector of θ, with simultaneous Hf cation motion in the opposite direction - SS(C) - whose angular frequency is given by[10]:

$$\omega^2_{SS(C)} = \frac{1}{\mu} [k_2(1+\cos\theta) + 4k_3(1-\cos\theta)] \tag{3}$$

where k_3 is the *bending* force constant. For $\bar{\nu}_{SS(C)} = 235\ cm^{-1}$ and $\theta = 172°$, one finally has $k_3 = 7\ Nm^{-1}$. The reason why the IR absorption edge of fluorohafnate glasses is only shifted to lower frequencies, relatively to fluorozirconate glasses, by a few wavenumbers, follows directly from eq. (2), given the negligible difference in reduced mass. On the other hand, the value of the force constant ratio $k_1/k_2 \sim 2.9$ suggests[11] that the *bridging/non-bridging* bond length ratio is relatively large (~ 1.4). This is consistent with a bridging bond order less than one, corresponding essentially to *one-electron* bonds between Hf(Zr) and F_b.

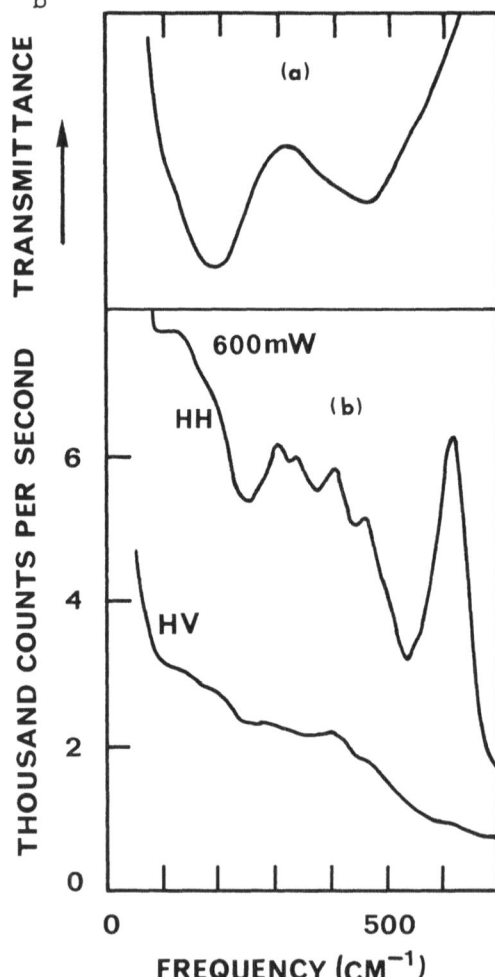

Fig. 3. a) IR transmission and b) polarized
Raman spectra of $43\,HfF_4 - 57\,ThF_4$ glass.

Fig. 3 shows the infrared absorption and polarized Raman
spectra of thorium-fluorohafnate glass B. This glass, which
was considerably less stable than HfF_4 - PbF_2 glasses, was representative of *non-modified* glasses where both HfF_4 and ThF_4
appear to behave as network-forming compounds[3]. A comparison
with the spectra of thorium fluorozirconate glasses[12] showed
that the high frequency completely polarized Raman band involved no appreciable Hf cation motion. The intensity of this
band relative to the intermediate frequency region of the
Raman spectrum was smaller than in lead fluorohafnate glasses,
but it increased with the HfF_4 content in the binary thorium-
-fluorohafnate system[3], showing that it was also due to a symmetric stretch of $Hf-F_{nb}$ bonds with fixed Hf atoms, designated
by $SS(F_{nb}^{Hf})$. The 45 cm^{-1} frequency increase of this mode relative to glass A could be due either to (a) a decrease in Hf coordination number, which is unlikely since fivefold coordinated
Hf is not known to occur, or to (b) an increase in the degree
of bridging of the glass network, which is likely to be the
case. Therefore, the Th-containing glass must have a considerably lower concentration of F_{nb} atoms than the Pb-containing
glass, but the Hf atoms will still be mostly sixfold coordinated as in modified fluorohafnate glasses.

Here, the *critical bridging angle* θ_c will be equal to 94°
and 93° for Hf-F-Hf and Th-F-Th bridges, respectively, with a
similar value for possible Hf-F-Th bridges. Therefore, given
the much larger values of the average bridging angle θ which
are predicted below, again solid-state effects will predominate.

Using the same generalized central force network model as
before, eq. (1) gives the Hf *non-bridging stretching* force
constant k_1^{Hf} at 430 Nm^{-1}. The increase, relatively to glass A,
is due to the increase in the degree of bridging for the network of glass B. The polarized Raman band at 464 cm^{-1}, which
does not involve Hf cation motion[12], can be attributed[3] to a
symmetric stretch of $Th-F_{nb}$ bonds with fixed Th atoms, designated
by $SS(F_{nb}^{Th})$. Eq. (1) yields, for the Th *non-bridging stretching*
force constant k_1^{Th}, a value of 241 Nm^{-1}. It is possible that
this band had a small contribution due to weak Raman activity
of the strong infrared mode at 465 cm^{-1}, which can be assigned
to an antisymmetric stretch of F_b atoms perpendicular to the
bissector of θ^{Hf}, with simultaneous Hf cation motion, designated by $AS(C^{Hf})$. The other strong, partially resolved infrared
band at 374 cm^{-1}, which occurred near the strongest IR mode of
crystalline ThF_4 at 368 cm^{-1} [12] and whose intensity increased
with the ThF_4 content of the glasses[3], can be attributed to a
similar vibration involving Th-F-Th bridges, designated by
$AS(C^{Th})$. The presence of Hf-F-Th bridges was not unambiguous-
ly revealed by the IR spectra and the origin of the 190 cm^{-1}
IR mode is not known at present.

The value of the average bridging angle is not known in the
case of thorium fluorohafnate glasses, nor are the values of
k_2^{Hf} or k_2^{Th}. However, one may assume that these two force constants scale approximately as the *non-bridging stretching* force
constants and, therefore, a system of two equations may be set
up based on eq. (2), taking the reduced masses μ as equal,

since both will be very close to the mass of the F atom. One then obtains cos θ^{Hf} as a function of cos θ^{Th}. Since crystalline ThF$_4$ is isomorphic with ZrF$_4$[13], for which the average angle $\theta = 164.2°$[14] and given the fact that the strong, high frequency IR mode of glass B (374 cm^{-1}) virtually coincides with that of crystalline ThF$_4$ (368 cm^{-1}), as opposed to the case of crystalline HfF$_4$ whose high frequency IR mode[3] (498 cm^{-1}) does not coincide with the highest frequency band of glass B (465 cm^{-1}), one can use the value of 164.2° for θ^{Th} to obtain a semi-quantitative estimate of θ^{Hf} at 134°. This shows that θ^{Hf} is considerably smaller than θ^{Th}, but still substantially larger than θ_c for both Hf -F- Hf and Th-F-Th bridges ($\sim 93°$). The assignment of the lower frequency Raman-active modes is difficult at present and is not highly relevant for the structural analysis.

The spectral characteristics of thorium fluorohafnate glasses and the probable values of the bridging angles suggest that Th atoms will predominantly be surrounded by eight F atoms, a majority of which being of the bridging type. The average bridging angle θ^{Th} is probably close to the value found in fully bridged crystalline ThF$_4$ ($\sim 164°$). Most Hf atoms appear to be sixfold coordinated as in lead fluorohafnate glasses. Given the overall anion/cation ratio of 4 in these non--modified glasses, it is also likely that the majority of the F atoms around each Hf will be of the bridging type, corresponding to a larger degree of bridging than in modified fluorohafnate glasses, which is consistent with a drop in θ^{Hf}. The proposed structure has an overall three-dimensional character, in which both Hf and Th appear to behave as network-forming cations. When compared with the essentially one-dimensional modified fluorohafnates, one concludes that the three-dimensional character of the non-modified glasses led to a loss of orientational freedom, which decreased the glass-forming ability.

3.2 Chloro-bromo-iodide glasses

Fig. 4 shows the Raman spectrum of chloro-iodide glass C, which is a modified chlorozincate glass; other glasses can also be prepared with KBr or KCl replacing KI, although with lower thermal stability. The spectrum of Fig. 4 was dominated by a highly polarized high frequency band at 294 cm^{-1} plus a highly polarized intermediate frequency region around ~ 150cm^{-1}. For another glass with 50 ZnCl$_2$ - 50 KBr, for example, the high frequency Raman band was unchanged, but the intermediate frequency region was somewhat modified and shifted to higher frequencies. On the other hand, in the Raman spectrum of pure vitreous ZnCl$_2$[15], a dominant band occurred at 228 cm^{-1} and it could be ascribed to a symmetric stretch of bridging chlorine atoms Cl$_b$ along the bissector of the Zn-Cl-Zn bridging angle. A comparison of the two spectra showed that *modification* introduced a polarized high frequency band which can be assigned to a *symmetric stretch* of non-bridging chlorine atoms Cl$_{nb}$, a species which is essentially absent in v-ZnCl$_2$. The high intensity intermediate frequency region of Fig. 4, also highly polarized, could in principle be due to symmetric stretches of

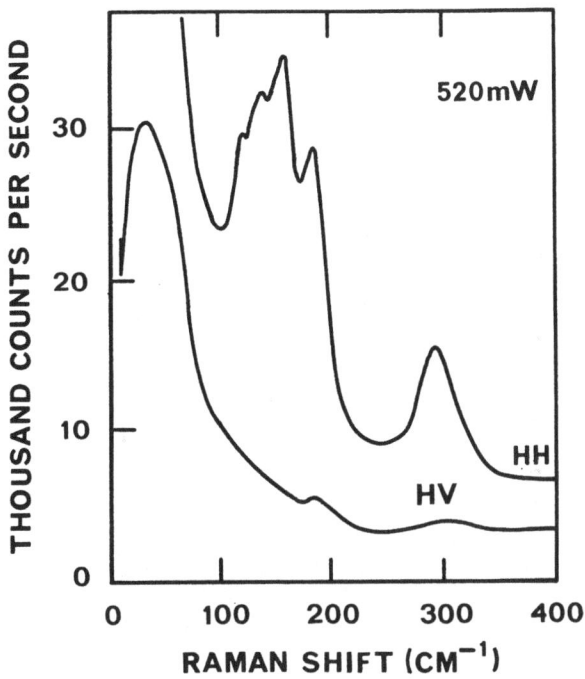

Fig. 4. Polarized Raman spectrum of
50 ZnCl$_2$ - 50 KI glass.

bridging and non-bridging iodine atoms or Cl$_b$ atoms with a
larger bridging angle than v-ZnCl$_2$.

Before attempting a more complete assignment of the vibrational modes of Fig. 4, it is useful to examine the Raman
spectrum of glass E with 50 ZnBr$_2$ - 50 KI, shown in Fig. 5,
which was a modified bromozincate glass. The appearance of
this spectrum was rather similar to that of v-ZnCl$_2$[15], although there is in principle no connection between the two.
In fact, the spectrum of Fig. 5 is also dominated by a completely polarized band, although at a frequency (139 cm^{-1}) lower
than in v-ZnCl$_2$ (228 cm^{-1}).

The totally symmetric (A$_1$) mode of ZnX$_4^{2-}$ tetrahedra (X = Cl,
Br, I) occurs at 276 cm^{-1}, 172 cm^{-1} and 122 cm^{-1} for ZnCl$_4^{2-}$,
ZnBr$_4^{2-}$ and ZnI$_4^{2-}$ ions, respectively[16]. Therefore, in addition
to the previously assigned symmetric stretch of Cl atoms at
294 cm^{-1} and taking into account the effects of bridging in
the glasses, one might tentatively associate the polarized
139 cm^{-1} mode of Fig. 4 with a symmetric stretch of I$_{nb}$ atoms
and the polarized 178 cm^{-1} mode of Fig. 5 with a symmetric
stretch of Br$_{nb}$ atoms. The polarized 185 cm^{-1} mode of the

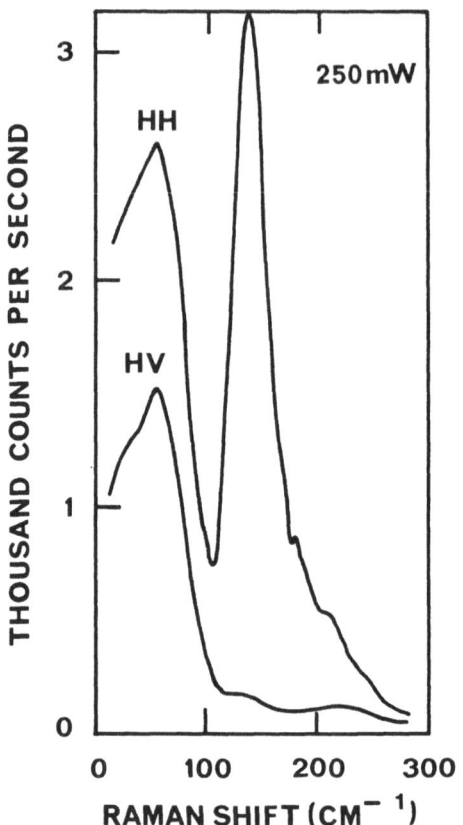

Fig. 5. Polarized Raman spectrum of
50 ZnBr$_2$ - 50 KI glass.

chloro-iodide glass may be due to a symmetric stretch of Cl$_b$
atoms (with a lower frequency than in v-ZnCl$_2$ due perhaps to a
larger bridging angle in the modified glass) and the dominant,
highly polarized 139 cm^{-1} mode of the bromo-iodide glass may
be due to a symmetric stretch of Br$_b$ atoms, which occurs at
150 cm^{-1} in liquid ZnBr$_2$[17].

The bromine and iodine-containing glasses are peculiar in
the sense that the network-forming cations are lighter than the
anions. This has important consequences with respect to the
values of the critical bridging angles, which are calculated
at 111° for Zn-Cl-Zn bridges, 145° for Zn-Br-Zn bridges
and >180° (physically meaningless) for Zn-I-Zn bridges. The
average bridging angle has been estimated at ∿110° for v-ZnCl$_2$
and it has a value near 106° for crystalline ZnBr$_2$, in which
the Br atoms form a cubic close-packed lattice with fourfold
coordinated zinc[18]. It is likely that the vitreous form of
this compound also has an average θ near the tetrahedral value.

Even if the average bridging angle increased somewhat with the structural breakdown in the modified glasses, it is likely that the vibrational spectra of these zinc halide-based glasses are dominated by *molecular*-like effects, with the relati vely uncoupled modes, particularly those involving Br and I atoms, resembling those of distorted ZnX_mY_n tetrahedra. Such reasoning is supported by the observation of sharp bands in the spectra of Figs. 4 and 5. The occurrence of mixed $ZnCl_mI_n$ and $ZnBr_mI_n$ structural units, however, complicates an attempt to such a type of analysis, given the fact that the vibrational frequencies of such ions, to the best of the author's knowledge, are not available.

Some force constant and frequency calculations, based on the generalized central force network model, will help in the assignment of the spectra. Starting with the spectrum of v-$ZnCl_2$, the frequency of the $SS(Cl_b)$ mode at 228 cm^{-1} may be given by[10]:

$$\omega^2_{SS(Cl_b)} = \frac{1}{m_{Cl}} [k_2(1+\cos\theta)+2k_3(1-\cos\theta)] \qquad (4)$$

If one takes the average θ at 110°, after Wright et al[4] and if one assumes $k_3/k_2 = 0.056$, as in modified fluorohafnate glas ses, one finds $k_2^{Cl} = 134$ Nm^{-1} and $k_3^{Cl} = 8$ Nm^{-1}. Also, ap plying eq. (1) to Cl_{nb} atoms, the $SS(Cl_{nb})$ frequency of glass C at 294 cm^{-1} yields $k_1^{Cl} = 181$ Nm^{-1}.

Before making a final assignment of the Raman spectrum of the chloro-iodide glass, it is useful to analyse the spectrum of the bromo-iodide glass E, for which the assignment is more straightforward. The non-bridging stretching force constant of Zn-Br_{nb} bonds can now be estimated by assuming that k_1^{Br}/m_{Br} and k_1^{Cl}/m_{Cl} scale approximately as the square of the fre quencies of the A_1 modes of $ZnBr_4^{2-}$ and $ZnCl_4^{2-}$ complex ions[16] at 172 cm^{-1} and 276 cm^{-1}, respectively. This yields $k_1^{Br} = 158$ Nm^{-1} which, when introduced into eq. (1), gives an $SS(Br_{nb})$ frequency of 183 cm^{-1}. This confirms that the 179 cm^{-1} shoulder in the spectrum of Fig. 5 was indeed due to the symmetric stretch of Br_{nb} atoms about fixed Zn atoms, as suggested above. If one now assumes that the k_1/k_2 ratio for Zn-Br bonds is the same as for Zn-Cl bonds (181 Nm^{-1}/134 Nm^{-1} = 1.35), the bridging stretching force constant for Zn-Br-Zn bridges is readily calculated at $k_2^{Br} = 117$ Nm^{-1}, from which the corresponding value of $k_3^{Br} \sim 0.056$ k_2^{Br} is estimated at 7 Nm^{-1}. Then, taking $\theta = 110^o$ as in v-$ZnCl_2$ ($\theta = 106^o$ for crystalline $ZnBr_2$), the application of eq. (4) to Br atoms yields the $SS(Br_b)$ frequency at 142 cm^{-1}, remarkably close to the frequen cy of the dominant Raman band of glass E (139 cm^{-1}). Converse ly, taking 139 cm^{-1} as the $SS(Br_b)$ frequency, one obtains a better estimate of the average bridging angle θ at 112°. The very high intensity of the 139 cm^{-1} Raman band suggests the existence of a large concentration of Zn-Br-Zn bridges. Gi ven the stoichiometry of glass E($KZnBr_2I$), its network should be formed by $(ZnBr_2I)_n^{n-}$ anionic units; if one further assumes that Zn is always 4-fold coordinated (as in $ZnCl_2$, $ZnBr_2$ and

ZnI$_2$), then a *chain-like* structure naturally follows, analogous to that of a metassilicate. The 4-fold coordination for Zn in glass C is evident from the fact that the SS(Cl$_{nb}$) frequency of glass C was even higher than in tetrahedral ZnCl$_4^{2-}$ complexes. Since the present results strongly favor bridging via Br$_b$ atoms, the glass structure will probably include *chains* of distorted [ZnBr$_3$I] tetrahedra, most of which with two Br$_b$, one Br$_{nb}$ and one I$_{nb}$ atom each, cross-linked by K-Br and K-I ionic bonds, as previously proposed[19]. Although the spectrum of Fig. 5 was similar to those of molten ZnBr$_2$ and v-ZnCl$_2$, it is not possible at present to exclude the occurrence of a small fraction of I$_b$ atoms as well. This relatively simple model is depicted in Fig. 6. For such a structure, one

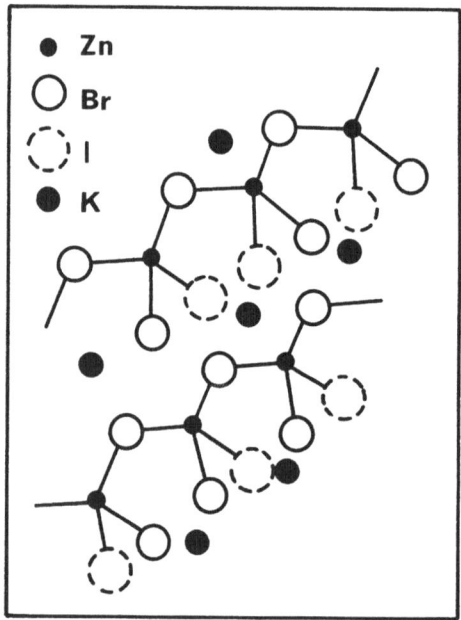

Fig. 6. Schematic 2-dimensional drawing of the
structure of 50 ZnBr$_2$ - 50 KI glass.

might expect to see a weakly Raman-active antisymmetric stretch of Br$_b$ atoms perpendicularly to the bissector of the Zn-Br-Zn angle about fixed Zn atoms, with a frequency given by[10]:

$$\omega_{AS}^2 = \frac{k_2}{m_{Br}} (1 - \cos \theta)$$

(5)

at 185 cm^{-1}, which may perhaps exist in Fig. 5. Finally, the symmetric stretch of I_{nb} atoms was also expected to occur, at a frequency slightly higher than in $ZnI_4{}^{2-}$ complex ions[16] (122 cm^{-1}), as was the case with $SS(Cl_{nb})$. Such frequency may be furnished by the spectrum of glass C.

The Raman spectrum of the chloro-iodide glass (Fig. 4) was considerably more complicated than that of the bromo-iodide glass, suggesting a more complex structure for the former. When that spectrum was compared with the Raman spectrum of v-$ZnCl_2$, the absence of extensive Cl atom bridging became apparent. This was confirmed by the spectrum of glass D, which exhibited two closely spaced peaks at 202 cm^{-1} and 217 cm^{-1}, not far from the $SS(Cl_b)$ of v-$ZnCl_2$ at 228 cm^{-1}. Therefore, the 50 $ZnCl_2$ - 50 KI glass has less Cl atom bridging than the amount of Br atom bridging in 50 $ZnBr_2$ - 50 KI glass, compensated by a significant fraction of I atom bridging. In the Raman spectrum of Fig. 4, after the assigment of the 294 cm^{-1} band, one has to look mainly for symmetric stretches of *bridging* Cl and I atoms, as well as the symmetric stretch of *non-bridging* I atoms. The $SS(Cl_b)$ mode is most likely associated with the sharp 185 cm^{-1} peak; the decrease in frequency from the 228 cm^{-1} value of v-$ZnCl_2$ may be due to an increase in the average Zn-Cl-Zn bridging angle θ^{Cl}. From eq. (3) and the previously calculated k_2^{Cl} and k_3^{Cl}, the 185 cm^{-1} frequency yields $\theta^{Cl} = 131°$. Using eq. (5) for Cl atoms with this value of θ, one calculates the frequency of $AS(Cl_b)$ at 326 cm^{-1}, which might be hidden under the 294 cm^{-1} band, given its weak activity. For the iodine-related modes, the non-bridging stretching force constant of $Zn-I_{nb}$ again can be estimated by assuming that k_1^{I}/m_I and k_1^{Cl}/m_{Cl} scale approximately as the square of the frequencies of the totally symmetric modes of ZnI_4^{2-} and $ZnCl_4^{2-}$ complex ions[16] at 122 cm^{-1} and 276 cm^{-1}, respectively. This yields $k_1^{I} = 126$ Nm^{-1}. When eq. (1) is applied to iodine atoms with this k_1 value, a frequency of 130 cm^{-1} is obtained for the $SS(I_{nb})$ mode. A similar reasoning to that used above for bromine atoms will now lead to $k_2^{I} = 94$ Nm^{-1} and $k_3^{I} = 5$ Nm^{-1}. Using $\theta^{I} = \theta^{Cl} = 131°$ as a first estimate, one would obtain $AS(I_b)$ at 144 cm^{-1} and $SS(I_b)$ at 82 cm^{-1}, two results which do not quite match the experimental spectra. However, given the semi-quantitative nature of several of the estimates, the 158 cm^{-1} and 139 cm^{-1} modes will tentatively be assigned to $AS(I_b)$ and $SS(I_{nb})$, respectively. Therefore, the 139 cm^{-1} band of Fig. 5 is expected to contain also a contribution from $SS(I_{nb})$ vibrations.

The above results are compatible with the structure shown in Fig. 7, for which the concentration of $Zn-Cl_{nb}$ bonds is expected to be substantially higher than the concentration of $Zn-I_{nb}$ bonds, although the relative values cannot be estimated at present. The intensities of the 294 cm^{-1} and 139 cm^{-1} bands of Fig. 4 cannot be compared and thus the structural model does not imply that the number of Cl and I bridges are equal. Although one would expect the 158 cm^{-1} band to be depolarized, the overlap with the 185 cm^{-1} and 139 cm^{-1} bands prevents an accurate evaluation of the depolarization ratio at 158 cm^{-1}.

Finally, the depolarized 53 cm^{-1} band of Fig. 5 may include

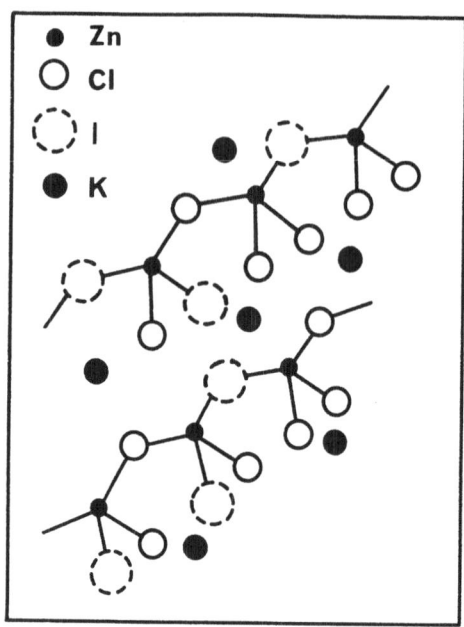

Fig. 7. Schematic 2-dimensional drawing of the
structure of 50 ZnCl$_2$ - 50 KI glass.

contributions from Br rocking vibrations and the 32 cm^{-1} (probably) depolarized band of Fig. 4 may in part be due to rocking motions of I$_b$ atoms. Considerably more work needs to be
done on zinc halide-based systems, however, before a more definitive vibrational assignment is made.

3.3 Structural considerations

In the case of modified fluorohafnate and fluorozirconate
glasses, it has been found that the *dihafnate* or *dizirconate*
compositions, for which a chain-like structure has been proposed, are the most stable, i.e. they have the highest glass-for
ming tendency. Although pure HfF$_4$ or ZrF$_4$ may not form glasses
simply because they sublime at ambient pressure, an increase
in the HfF$_4$ or ZrF$_4$ concentrations above that of the dihafnate
or dizirconate compositions progressively reduced the glass-
-forming tendency.

Thorium-fluorohafnate (or fluorozirconate) glasses have an
overall three-dimensional character, but their glass-forming
ability is much lower than that of modified lead dihafnate
glass, for example.

In the case of zinc halide-based glasses, although zinc chloride itself can form a glass when stringent precautions are taken in order to eliminate residual water, it is clear that modified zinc chloride or zinc bromide-based glasses are much more stable than the pure zinc compounds. In particular, the 50 - 50 (mol %) compositions appeared to have the best glass-forming ability and their structure is chain-like, similar to a metassilicate. Although these glasses are highly covalent, their glass-forming ability was not appreciably higher than that of multicomponent fluorohafnate or fluorozirconate glasses such as ZBLA (57 ZrF_4-36 BaF_2-3 LaF_3-4 AlF_3), where the concentration of modifier (BaF_2) is the same as in a dizirconate glass.

Therefore, it appears that the degree of covalency of the glass network is not particularly important. On the other hand, the addition of a modifying halide to a network-forming halide appears to drastically increase its glass-forming ability. A chain-like structure, where the orientational freedom of the network is highest, seems to be the most favorable structural configuration for halide systems, with the possible exception of v-BeF_2, which has a weakened version of the structure of v-SiO_2.

4. CONCLUSIONS

The general state of the different studies of halide glass structure by vibrational spectroscopy is fairly good, at present. Although a great deal of work still needs to be done, a reasonable understanding of the nature of the vibrational modes and the short range structures (coordination numbers and interbond angles) has now been achieved.

During the next five years, it is desirable that additional compositions and systems will be studied by infrared (absorption and reflectivity) and Raman spectroscopies, particularly in the case of ThF_4 - based and zinc halide-based glasses. It is also important that inelastic neutron scattering studies will be performed for some glasses in order to obtain a measure of the density of vibrational states (DOVS), particularly for fluorozirconate-type and ThF_4 - based glasses.

Finally, fruitful collaboration may perhaps be carried out between Portugal and France or U.S.A., in the area of vibrational studies of zirconium-free fluoride glasses and between Portugal and the United Kingdom, in the area of inelastic neutron scattering studies of different halide glasses.

ACKNOWLEDGMENTS

I would like to thank the Instituto Nacional de Investigação Científica and the Instituto de Engenharia de Sistemas e Computadores for partial support of the present work and Mrs. M.C. Silva for preparing some of the glass samples. I would also like to thank Prof. J.J.C. Teixeira Dias for the use of the Raman spectrometer.

REFERENCES

1. Baldwin, C.M., Almeida, R.M. and Mackenzie, J.D., J. Non--Crystalline Solids, 43, 309 (1981).
2. Etherington, G., Keller, L., Lee, A., Wagner, C.N.J. and Almeida, R.M., J. Non-Crystalline Solids 69, 69 (1984).
3. Almeida, R.M. and Mackenzie, J.D., J. Non-Crystalline Solids 68, 203 (1984).
4. Desa, J.A.E., Wright, A.C., Wong, J. and Sinclair, R.N., AERE-R 10186 Report, Harwell, United Kingdom (1981).
5. Almeida, R.M. and Mackenzie, J.D., J. de Physiqué Colloque C8, suppl. no. 12, 46, 75 (1985).
6. Toth, L.M., Quist, A.S. and Boyd, G.E., J. Phys. Chem. 77, 1384 (1973).
7. Almeida, R.M. and Mackenzie, J.D., J. Chem. Phys. 78, 6502 (1983).
8. Etherington, G., Wagner, C.N.J., Almeida, R.M. and Faber Jr., J., Repts. Hahn-Meither Institute B 411, 64 (1984).
9. Sen, P.N. and Thorpe, M.F., Phys. Rev. B15, 4030 (1977).
10. Almeida, R.M. (to be published).
11. Badger, R.M., J. Chem. Phys. 2, 128 (1934); ibid. 3, 710 (1935).
12. Almeida, R.M. and Mackenzie, J.D., Glastechn. Ber. 56 K, 850 (1983).
13. Zachariasen, W.H., Acta Cryst. 2, 388 (1949).
14. Burbank, R.D. and Bensey, F.N., Union Carbide Nuclear Company Report K-1280 (Oct. 31, 1956).
15. Galeener, F.L. et al., J. Non-Crystalline Solids 42, 23 (1980).
16. Nakamoto, K., Infrared and Raman Spectra of Inorganic and Coordination Compounds, 3rd ed., John Wiley, New York (1978).
17. Ellis, R.B., J. Electrochem. Soc. 113, 485 (1966).
18. Chieh, C. and White, M.A., Z. Kristal. 166, 189 (1984).
19. Almeida, R.M., Mater. Sci. Forum 6, 427 (1985).

DISCUSSION

Q: C. Pantano. Does your analysis based on a random close-
-packing really contradict your structural model? Your calculations used extreme assumptions and the results bracket
the experimental densities.

A: R. Almeida. No. The analysis does not exclude the possibility of random close-packing, but it certainly does not sup
port it.

Q: G. de Leede. What was your measure of the degree of covalency?

A: R. Almeida. I just used Pauling's model, based on the differences of electronegativities.

Q: G. de Leede. Why do you think that the degree of covalency
of the glass network is not particularly important for
glass formation?

A: R. Almeida. Because glass-forming systems of widely different degrees of covalency have similar glass-forming
tendencies. For example ZrF_4-based and $ZnBr_2$-based glasses.

DIFFRACTION STUDIES OF HALIDE GLASSES

Adrian C. WRIGHT

J. J. Thomson Physical Laboratory, Whiteknights, Reading, RG6 2AF, U.K.

ABSTRACT

A review is given of X-ray and neutron diffraction studies of halide and oxyhalide glasses with special emphasis on experimental techniques for the separation, or partial separation, of individual component correlation functions. These are particularly important in structural investigations of the complex multicomponent fluorozirconate and other heavy metal fluoride glasses which are the subject of the present workshop. The structures of heavy metal fluoride glasses are discussed in terms of the concepts of radius ratio and hole filling and a first order structural model is proposed based on a locally ordered random close packing. Several chain models have been suggested for these glasses but the extent of chain formation in real systems remains to be established.

1. INTRODUCTION

For many years the only known halide glasses were those based on BeF_2 and $ZnCl_2$ and these were regarded more as a scientific curiosity than as technologically useful materials. Recently, however, there has been a tremendous increase in interest in halide glasses stimulated by the discovery of many new glass-forming systems and also the possible applications of fluoroberyllate glasses in high-powered lasers for fusion reactors and of fluorozirconate glasses in optical fibres, the latter being the raison d'etre of the present workshop.

An important stage in the full understanding and utilisation of the properties of any material is a knowledge of its structure, but for many of the newer halide glass systems there is still controversy in the literature as to the co-ordination polyhedra present (short range order), let alone the way in which these combine together to give the so-called intermediate range order. In the following sections a summary is given of diffraction studies of halide glass structure, with particular emphasis on fluorozirconate glasses, together with a brief outline of the diffraction technique and a discussion of its major strengths and limitations.

Before describing any particular halide glass system in detail it is instructive to outline the two relevant first-order models of glass structure, viz. the random network and the random close packing. While in some ways these models are idealistic they nevertheless form a good foundation from which to discuss the structure of real glasses. It is also important to establish general principles as many of the technologically important halide glasses are complex multicomponent materials, a factor which will severely limit the structural information obtainable from these specific systems, particularly in respect of their intermediate range order.

1.1 Basic Theories of Glass Structure

The structure of covalent glasses with directional bonding is usually discussed in terms of the random network theory [1] in which well-defined structural units are connected together randomly to form a continuous network which lacks symmetry, periodicity and long range order, as

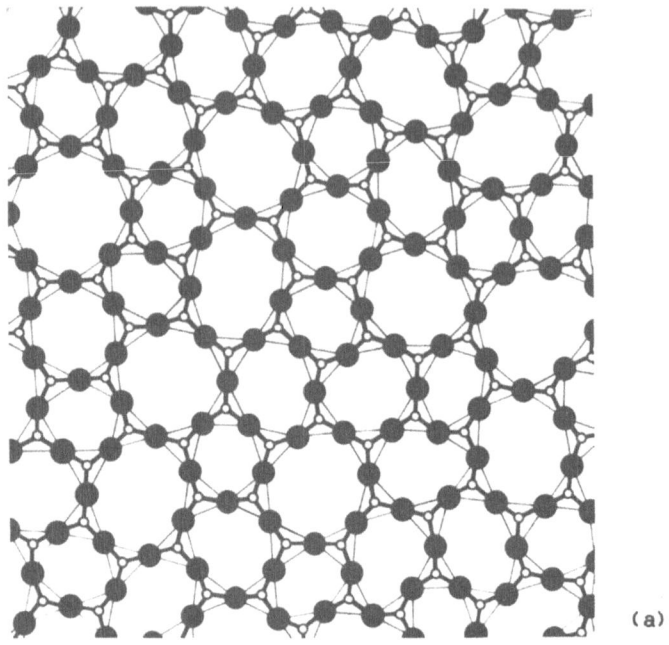

(a)

FIGURE 1 Schematic representation of (a) A_2X_3 random network (b) $MX-A_2X_3$ random network (c) invert glass.

○　　A Atom

●　　X Atom

AX_3 triangle (structural unit)

◐　　M^{2+} ion

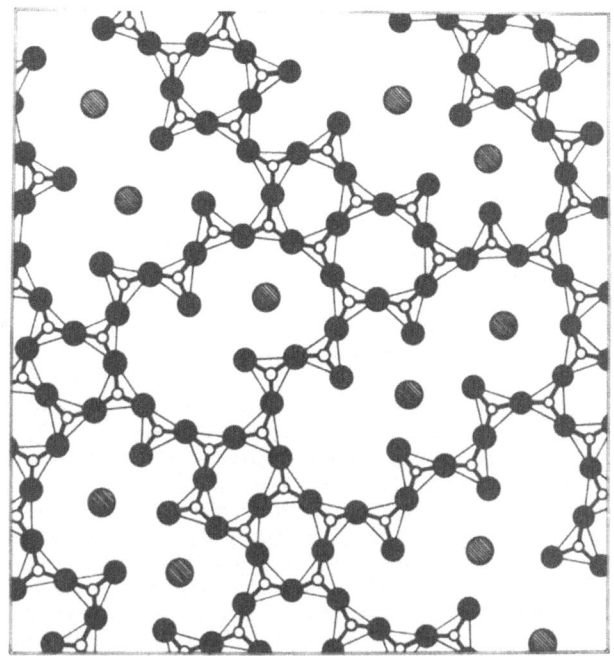

(b)

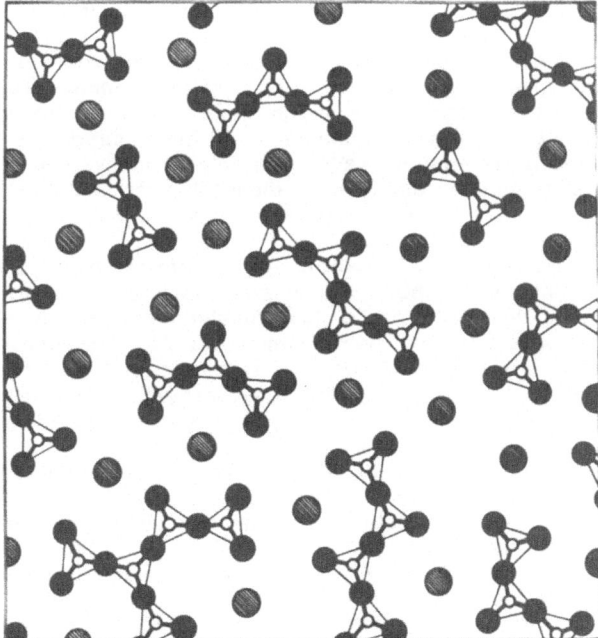

(c)

Indicated schematically in figure 1(a) for a glass of composition A_2X_3. In three dimensions disorder is introduced into such a structure through torsional rotations about A-X bonds and a distribution of A-X-A bond angles, any distortion of the structural units (AX_3 triangles in figure 1) themselves usually being comparable to that in the corresponding crystalline material. Zachariasen's 'rules' for glass-formation [1] suggest that glass-formation in such systems is favoured by structural units AX_n with a relatively small co-ordination number $n_{A(X)}$ of X around A ($n = 3$ or 4), which share corners rather than edges or faces, and also that the formation of a three dimensional network requires that at least 3 corners of each unit must be shared.

The addition of a network modifier MX to the glass of figure 1(a) results in the formation of non-bridging X atoms

$$A - X - A \quad + \quad MX \quad \longrightarrow \quad 2 \left[A - X^- \right] + M^{2+}$$

Traditionally the network modifying cation M^{2+} has been pictured as occupying holes in the resulting network close to the non-bridging X atoms as shown in figure 1(b), but modern diffraction and EXAFS [2] evidence indicates that the M^{2+} modifies its local environment to obtain its own desired co-ordination polyhedron as found in related crystalline materials. It should be noted, however, that the fact that a particular ion modifies its local environment in this way does not necessarily mean the introduction of regions of crystalline order (crystallites) but is probably simply the consequence of the appropriate radius ratio as discussed below for ionic systems. At very high network modifier concentrations the network is completely destroyed and the glass consists of a mixture of polyanions and cations as illustrated by the structure for invert glasses [3] in figure 1(c), although again it is highly probable that each cation will modify its own local environment to suit itself.

For metallic and ionic systems where there is no well-defined directional bonding a structural description in terms of a random network and network modifiers is less appropriate, a much more relevant first-order model being a random close packing of spheres. In such structures the important concepts are those of radius ratio, packing density and hole filling. These ideas are frequently used in a discussion of metallic and ionic crystal structures, but are often neglected when considering the structures of the equivalent amorphous solids.

Consider first the coordination polyhedron AX_n. For stability the A atom should fill the cavity defined by the X atoms and there exists a critical radius ratio r_A/r_X, at which this criterion is just fulfilled with the A atom just in contact with all the X atoms and below which the AX_n polyhedron is unstable. Above this value the AX_n polyhedron will remain stable until the point is reached where extra X atoms can be incorporated. The critical radius ratio $(r_A/r_X)_c$ is given in table 1 for the common coordination polyhedra.

The space filling properties of a packing of equal-sized spheres is best described in terms of the dimensionless packing density, η. For crystalline close packing (face centred cubic or hexagonal close packing) η is 0.7405, but this value is reduced to 0.6366 [4] for a random close packing. In the case of crystalline close packing the remaining holes in the structure are either tetrahedral (2 per sphere/atom) or octahedral (1 per sphere/atom), as illustrated in figures 2(a) and 2(b), and various simple ionic and metallic alloy crystal structures (rocksalt, zinc blende, etc.) are generated by filling these holes with atoms/ions of the appropriate radius ratio. The unoccupied space in a random close packing on the other hand comprises a number of larger holes (c.f. figure 2(c)-(e)), in addition to the simple tetrahedra and octahedra, as first characterised by Bernal [5]. The presence of these larger holes is a reflection of the lower packing density of the random close packing and their relevance to the structure of halide glasses will become apparent in later sections of this paper.

The five idealised canonical or Bernal holes in figure 2 define polyhedra that are all representatives of the class known as convex deltahedra [6] in which every face is an equilateral triangle. They comprise: -

        (a)   a tetrahedron,

        (b)   an octahedron (tetragonal bipyramid or trigonal antiprism),

        (c)   a trigonal prism, with the three square faces capped with half octahedra,

        (d)   an Archimedean antiprism, capped with two half octahedra, and

        (e)   a dodecadeltahedron.

A convex structure is essential for the hole to be stable and the triangular faces are formed by three atoms in contact. There are a

TABLE 1 Critical radius ratios for common co-ordination polyhedra and Bernal Holes.

Polyhedron	Radius ratio	A-X		A-(2)X	
		$n_{A(X)}$	r_{A-X}/r_X	$n_{A(X)}$	r_{A-X}/r_X
Triangle	0.155	3	1.155	–	–
Tetrahedron	0.225	4	1.225	–	–
Dodecadeltahedron	0.353*	4	1.353	4	1.860
Octahedron	0.414	6	1.414	–	–
Trigonal Prism	0.528	6	1.528	$3^†$	1.992
Archimedean Antiprism	0.645	8	1.645	$2^†$	2.255
Cube	0.732	8	1.732	–	–
Icosahedron	0.902	12	1.902	–	–

*   for atom at centre. A larger atom can be inserted off-centre.

†   refers to half octahedron cap.

total of 8 possible convex deltahedra, the remaining three being
        (f)   the trigonal bipyramid,
        (g)   the pentagonal bipyramid   and
        (h)   the icosahedron (pentagonal antiprism capped with two
                   pentagonal pyramids).
The hole in the centre of a trigonal bipyramid, which consists of two
tetrahedra sharing a common face, corresponds to triangular co-ordination
for which the radius ratio is less (c.f.   table 1) than if the included atom
is displaced into the centre of one of the tetrahedra.   The trigonal
bipyramid is thus best considered as a pair of tetrahedral holes.   The
pentagonal bipyramid and the icosahedron both have five-fold rings of atoms
around the hole centre and it is likely that these become unstable during
the compaction stage of the formation of a random close packing and

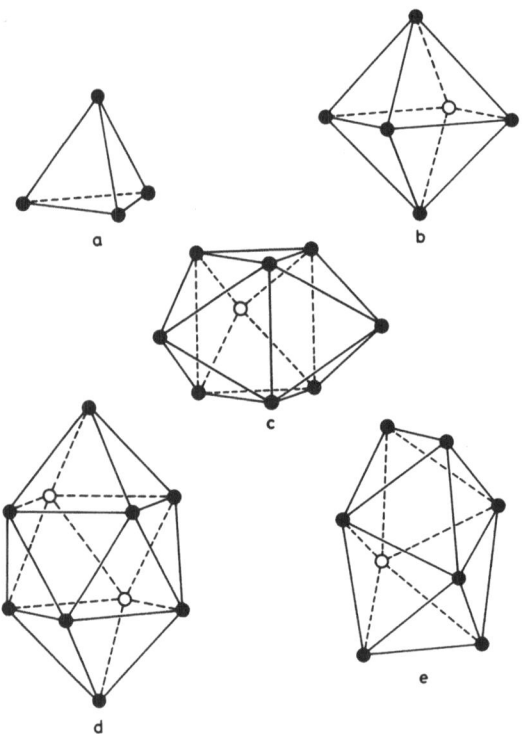

FIGURE 2   The five canonical or Bernal holes in a random close packing
[5]. (a) Tetrahedron, (b) Octahedron, (c) Trigonal Prism (capped with
three half octahedra), (d) Archimedean Antiprism (capped with two half
octahedra) and (e) Dodecadeltahedron.

decompose into two or more small holes.   The radius ratio for an atom at the centre of the pentagonal biprism is in any case very small. Note, however, the relationship between the pentagonal biprism and the dodecadeltahedron, which is present and may be considered as two distorted pentagonal pyramids joined together by three of their base edges with a wedge occupying the skew quadrilateral formed by the four remaining base edges.   Critical radius ratios for six of the eight convex deltahedra are included in table 1 together with details of the first two co-ordination shells of the included atom.

2 DIFFRACTION TECHNIQUES

A full discussion of the theory and practice of the diffraction method is beyond the scope of this review and so only a limited account will be presented here, with particular emphasis on the techniques for separating individual component correlation functions.   Amorphous solids are in general isotropic and hence a diffraction experiment yields a scattered intensity $I(Q)$

T(r) from a Neutron Diffraction Experiment

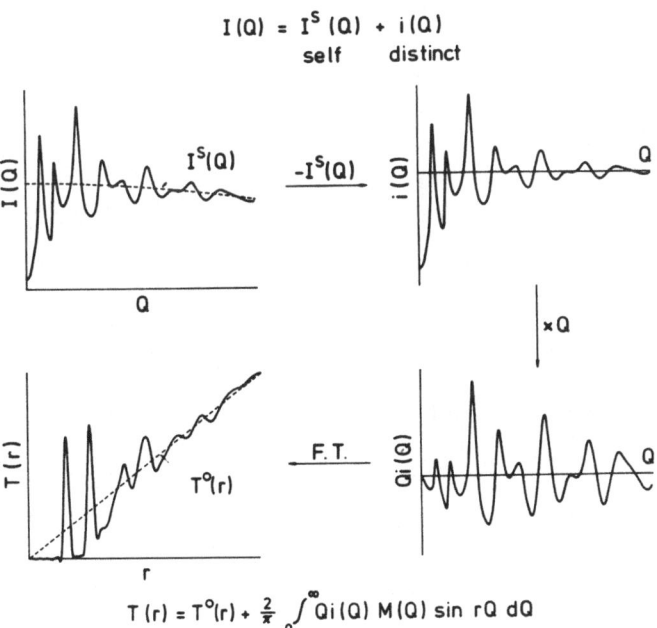

$$I(Q) = I^S(Q) + i(Q)$$
$$\text{self} \quad \text{distinct}$$

$$T(r) = T^0(r) + \frac{2}{\pi} \int_0^\infty Qi(Q) \, M(Q) \sin rQ \, dQ$$

FIGURE 3 The relationship between the corrected, normalised diffraction pattern, $I(Q)$, and the real space correlation function, $T(r)$.

which is a function only of the magnitude of the scattering vector

$$Q = \frac{4\pi}{\lambda} \sin\theta \qquad (1)$$

and not its direction.   As discussed elsewhere [7], the consequent one-dimensional nature of such diffraction data means that it is impossible to uniquely determine the structure of an amorphous solid and is the reason why modelling plays such an important role in amorphographic studies.

The relationship between the corrected, normalised diffraction pattern I(Q) and the real space correlation function T(r) is outlined diagrammatically in figure 3 using neutron data.   For a sample containing n elements the correlation function T(r) from a single diffraction experiment is a weighted sum of n(n+1)/2 independent components $t_{jk}(r)$ each convoluted with the appropriate peak function $P_{jk}(r)$ which defines the

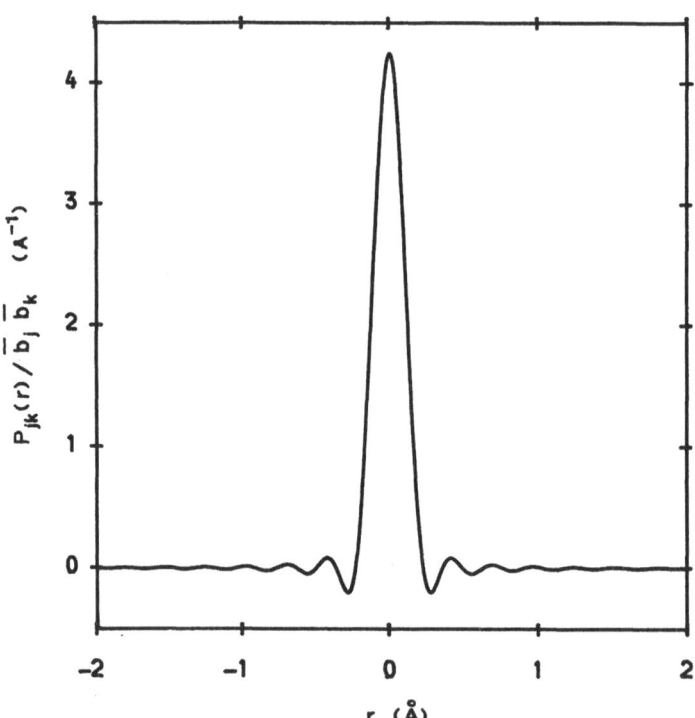

FIGURE 4   The reduced neutron peak function $P_{jk}(r)/\overline{b}_j\overline{b}_k$ (Q_{max} = 22.64Å^{-1}) used to generate figure 10.

experimental resolution in real space and is the Fourier cosine transform of the modification function $M(Q)$.

$$T(r) = \sum_j \sum_k \int_0^\infty t_{jk}(r') \left[P_{jk}(r-r') - P_{jk}(r+r') \right] dr' \qquad (2)$$

r' is a dummy convolution variable and the j summation is taken over the atoms in one composition unit, while that for k is over atom types (elements). For X-rays the shape of $P_{jk}(r)$ changes for the different j-k components and its area is proportional to the atomic/ionic electron product $K_j K_k$ while for neutrons the shape of $P_{jk}(r)$ is constant (c.f. figure 4) and the magnitude varies as $\overline{b}_j \overline{b}_k$. $\overline{b}_j$ being the neutron scattering length of atom j. The width of $P_{jk}(r)$ is inversely proportional to Q_{max} ($\Delta r_{FWHM} = 5.437/Q_{max}$ for neutrons with the Lorch modification function). The effective number of electrons on any given atom/ion K_j clearly depends on the degree of ionicity and a lack of knowledge of the exact electron distribution can lead to significant errors in X-ray co-ordination numbers involving elements of low atomic number Z. Formally, the product $K_j K_k$ is given by [8]

$$K_j K_k = \frac{f_j(0) f_k^*(0)}{\overline{f_e(0)}^2} \qquad (3)$$

in which $f_j(0)$ is the atomic form (scattering) factor of atom j at $Q=0$, including any dispersion corrections, and $\overline{f_e(0)}$ is the average form factor per electron for the sample in question. For BeF_2, in the absence of dispersion corrections, $K_j K_k = 36e^2$ for free atoms and $20e^2$ for Be^{2+} and F^- ions, while the corresponding variation for heavy metal fluoride glasses is $\leq 10\%$. The magnitude of the effect for vitreous BeF_2 may be seen from figure 5, which shows the X-ray correlation functions for Ag K_α X-rays ($Q_{max} = 20\text{Å}^{-1}$) calculated from a Molecular Dynamics simulation of Brawer [9] using free atom and ionic form factors. Note in particular that, although there is a (relatively small) change in height of the first Be-F peak, the difference in area mainly occurs away from the central maximum and is a reflection of the fact that it is the electrons in outer orbitals which are involved in the transition from free atoms to free ions. Great care is thus required in the interpretation of peak areas (co-ordination numbers) under any correlation function extracted from an X-ray diffraction experiment on fluoride glasses. It is also important to state the form factors used in the data analysis and to clearly specify the values of K assumed in the calculation of co-ordination numbers.

Structural studies of glasses rapidly get more complicated as the number of elements increases. This is an important factor in explaining the present very limited knowledge concerning the structure of heavy metal fluoride glasses since these contain a minimum of 3 elements (6 independent components) and in the case of the commercially important glasses typically 5 or 6 (15 or 21 independent components respectively). In studying such materials good real space resolution is imperative and this means making accurate measurements to high Q_{max}. Where possible advantage should be taken of the various methods which exist for the separation of individual component correlation functions, $t_{jk}(r)$, based on the variation of the component weighting factors $\overline{b}_j \overline{b}_k$ or $K_j K_k$. In order to achieve a complete separation of the component correlation functions for an n-element sample it is necessary to perform a minimum of $n(n+1)/2$ distinct scattering experiments in which the scattering amplitude of at least n-1 elements is varied. For complex multielement systems a complete separation is rarely possible and in any case is subject to large uncertainties. Nevertheless a partial separation involving a variation of the scattering amplitude of a single element can often be used with great effect to investigate one particular aspect of the structure. An example of this technique – a study of the Dy^{3+} environment in vitreous $NaF-DyF_3-BeF_2$ – is described in section 5.1.

If two experiments are performed in which the scattering amplitude of element A is changed then it is possible to separate linear combinations of any two of the three contributions A–A, A–X and X–X (where X is any

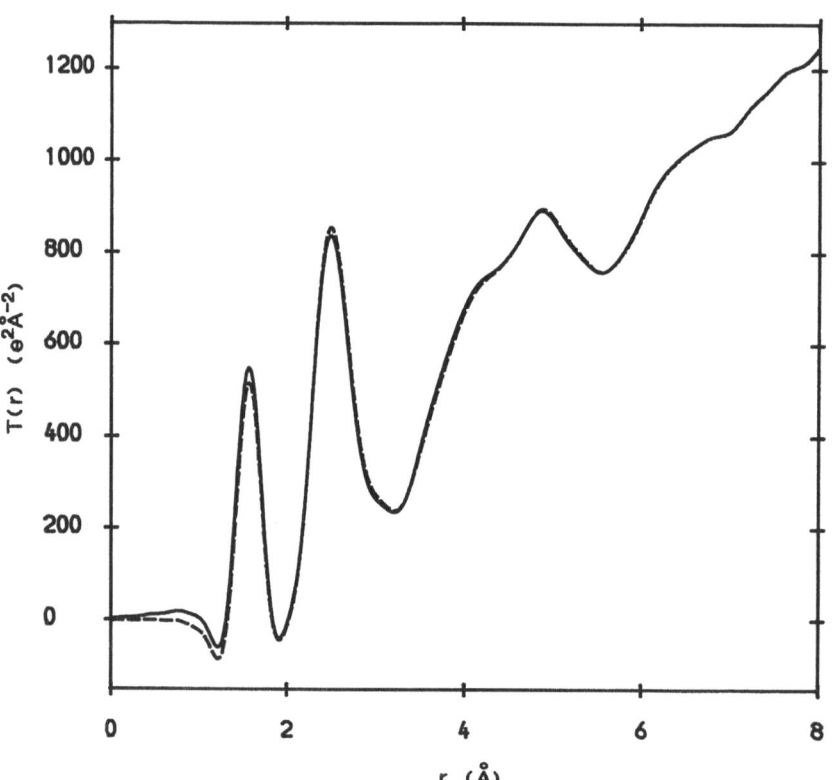

FIGURE 5 The X-ray correlation function for vitreous BeF_2 calculated from a Molecular Dynamics simulation of Brawer [9], using free atom (————) and ionic (------) form factors ($Q_{max} = 20Å^{-1}$).

element other than A), the two most useful being A-X + A-A and X-X - A-A [10].   A further variation for element A allows a complete separation of A-A, A-X and X-X.   Note that if A is present in low concentrations (e.g. as a minority constituent in a multicomponent glass) the contribution from the A-A component is very small and can sometimes be neglected, particularly if the relative spacing of A atoms is random and large.   The techniques for varying scattering amplitudes are outlined in sections 2.1 to 2.3, together with some approximate methods (section 2.4 to 2.6) which can often yield valuable information providing they are used with the necessary caution. The latter are particularly useful for samples where the exact techniques of sections 2.1 to 2.3 are not applicable, which is frequently the case for the systems of greatest interest.

The important complementary role played by EXAFS studies should also be stressed.   The diffraction and EXAFS techniques have been compared in detail elsewhere [11] and so only a few brief comments will be made here.   A measurement at an absorption edge of element A yields information concerning the environment of A atoms directly (A-X + A-A components) and is thus equivalent to the difference function derived from two diffraction experiments in which the scattering amplitude of A is varied. For multicomponent systems the environment around each element in turn can be investigated and the fact that the EXAFS signal is obtained by direct measurement, rather than as a difference, means that EXAFS spectroscopy can be used at very much lower concentrations.   The major disadvantage of EXAFS spectroscopy is the lack of data at low wavevector k which limits the accuracy with which co-ordination numbers can be determined and means that the technique is confined to the study of relatively sharp peaks in real space (short range order).

2.1 Isotopic Substitution

In the absence of magnetic scattering neutrons are scattered by the nucleus and in general each isotope i has a different neutron scattering length $\overline{b}_i$.   The scattering length for element A in any sample in which its various isotopes are distributed randomly between the different A atom sites is simply the average value calculated using the appropriate isotopic abundances ω_i

$$\overline{b}_A = \sum_i \omega_i \overline{b}_i \qquad (4)$$

Different samples may thus be prepared for which the isotopic composition of element A and hence its average scattering length $\overline{b}_A$ is varied.   Note, however, that the reliability of the isotopic substitution technique depends critically on the manufacture of samples which are identical in every respect except for the isotopic composition of element A and also that it is imperative that the distribution of the various isotopes over the A sites is indeed random.   In this connection it is important to remember that a change in mass can effect chemical kinetics, particularly for elements of low atomic number such as in H/D substitution.

For a few elements (H, Li, Ti, Cr, Ni, Sm, Dy, W) isotopes exist with negative scattering lengths, corresponding to scattering without the normal phase change of π.   This gives rise to a special form of isotopic substitution known as the null technique [10] in which a combination of isotopes is used such that element A has a zero scattering length and hence does not contribute to the measured interference or correlation functions. A measurement of the scattering from such a sample thus yields the X–X component directly and a subtraction of the scattering from that for a sample containing natural A (or A with any other non-zero $\bar{b}_A$) yields the A–X + A–A contribution. For a binary system where both elements can be obtained with a zero scattering length the use of the double null technique [10] yields both like-atom component correlation functions directly. An alternative approach is that of balanced scattering lengths [10] in which two samples are prepared with $\bar{b}_A$ equal in magnitude but of opposite sign. In this case the difference between the two measurements yields the A–X component without interference from A–A interactions.

2.2 Anomalous Dispersion

Near an absorption edge the X-ray atomic form factor becomes complex

$$f(Q) = f_o(Q) + \Delta f'(\lambda) + i\Delta f''(\lambda) \tag{5}$$

and thus the form factor for element A can be varied by performing measurements at two or more different wavelengths [12]. To be really effective X-rays are required with a continuously variable incident wavelength and hence the X-ray anomalous dispersion technique has greatly benefitted from the development of X-ray scattering facilities at intense synchrotron radiation sources. The advantage over isotopic substitution is that all the measurements are performed on the same sample but at present accuracy is limited due to lack of knowledge concerning the exact wavelength dependence of $\Delta f'(\lambda)$ and $\Delta f''(\lambda)$. The change in form factor also tends to be relatively small amounting to only ≤ 10–20%.

Anomalous dispersion experiments are also possible with neutrons [13]. In this case the variation in scattering length occurs close to an absorption resonance and is much larger (a factor of ~ 5–10) for the isotope in question. At present the neutron anomalous dispersion technique is limited to very few elements but this number should increase with the development of the next generation of amorphous diffractometers on spallation pulsed neutron sources which make measurements at smaller scattering angles and much shorter neutron wavelengths [14]. Possible elements will then include [113]Cd, [149]Sm, [151]Eu, [167]Er, [176]Lu, [237]Np and [239]Pu.

2.3 Magnetic Neutron Diffraction

The role of neutron scattering techniques in the study of amorphous solids containing magnetic ions has been reviewed elsewhere [15]. As with crystalline materials neutron magnetic diffraction can provide unique information concerning the structural arrangement of magnetic ions within a glass. Measurements on magnetic systems are in general limited to $Q \leq 10 \text{Å}^{-1}$ due to the fall-off in the magnetic form factor $f_M(Q)$ with increasing Q, which limits real space resolution, but on the other hand the correlation functions obtained are simpler since they refer only to the magnetic ions.

For a paramagnetic sample containing a single magnetic species, in which there is no correlation in the orientation of adjacent magnetic moments, a measurement of the magnetic component of the diffraction pattern yields the square of the magnetic form factor, $f_M^2(Q)$. If, however, the atomic magnetic moments align on cooling the sample to lower temperatures then, by measuring the magnetic diffraction pattern as a function of temperature through the magnetic transition region, it is possible to extract the magnetic interference function $Qi_M(Q)$. The usual assumption made is that the nuclear scattering is unchanged through the magnetic transition region so that $i_M(Q)$ may be obtained by simply subtracting the total diffraction pattern (magnetic + nuclear) for the paramagnetic state from that for the magnetically ordered state. If the temperature difference is too great, however, problems may occur due to changes in the nuclear scattering Debye-Waller factor as discussed by Le Ball et al. [16].

For a sample in which there is no correlation between the magnetic moment directions and the radius vector r, $Qi_M(Q)$ may be Fourier transformed to give the magnetic correlation function, which with a single magnetic species takes the form [15]

$$d_M(r) = \frac{2}{\pi} \int_0^\infty Qi_M(Q) M(Q) \, \sin rQ \, dQ = 4\pi r \left[\rho_M(r) - \rho_M^0 \right] \qquad (6)$$

where

$$\rho_M(r) = \rho(r) \langle \mu(0).\mu(r) \rangle / \mu_p^2 \qquad (7)$$

and ρ_M^0 is the corresponding average density. $\mu(r)$ is the atomic magnetic moment and μ_p the appropriate paramagnetic moment. Note that $\rho_M(r)$ depends on both the relative positions and the magnetic moments of the magnetic ions and hence it is possible to extract information concerning both the atomic and magnetic structure. The former application is particularly useful when studying complex systems containing a single magnetic species A, since it gives the A-A distances alone, and an example is given later in section 5.3. For samples where no magnetic ordering occurs at low temperatures, or where devitrification occurs below or within the magnetic ordering transition range, an alternative approach is to align the atomic magnetic moments at low temperature with the aid of a high magnetic field, the optimum arrangement being to measure diffraction patterns with the field perpendicular (maximum magnetic scattering) and parallel (zero magnetic scattering) to Q.

A much more elegant method of investigating amorphous solids containing magnetic ions is to employ polarised neutrons, which allow a complete separation of the A-A, A-X and X-X components. Polarisation techniques are particularly powerful when a polarisation analyser is also used in the diffracted beam so that it is possible to distinguish scattering events involving spin-flip and non-spin-flip processes. The optimum configuration for this type of experiment is to align the incident neutron polarisation parallel or anti-parallel to Q in which case the coherent nuclear and magnetic scattering are completely separated.

2.4 Isomorphous Substitution

Isomorphous substitution involves the replacement of one element with another of different scattering amplitude along with the assumption that the structure is otherwise unchanged. In order for this to be the case the chemistry of the two elements must be identical for the system in question and also the bonding to the other elements present, particularly in respect of bond lengths and angles. In general isomorphous substitution, like doping, is not nearly so useful in glasses as in crystals since there is no longer the constraint of a lattice and hence an atom/ion has a much greater tendency to modify its immediate environment to give the optimum (minimum energy) local structure. The most likely candidates for isomorphous substitution in halide glasses are the Lanthanide rare earth ions, since their chemistry and ionic radius are relatively insensitive to changes in the 4f electron configuration. Other possibilities include transition metal ions such as the substitution of Hf^{4+} for Zr^{4+} in fluorozirconate glasses (c.f. section 5.3). In performing isomorphous substitution experiments it is preferable to perform measurements not only on the end members of the substitution series but also on a few intermediate partially substituted samples to ensure that a consistent answer is obtained as the substitution proceeds.

2.5 The Difference Technique

In any multicomponent system the information obtainable from diffraction experiments is greatly increased if a systemmatic study is carried out as a function of composition. A common approach in such cases is to use the difference technique in which the correlation function for the base glass is subtracted from those for the more complex glasses and the assumption made that the resulting difference correlation functions apply only to the added component(s). Unless the concentration of the added component(s) is small, however, this assumption is unlikely to be valid except for the first co-ordination shell. As an example consider the system $NaF-BeF_2$ and the difference curve obtained by subtracting the correlation function for pure vitreous BeF_2. The added NaF acts as a network modifier (c.f. section 1.1) breaking up the network with the formation of non-bridging F atoms and hence it is totally unreasonable to assume that the scattering from the fluoroberyllate network is unchanged. The difference technique may, however, give information concerning the first Na-F co-ordination shell providing the Be-F bond length is similar for bridging and non-bridging F atoms. In interpreting difference curves there is also a tendency to forget cross-correlations between the two components (e.g. Na-Be interactions) which are particularly important when the concentration of the added component is small.

2.6 A Combination of Diffraction Techniques

As discussed above the component weighting factors for X-rays and neutrons are in general different and hence considerably more information can be obtained by combining the two techniques. It has been suggested that with the addition of neutron magnetic diffraction it would be possible to extract the three component correlation functions for a binary system. Unfortunately, however, this is not straightforward as the shapes of the peak functions $P_{jk}(r)$ are different for X-rays and neutrons. Nevertheless the requirement for a structural model to reproduce both the X-ray and neutron correlation function is a much more stringent test than the comparison with either correlation function alone. As with crystalline materials the fact that $\bar{b}$ does not vary systematically with Z means that neutrons are particularly suited to studying low-Z atoms in a higher-Z matrix (e.g. neutrons are very much more sensitive than X-rays to the F^- ions in a heavy metal fluoride glass) or to differentiating between atoms of neighbouring high Z as in isomorphous substitution experiments.

A further example of the utilisation of the variation of the scattering amplitude with atomic number Z is the X-ray heavy element technique, which has been used with great success in the study of crystalline materials. In glasses containing a heavy metal A it is very tempting to identify the most prominent maxima in the correlation function with A-A interactions. Prins [17] has pointed out that this is by no means always correct as can be seen by considering the case of a glass containing discrete pairs of A ions, each ion being surrounded by $n_{A(F)}$ fluorine ions. The area under the A-A peak is approximately proportional to $n_{A(A)}K_A^2$ whereas for the A-F spacing the equivalent quantity is $2n_{A(F)}K_AK_F$. Hence for equal areas

$$n_{A(A)}K_A = 2n_{A(F)}K_F \qquad (8)$$

Taking $n_{A(A)} = 1$, $n_{A(F)} = 8$ and $K_F = 10$ gives $K_A = 160$ showing that in this case the A-F peak is always greater in area irrespective of the atomic number of A.

3. SMALL Q SCATTERING

Although long range order is absent in glasses, long range fluctuations in the average scattering density may still occur, for example as a result of phase separation or the onset of devitrification [18], and give rise to scattering at small Q (often termed small angle scattering despite the fact that with neutrons measurements are normally made at long wavelengths $(\lambda \approx 5-15 \text{Å})$ and frequently extend to angles in excess of $10°$). The phenomena which give rise to small Q scattering are particularly sensitive to preparation conditions and heat treatment and are the same as those responsible for the light scattering which ultimately limits the performance of a glass as a light transmission medium. Note that the scattering vector Q for light scattering is typically of the order of 0.001Å^{-1}. Small Q scattering measurements can also be employed to study surfaces [18] and even for a bulk, single-phase sample there is interest in the scattering at low Q since the zero Q limit, $I(0)$, is determined by bulk thermodynamic fluctuations which for a melt-quenched glass are related to the isothermal compressibility at T_g [8, 19]. The correlation function from a conventional diffraction experiment is a gross average over the whole irradiated volume and as such is relatively insensitive to the effects which give rise to small Q scattering, especially since any small Q scattering is usually removed before Fourier transformation by extrapolating the low Q side of the first diffraction peak to a low value of $I(0)$.

Since the momentum transfers involved in small Q scattering experiments extend to much higher values than for light scattering, X-ray and neutron small Q scattering are sensitive to inhomogeneities of considerably smaller dimension (typically $\sim 10-1000 \text{Å}$) than is the case for light scattering and hence can detect phase separation and devitrification at a much earlier stage. Neutrons have several distinct advantages over X-rays for these experiments. Absorption cross-sections for neutrons are in general much lower than for X-rays, particularly at longer wavelengths, which means that it is possible to work with larger samples, to greatly reduce unwanted surface effects, and at longer wavelengths to avoid making measurements at extremely small scattering angles with their attendant collimation, resolution and background problems. For samples which have very little scattering at low Q (e.g. $I(0)$ measurements on bulk homogeneous glasses) the use of long wavelength neutrons has the added advantage that most of the conventional diffraction range, where the scattering is much higher, is beyond the Q cut-off $(4\pi/\lambda)$ and therefore cannot contribute to multiple scattering in the small Q region.

Roth and Zarzycki [20] have pointed out that there are two types of long range fluctuations in glasses, composition fluctuations and density fluctuations and that it should be possible to separate these by combining X-ray and neutron results for which the relative scattering amplitudes are different. As with diffraction studies neutrons can be used to investigate the distribution of low-Z atoms or to distinguish between atoms of neighbouring high Z in complex systems. It is also possible to use isotopic or isomorphous substitution, while for samples containing magnetic species small Q neutron magnetic scattering can be used to probe longer range fluctuations in magnetic moment such as the variation in magnetic correlation length around the magnetic critical temperature T_c.

To date the only small Q scattering studies of halide glasses known to the author are limited to two neutron investigations of fluorozirconate glasses [21,22]. Data for vitreous $(NaF)_{0.20}(BaF_2)_{0.20}(AlF_3)_{0.03}(LaF_3)_{0.05}(ZrF_4)_{0.52}$ [22] are shown in figure 6, before and after heat treatment at 295°C for 91 mins. The arrows indicate the incoherent macroscopic differential scattering cross-section while the corresponding self scattering level, which represents the average scattering in the diffraction region, is off scale at $0.0255 cm^{-1}st^{-1}$. It is thus apparent that the level of coherent small Q scattering is extremely small indicating the absence of any detectable phase separation on a scale of ~600Å or less. The slight diminution in scattering on annealing can be

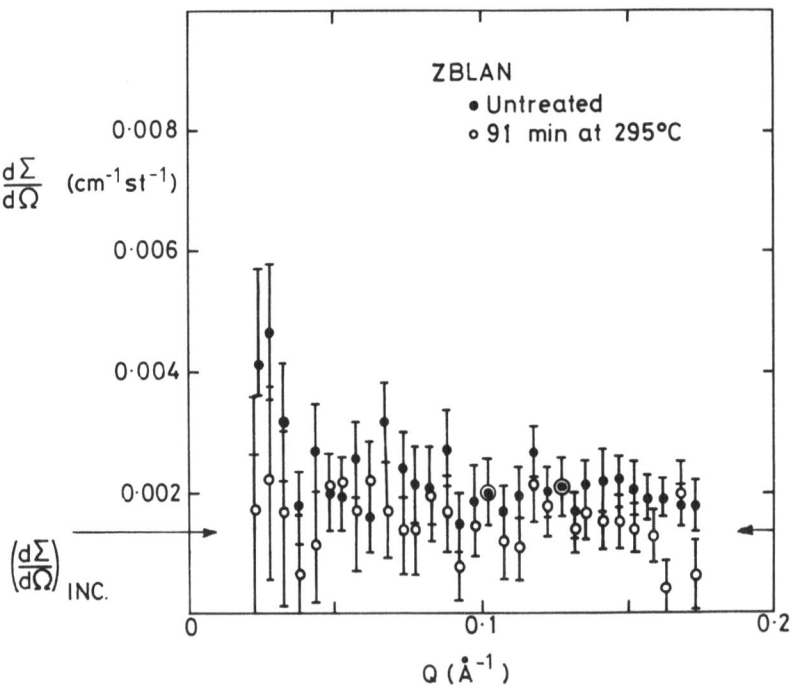

FIGURE 6 Small Q scattering from vitreous $(NaF)_{0.20}(BaF_2)_{0.20}(AlF_3)_{0.03}(LaF_3)_{0.05}(ZrF_4)_{0.52}$ [22].

explained by a reduction in the fictive temperature and hence in the compressibility limit. Santini et al. [21] also find very little small Q scattering for a glass in the system $BaF_2-AlF_3-LaF_3-ZrF_4$.

4   INTERPRETATION OF DIFFRACTION DATA

The interpretation of diffraction data can conveniently be discussed under two headings; short range order involving the order within a single structural unit and its connection to its immediate neighbours, or in systems where there is no well defined structural unit the local configuration (first co-ordination shell) around each atom/ion, and intermediate range order describing the way in which the structural units or local configurations combine together to produce the structure on a scale up to ~10-20Å, beyond which the differential correlation function D(r) is essentially zero. Before discussing these two regimes individually, however, it is first useful to make a few comments concerning the intercomparison of models and experiment, which are common to both short and intermediate range order.

When comparing a calculation of T(r) with experiment it is imperative to correctly include the effects of the finite range of Q over which measurements have been performed, otherwise the comparison is meaningless. In real space this takes the form of a convolution of the true component correlation functions

$$t_{jk}(r) = 4\pi r \rho_{jk}(r) \qquad (9)$$

with the appropriate component peak functions $P_{jk}(r)$ (equation 2). Note that $P_{jk}(r)$ has satellite features on either side of the central maximum which can themselves generate features in T(r) so that a Gaussian approximation to $P_{jk}(r)$ is simply not adequate, especially when the width is arbitrarily chosen to give the best agreement with experiment. For X-rays each component peak function has a unique shape due to the different Q dependence of the relevant form factors, as illustrated for BeF_2 in figure 7. It is also important to realise that the real and reciprocal space functions T(r) and Qi(Q) emphasise different aspects of any given structure and hence it is necessary to examine any discrepancies between theory and experiment in both spaces. Good agreement with experiment does not just mean getting peaks in either real or reciprocal space in the correct position (only very poor models fail to do this) but also getting peak widths and areas correct and doing this in *both* cases. The problem with diffraction data is one of uniqueness in that even if a perfect fit is obtained to experiment this is no guarantee that there are not other models which will also fit equally well. Agreement with diffraction data is thus *a necessary but not sufficient criterion* for any valid structural model.

4.1 Short Range Order

A specification of the short range order within a glass includes not only an identification of the basic structural unit(s) present, if any, but also quantitative parameters such as co-ordination numbers and the detailed distribution of bond lengths and angles. Too often interpretation is limited to the derivation of bond lengths from peak positions and co-ordination numbers from peak areas, the latter being obtained by assuming a peak is symmetric about its maximum in the radial distribution function rT(r). This procedure can, however, lead to considerable errors, particularly if a peak is asymmetric and/or only partially resolved as is frequently the case in multicomponent systems. Note that the experimental broadening is

92

asymmetric in rT(r) (symmetric in T(r) – c.f. equation 2) and also that in order to obtain the true area under any peak it is necessary to include the area under all the termination ripples on either side of the central maximum of $P_{jk}(r)$ (c.f. figure 5). In general, therefore, it is much better to extract short range order parameters using peak fitting techniques. Usually it is assumed that the true peak shape in $t_{jk}(r)$ is Gaussian [8] and values

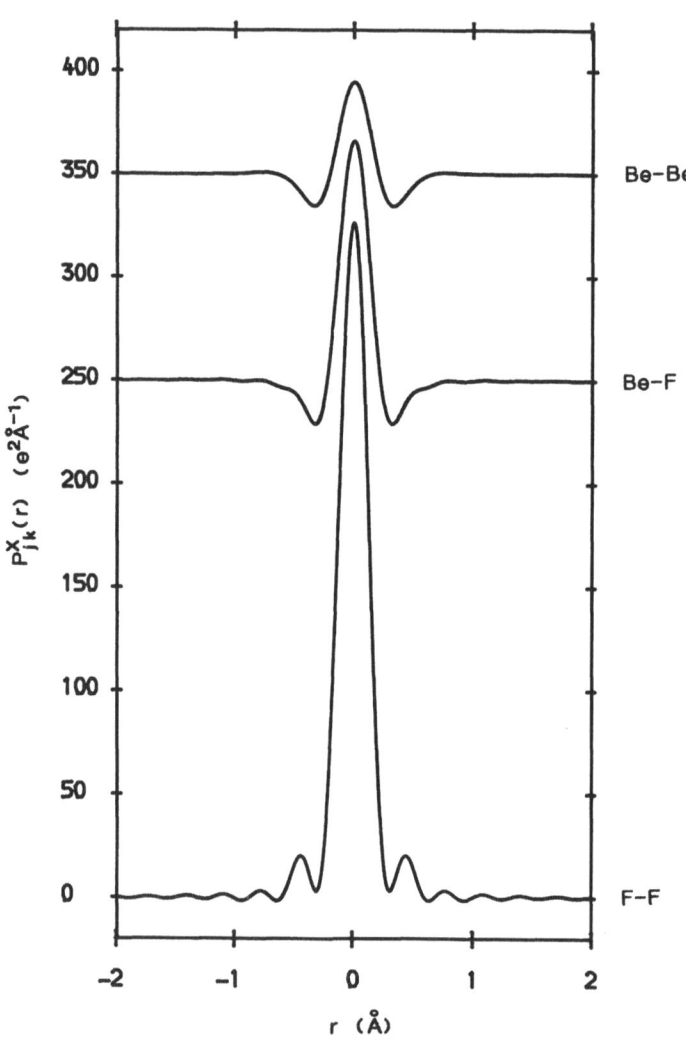

FIGURE 7 X-ray component peak functions, $P_{jk}(r)$, for vitreous BeF_2 (ionic form factors; $Q_{max} = 20Å^{-1}$).

are extracted for the peak area (co-ordination number), position and width
(r.m.s. bond length variation) and an example of such a fit for vitreous
$ZnCl_2$ is shown in figure 8 [23], but more sophisticated (asymmetric) peak
shapes can be used such as that proposed by Da Silva ot al. [24].
The A-X-A bond angle may be obtained from a knowledge of the
distribution of A-X and A-A distances but it should be stressed that the
most probable bond angle is not simply obtained from the A-X and A-A
peak maxima, as discussed by Coombs et al. [25].

The data for vitreous $ZnCl_2$ in figure 8 were obtained with the D4
diffractometer at the Institut Laue-Langevin at an incident wavelength of

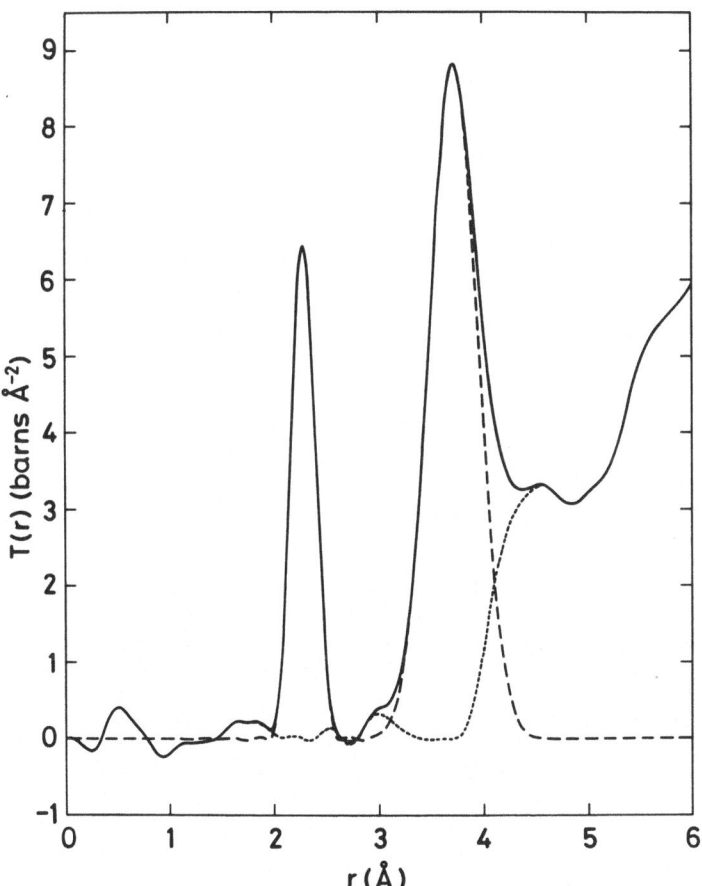

FIGURE 8 Peak fit for vitreous $ZnCl_2$ [23]. ———— experiment, ----- fit
and .... residual.

0.5Å (Q_{max} =23.6Å^{-1}). The width of the first peak in T(r) corresponds to an r.m.s. bond length variation of 0.075Å, showing that there is still appreciable experimental broadening even at this relatively high value of Q_{max}. The peak fit indicates that the basic structural unit in vitreous $ZnCl_2$ is the $ZnCl_4$ tetrahedron with a well defined Zn-Cl bond length of 2.29Å.

4.2 Intermediate Range Order

Intermediate range order in glasses is usually investigated by means of modelling studies, the most useful approaches for halide glasses being random network models, random close packings of appropriately sized hard spheres, Monte Carlo and Molecular Dynamics computer simulations, crystal based models and random displacement iterative techniques. Random network models are relevant for vitreous BeF_2 and the fluoroberyllate glasses and perhaps vitreous $ZnCl_2$. The great similarity between BeF_2 and SiO_2 (c.f. section 5.1) means that random network models for the latter are immediately applicable to the former. The majority of modelling studies of the heavy metal fluoride glasses have taken the form of Molecular Dynamics simulations and this technique is the subject of the paper by Parker [26]. Molecular Dynamics simulations have also been performed for vitreous BeF_2 and vitreous $ZnCl_2$ and are discussed in section 5.

It is important to realise that frequently the correlation function for a model can only be calculated for distances less than some maximum value r_{max} and that exactly the same considerations apply when such data are transformed to reciprocal (intensity) space as pertain to the transformation of experimental data to real space i.e. the model interference function will be broadened by convolution with a peak function P(Q) which reflects the limited data range in real space, the width of P(Q) being inversely proportional to r_{max}. Thus before the model can be compared with experiment in reciprocal space it is essential to convolve the *experimental* interference function with P(Q). The exact form of P(Q) will depend on whether the differential correlation function, D(r), for the model is truncated by a step function or whether a more gradual cut-off is present due to some model shape function F(r). Problems also exist with any model incorporating periodic boundary conditions [7] in ensuring that the factors which limit the range of order, which are the very essence of the amorphous state, are indeed structural and are not just an artifact of the method of calculation.

In generating structural models of glasses it is common to relax the model to its minimum potential energy configuration using an appropriate interatomic potential. The model co-ordinates then represent the equilibrium atomic positions and consequently it is necessary to include thermal vibration before the model can be compared with experiment. This is usually done by assuming a Gaussian distribution in real space or employing a Debye-Waller factor in reciprocal space. On the other hand, correlation functions extracted from Molecular Dynamics simulations, as an average over a series of frames at the required temperature, already include thermal effects and only need convolution with the appropriate component peak functions before comparison with experiment.

Crystal based models may range from a simple comparison of the glass with its crystalline analogues to more sophisticated models in which the crystalline order is limited to give a diffraction pattern characteristic of a glass, usually by multiplying the differential correlation function D(r) by some factor F(r) which tends to zero at higher r. For example in the quasi-crystalline model [27] the form of F(r) is appropriate to a spherical crystallite embedded in a structureless medium of the same average scattering density. The polycrystalline (powder) diffraction pattern (a series of δ-functions, which comprise the Bragg peaks, + thermal diffuse scattering) is thus broadened by the Fourier cosine transform of $F(r)$. An interesting extension of this technique has recently been described by Le Bail et al. [28] which makes use of the Rietveld method [29] for the refinement of crystal structures from powder diffraction data, modified to

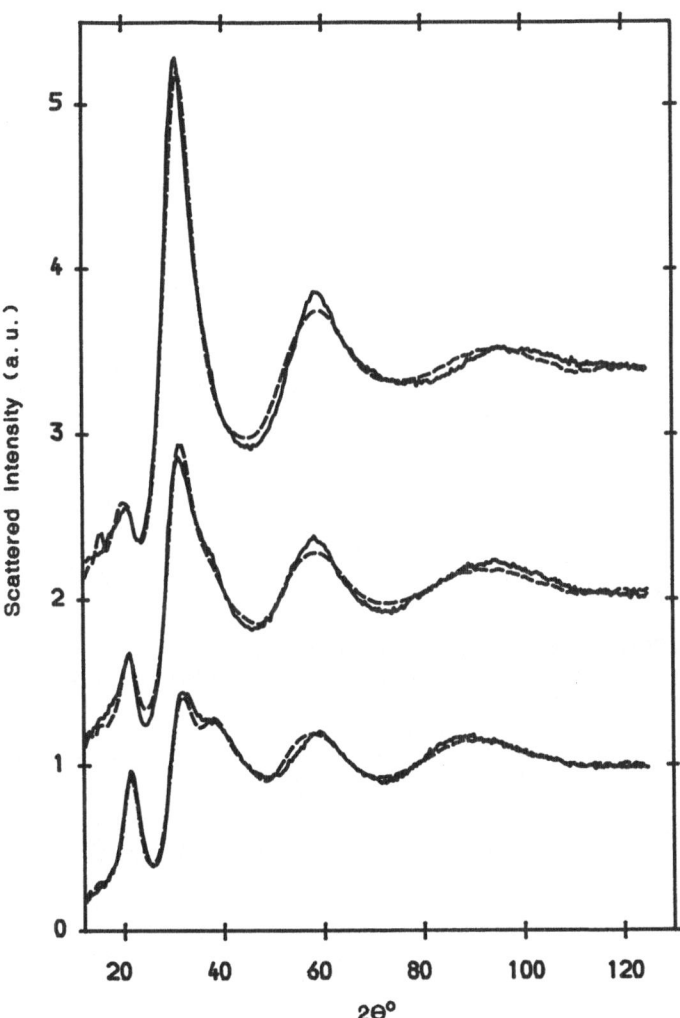

FIGURE 9 Modified Rietveld fit to $2PbF_2 \cdot TF_2 \cdot MF_3$ glasses [28,30]. Top $2PbF_2 \cdot MnF_2 \cdot VF_3$, middle $2PbF_2 \cdot MnF_2 \cdot FeF_3$ and bottom $2PbF_2 \cdot ZnF_2 \cdot FeF_3$. _____ experiment and ----- fit.

include strain broadening. The atomic co-ordinates within the quasi-crystalline starting model are refined using least-squares techniques in the normal way together with parameters describing the strain distortion which limits long range order and leads to a broadening of the Bragg peaks from the ideal (unstrained) crystal. An example [28,30] of this technique is given in figure 9 which shows the result of refining the structure of a series of $2PbF_2 \cdot TF_2 \cdot MF_3$ glasses by comparison with experimental data obtained using isomorphous substitution for T^{2+} and M^{3+}. The reliability factors (R factors) for the model fit to the three data sets

$$R = 100 \left[\frac{\sum \left[I_{exp}(2\theta) - I_{fit}(2\theta) \right]^2}{\sum I_{exp}^2(2\theta)} \right]^{\frac{1}{2}} \quad (10)$$

lie between 3.2 and 5.2%.

As indicated in the previous paragraph an alternative approach to minimising the potential energy for a predetermined model is to move the atoms in the model according to some specific algorithm in such a way as to improve the agreement between model and experiment in either real or reciprocal space. This is the basis of the random displacement technique first used by Averbach and co-workers [31-34]. The problem with this method was that the then available computer power limited the number of atoms that could be used (N=100-200) but with the very great increase in computer power during recent years the random displacement technique should now be capable of considerable improvement. The results obtained must, however, be subject to careful analysis, e.g. by comparing models generated from several very different starting configurations, since the number of parameters involved in the fit to experiment is very large (3N plus thermal factors – in comparison the cubic spline fit to the vitreous BeF_2 data used later in figure 10 only requires 231 parameters and has an R factor of 0.6%).

5 DIFFRACTION STUDIES OF HALIDE GLASSES

In this section a brief review is given of existing X-ray and neutron diffraction studies of halide glasses known to the author together with a few results for oxyhalide glasses. The amorphographic relationships between the various structures are discussed and also those with other non-halide amorphous solids.

5.1 Fluoroberyllate Glasses

The remarkable correspondence between the polymorphism of BeF_2 and that of SiO_2 is held [35] to be due to the similar radii and polarisabilities of the F^- and O^{2-} ions and the fact that for both materials the radius ratio is appropriate for tetrahedral bonding. The strength of the Be-F bond is, however, much less than that of the Si-O bond and hence BeF_2 may be considered a weakened analogue of SiO_2 having a lower melting point (540°C) and hardness and higher solubility and chemical reactivity [35]. Vitreous BeF_2 is the most extensively studied halide glass and a number of structural investigations have been performed using both X-rays [36-45] and neutrons [45,46]. These indicate that the structure of vitreous BeF_2 is indeed closely similar to that of vitreous SiO_2 comprising a network of corner sharing BeF_4 tetrahedral units with an average Be-F bond length of 1.55Å. Vitreous BeF_2 has an open structure with an anion packing density of 0.427 and the average Be-F-Be angle appears to be close to the Si-O-Si angle in vitreous SiO_2. Various crystal based models have been used in interpreting the structure of vitreous BeF_2. Narten [42] proposes a model based on the β-quartz structure with 6% random vacancies whereas Leadbetter and Wright [45] find that both X-ray and neutron data are best explained by a quasi-crystalline model based on the α-cristobalite structure with a correlation length of 11Å. In both these models it was necessary to use positional parameters from the corresponding silica polymorphs due to the lack of crystal structure data for any of the BeF_2 polymorphs. Recently, however, the structures of α and β-quartz BeF_2 have been determined [47] and the parameters obtained are remarkably similar to those of the corresponding SiO_2 polymorphs. Liquid BeF_2 has been studied by Vaslow and Narten at 700°C [43] and found to retain a corner-sharing tetrahedral network structure.

Several Monte Carlo and Molecular Dynamics simulations have been performed of liquid and vitreous BeF_2 [9,48-53] and predict the basic corner-sharing BeF_4 tetrahedral network. The simulation by Brawer [9] is compared to recent neutron diffraction data in figure 10. The Molecular Dynamics simulation accurately predicts the width of the first Be-F peak, but the distribution of F-F distances is much too broad. The mean Be-Be distance (Be-F-Be angle) is too high while the structure in the correlation function at higher r is much less than that found for the real material.

The structures of the alkali fluoride – beryllium fluoride glasses are thought to be similar to those of the vitreous alkaline earth silicates. Zarzycki [40] has investigated a series of glasses of composition $MF \cdot 2BeF_2$ (M = Li, Na or K) using X-ray diffraction and concluded that the glasses are phase separated probably into pure BeF_2 and $MF \cdot BeF_2$. A more detailed X-ray study of the $LiF-BeF_2$ system has been performed by Vaslow and Narten [41,43], including both vitreous $0.15LiF \cdot 0.85BeF_2$ and liquids of composition $LiF \cdot BeF_2$ (400°C), $2LiF \cdot BeF_2$(555°C) and $4LiF \cdot BeF_2$ (745°C). In the glass a basic tetrahedral network is retained whereas in the liquid with increasing LiF concentration the network becomes progressively distorted in such a way that only Be^{2+} ions retain their well-defined tetrahedral environment, while the Li^+ ions occur in local, instantaneous environments which are grossly distorted from average tetrahedrality. Vaslow and Narten conclude that their mean distances and coordination numbers are in good agreement with the results of a complementary Molecular Dynamics study by Rahman et al [48]. A similar study of liquids in the $NaF-BeF_2$ system has been reported by Umesaki et al [54,55], at compositions $NaF \cdot 2BeF_2$ (470°C), $NaF \cdot BeF_2$ (470°C) and $2NaF \cdot BeF_2$ (650°C). Again the basic tetrahedral network is gradually destroyed with the addition of the network modifier until at the highest NaF content the melt contains mainly monomeric BeF_4^{2-} anions. The authors suggest that short chain anions (eg $Be_2F_7^{3-}$ and/or $Be_3F_{10}^{4-}$) mainly occur in $NaF \cdot BeF_2$ (c.f. figure 1(c)) whereas $NaF \cdot 2BeF_2$ contains $(Be_nF_{3n})^{n-}$ cyclic anions (n = 3,4,5 and 6).

A study of the rare earth environment in vitreous $NaF-DyF_3-BeF_2$ has been performed by the present author and co-workers using the null technique [13,46] and the results compared with Molecular Dynamics simulations of the same glass by Brawer [9]. Data have also been obtained for vitreous $NaF-BeF_2$ with the same $NaF:BeF_2$ ratio and for pure vitreous BeF_2. A fit to the first Dy-F peak in the difference correlation function for the glass containing DyF_3 ($\bar{b}_{Dy}$ = 0.0 and 1.69 x 10^{-14}m) indicates that on average each Dy^{3+} ion is surrounded by 7.3 ± 0.5 F^- ions at a distance of 2.30Å with an r.m.s. bond length variation of 0.086Å

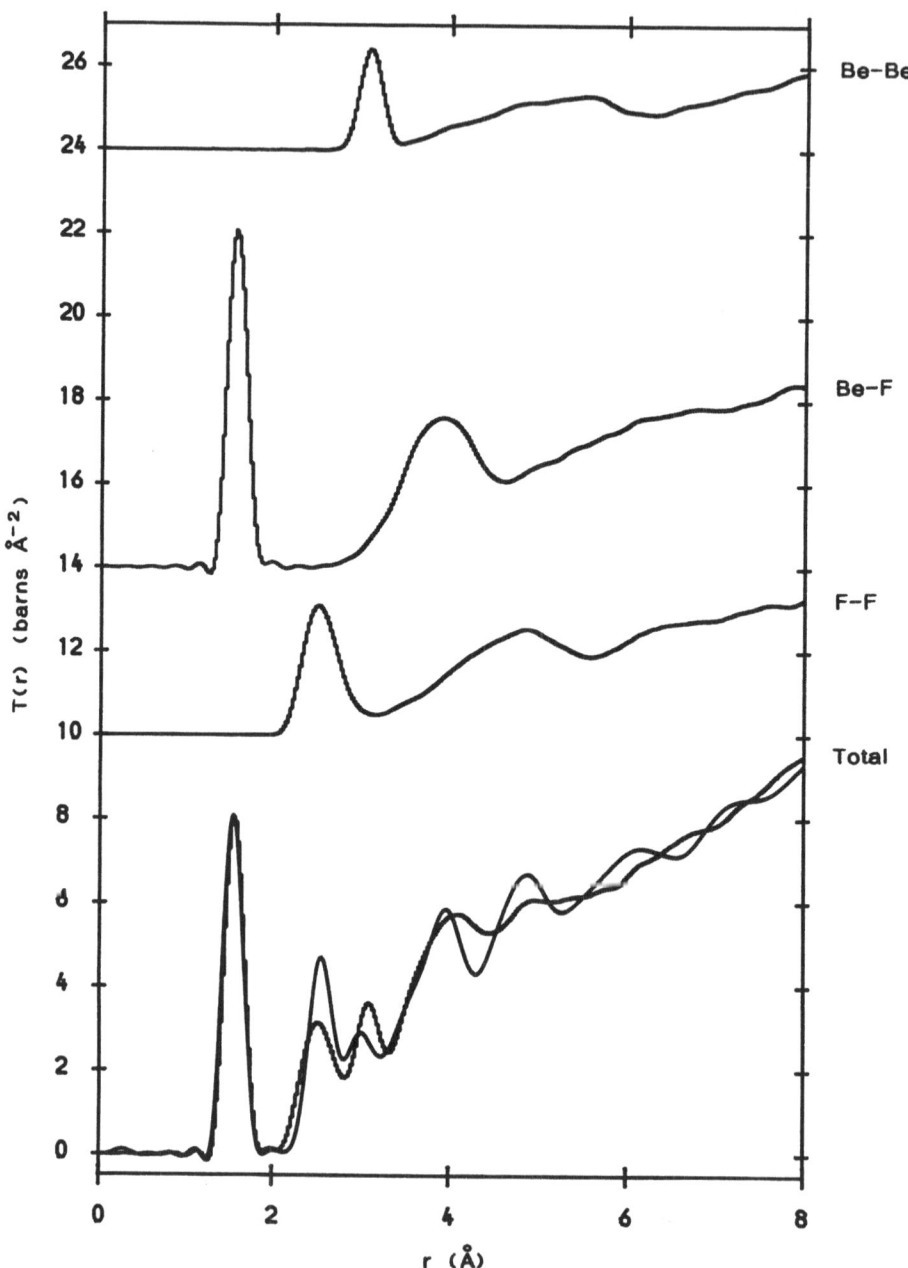

FIGURE 10 A comparison of the neutron correlation function for vitreous BeF$_2$ (solid line) with the Molecular Dynamics simulation of Brawer [9] (histogram).

[13], although the peak fit does exclude a slight asymmetry on the high r side of the peak which if included would slightly increase the co-ordination number. The first Dy(F) co-ordination shell is thus well defined and not simply the result of the Dy^{3+} ions occupying unmodified holes in the basic fluoroberyllate network. The Molecular Dynamics simulations suggest that the addition of NaF and DyF_3 to vitreous BeF_2 leads to the formation of

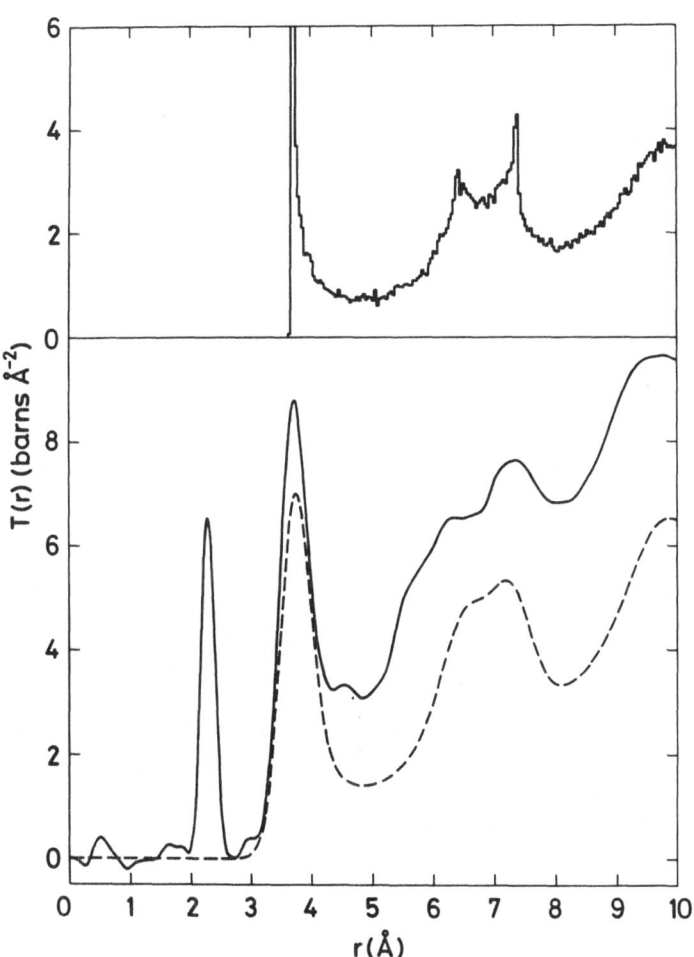

FIGURE 11 The component correlation function for a random close packed array of Cl^- ions (dashed curve) compared to the experimental correlation function for vitreous $ZnCl_2$ [23] (full line). The histogram shows $t_{Cl-Cl}(r)$ without thermal/disorder broadening.

some Be atoms five-fold co-ordinated by F and this point is currently being investigated in detail using data with much higher real space resolution. (The high resolution data for pure vitreous BeF_2 were used to construct figure 10.)

5.2 Vitreous $ZnCl_2$

The structure of vitreous $ZnCl_2$ has been investigated using both X-ray [56,57] and neutron diffraction [23,58], and that of the corresponding liquid at 327°C using neutron diffraction with isotopic substitution to obtain the three reduced component radial density functions [59]. All the known crystalline polymorphs of $ZnCl_2$ have close packed arrays of Cl^- ions and a calculation of the Cl^- ion packing density for vitreous $ZnCl_2$ (0.656) gives a value very close to that for a random close packing (0.6366 [4]). Figure 11 compares the component correlation function for a random close packed array of Cl^- ions with the neutron total correlation function $T(r)$ for vitreous $ZnCl_2$ and it can be seen that the $Cl-Cl$ component correlation function accounts for all the major features in $T(r)$ above 3A. Desa et al. [23] therefore conclude that the structure of vitreous $ZnCl_2$ is best described as a distorted random close packed array of Cl^- ions with the Zn^{2+} ions occupying tetrahedral holes in such a way as to maximise corner sharing of the resulting $ZnCl_4^{2-}$ tetrahedra at the expense of edge and face sharing. The $ZnCl_4^{2-}$ tetrahedra themselves are as regular as those found in the crystalline $\delta-ZnCl_2$ obtained on devitrificaiton. An alternative way of viewing the structure is that it is the result of reducing the $Zn-Cl-Zn$ angles in a silica-like random network, allowing bond rotations, until the anions become approximately randomly close packed. Vitreous $ZnCl_2$ is thus intermediate between BeF_2, with its open network structure, and the heavy metal fluoride glasses described in the next two sections, which also appear to have structures based on a random close packing. The $Zn-Cl$ bond length is 2.29Å and the bond is seen to have some covalent character as evidenced by the fact that Molecular Dynamics simulations using purely ionic potentials [49,60,61] predict a first $Zn-Zn$ distance and $Zn-Cl-Zn$ angle which are much too large (c.f. vitreous BeF_2). It is interesting to note that in the first Molecular Dynamics simulation of liquid $ZnCl_2$ Woodcock et al. [49] obtained a distribution of $Zn(Cl)$ co-ordination numbers from 3 to 6 with an average of 4.7. However, a tetrahedral structure for the glass at 300K was obtained [60] on reducing the ionic radius of Zn^{2+} to give the correct $Zn-Cl$ distance, together with a radius ratio appropriate to a tetrahedral $ZnCl_4$ unit. This result emphasises the role played by radius ratio in determining structure in these simulations which employ non-directional ionic potentials.

5.3 Heavy Metal Fluoride Glasses

Diffraction studies of fluorozirconate glasses [62-69] have concentrated on the BaF_2-ZrF_4 binary system which is the base glass for systems of commercial interest but has the advantage over the latter of only containing 3 elements. In addition Etherington et al [66,67] have assumed isomorphous substitution of Hf^{4+} for Zr^{4+} in an investigation of the corresponding $BaF_2 \cdot 2HfF_4$ fluorohafnate glass and Kawamoto and Horisaka [65] have investigated vitreous $PbF_2 \cdot ZrF_4$.

The glass-formation region in the BaF_2-ZrF_4 system is from 20 to 50% BaF_2 with the glass of composition $BaF_2 \cdot 2ZrF_4$ being particularly easy to prepare [70]. $xBaF_2 \cdot (1-x)ZrF_4$ glasses have been studied systemmatically as a function of composition using X-rays by Coupé and co-workers [62-64] (x = 0.25, 0.3, 0.35 and 0.4 together with devitrified $2BaF_2 \cdot 3ZrF_4$ (x =

0.4)). These authors find a mean Zr-F bond length of 2.09-2.11Å and identify two Ba-F distances of 2.61-2.66 and 3.07-3.19Å. With increasing BaF_2 content the Zr(F) co-ordination number decreases from 8.4 to 7.1 whereas the total Ba(F) co-ordination number increases from 8.7 to 10.7. These values are calculated using free atom form factors and are similar to

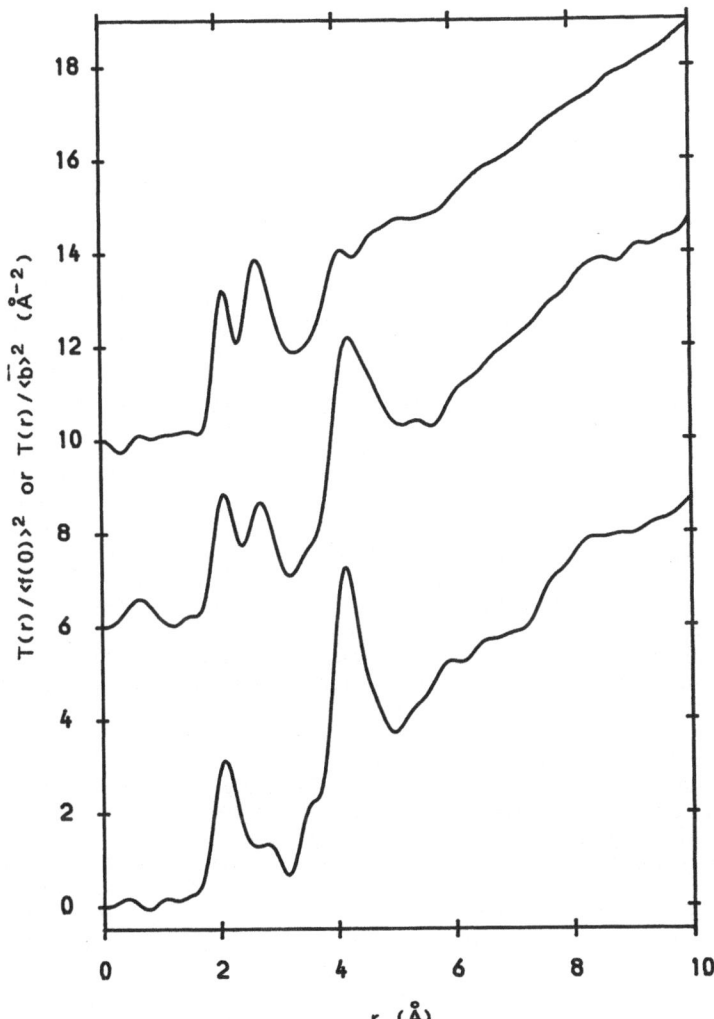

FIGURE 12 X-ray and neutron correlation functions for barium fluorozirconate and fluorohafnate glasses [66-68]. Top $BaF_2 \cdot 2ZrF_4$ neutron, middle $BaF_2 \cdot 2ZrF_4$ X-ray and bottom $BaF_2 \cdot 2HfF_4$ X-ray.

those $(Zr(F)=7$ or 8; $Ba(F)=10)$ found in related crystalline compounds [71]. The Zr-F interaction is, in fact, a special case in that the electron product $K_{Zr}K_F$ is the same $(360e^2)$ for both free atoms and free ions. Coupé et al. suggest that the average values they obtain for the Zr-F-Zr bond angle of 150 to 160° eliminate the possibility of edge sharing between adjacent ZrF_n polyhedra, but in a later paper Lucas et al [64] identify a second Zr-Zr distance and calculate that the ratio of corner sharing to edge sharing is about 3 to 1, a ratio which is almost independent of composition.

Kawamoto and Horisaka [65] interpret X-ray diffraction data for the glasses $BaF_2 \cdot ZrF_4$ and $PbF_2 \cdot ZrF_4$ on the basis of the crystal structures of β-$BaZrF_6$ and α-$PbZrF_6$ respectively. The radial distribution functions for the two glasses are very similar and Kawamoto and Horisaka propose a structure comprising edge-sharing ZrF_8 chains linked ionically by Ba^{2+} or Pb^{2+} ions.

The isomorphous substitution technique is used by Etherington et al. [66-68] for a glass of approximate composition $BaF_2 \cdot 2ZrF_4$, the Zr^{4+} being fully replaced by Hf^{4+} to give $BaF_2 \cdot 2HfF_4$. X-ray data have been obtained for both glasses and neutron data for $BaF_2 \cdot 2ZrF_4$ and the resulting reduced correlation functions $T(r)/(f(0))^2$ and $T(r)/(\bar{b})^2$ are reproduced in figure 12, the averages being taken over the atoms in one composition unit. The X-ray data indicate that the first Zr-F and Hf-F distances in the two glasses are identical (2.08Å) as are the Zr(F) and Hf(F) co-ordination numbers obtained from peak areas within experimental error $(7.4$ and 7.7 ± 0.5 respectively, giving an average value of 7.6). A subsequent reanalysis of the combined X-ray and neutron data for vitreous $BaF_2 \cdot 2ZrF_4$ using peak fitting techniques, however, leads to a lower mean Zr(F) co-ordination number of 6.7 and a mean Ba(F) co-ordination number of 15.5. The authors conclude that Hf does indeed replace Zr isomorphously in these glasses. The position and width of the first A-F peak (A=Zr or Hf) indicates that the basic AF_n units contain both bridging and non-bridging F^- ions and the mean A-F-A angle is found to be 170-180°. The results of Etherington et al. are in good agreement with those of Coupé et al. who for their glass of composition closest to $BaF_2 \cdot 2ZrF_4$ find a Zr-F distance of 2.09Å and a co-ordination number of 7.35. Etherington et al. find no evidence for a second Ba-F distance of ~3.2Å in either of their samples and suggest that this may be the result of termination effects, especially as Coupé et al. did not use a modification function when Fourier transforming their data.

Two structural models are discussed by Etherington et al. [66]. The first is based on ZrF_8 Archimedean antiprismatic units. In crystalline β-ZrF_4 [72] these units share all corners to form a three-dimensional array. In BaF_2-ZrF_4 glasses, however, Baldwin et al. [73] postulate that the fully connected network of ZrF_8 units is broken down by the addition of BaF_2 which acts as a conventional network modifier. If it is assumed that the BaF_2 becomes fully ionised then for $BaF_2 \cdot 2ZrF_4$ this would give rise to six bridging and two non-bridging F atoms per ZrF_8 unit. An alternative model has been proposed by Almeida and Mackenzie [74] as a result of Raman studies which suggest that the co-ordination number of F atoms **bonded** to Zr atoms is unlikely to be 8. Chains of ZrF_6 octahedral units are cross-linked by Ba-F ionic bonds as indicated schematically in figure 13 for $BaF_2 \cdot 2HfF_4$ [75]. Each ZrF_6 unit thus has two bridging and four non-bridging F atoms, while the total Zr(F) co-ordination number would be increased above 6 by one or two extra non-bridging fluorine neighbours from adjacent chains. Etherington et al. [66] conclude that their data are not inconsistent with either of these models although in their later paper describing the neutron data [67] they discuss only the second model.

A combined X-ray diffraction and Molecular Dynamics simulation of BaF_2-ZrF_4 glasses of two compositions, $BaF_2 \cdot 3ZrF_4$ and $2BaF_2 \cdot 3ZrF_4$, has been reported by Inoue et al. [69,76]. These authors perform an extensive analysis to elucidate the structural units present and compare their data with models containing Zr in 6,7 and 8-fold co-ordination. The optimum model for octahedral co-ordination comprises chains of octahedra sharing two corners similar to that proposed by Almeida and Mackenzie [74,75], whereas with 7-fold co-ordinated units the best fit is obtained with some polyhedra sharing two edges and others sharing one corner and one edge. Correlation functions are shown for 8-fold co-ordinated units sharing, corners, edges and a combination of corners and edges. Inoue et al. suggest that these model fits indicate that the Zr(F) co-ordination number is close to 8. The models were subsequently relaxed by Molecular Dynamics simulation at 300°C (close to T_g) whereupon the Zr(F)

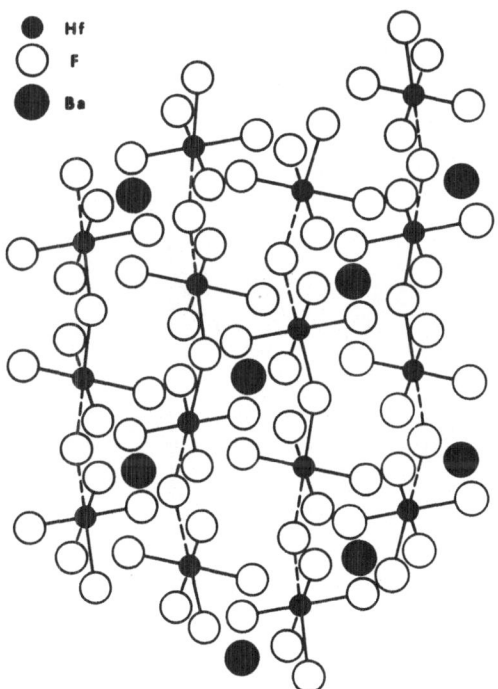

FIGURE 13 Two-dimensional schematic representation of the structural model for vitreous $BaF_2 \cdot 2HfF_4$ proposed by Almeida and Mackenzie [75].

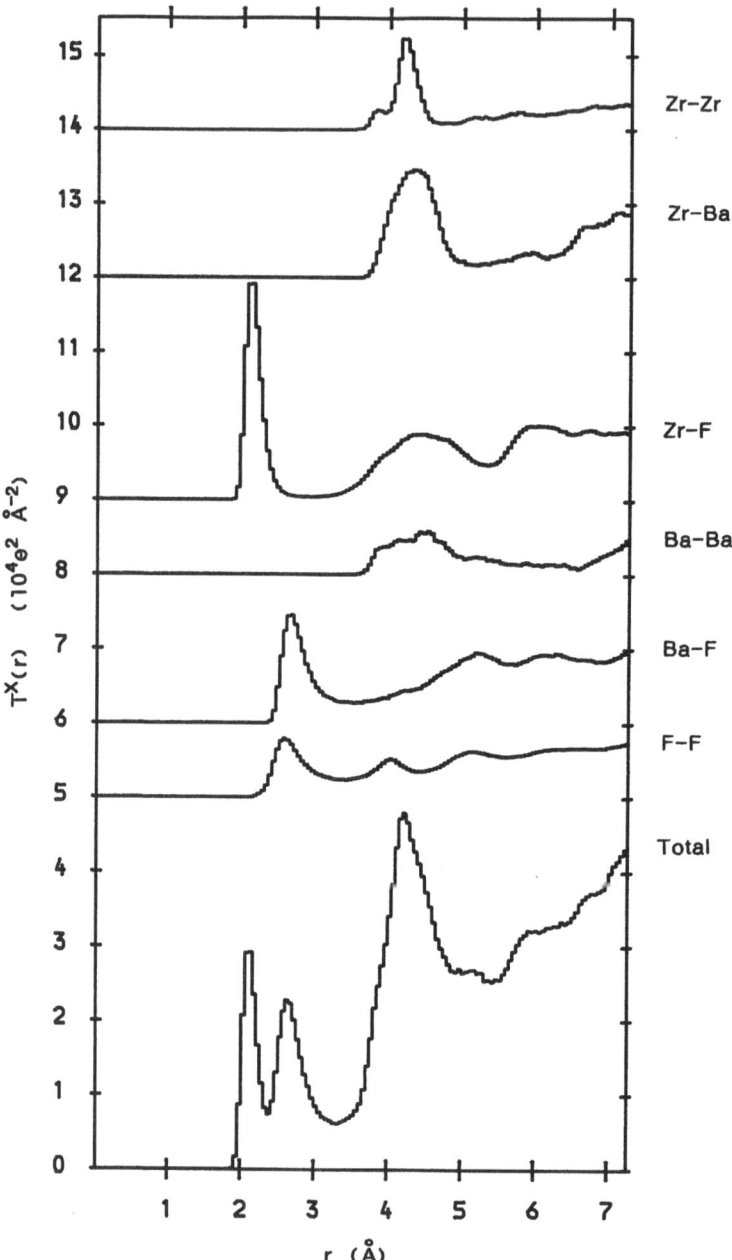

FIGURE 14 Unbroadened X-ray (ionic) correlation function and individual contributions for vitreous $2BaF_2 \cdot 3ZrF_4$ calculated from the Molecular Dynamics simulation of Hamill and Parker [78,81].

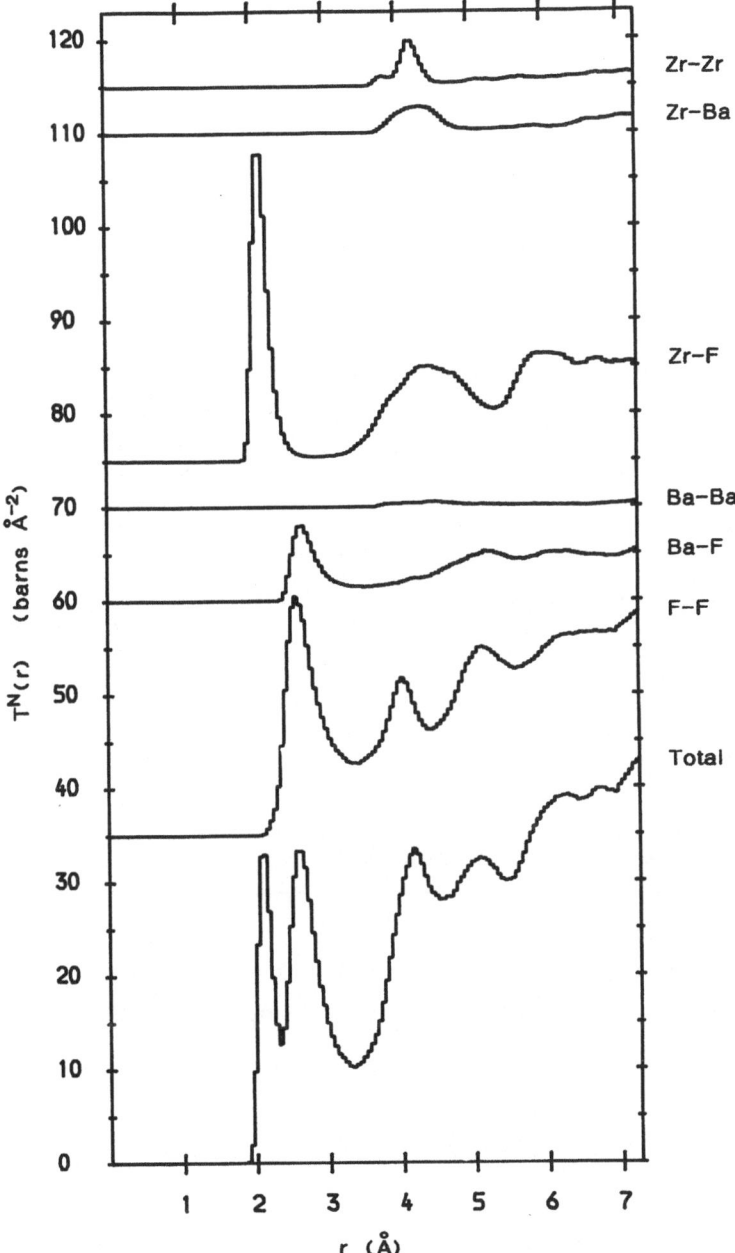

FIGURE 15 Unbroadened neutron correlation function and individual contributions for vitreous $2BaF_2 \cdot 3ZrF_4$ calculated from the Molecular Dynamics simulation of Hamill and Parker [78,81].

co-ordination number for the models containing ZrF_6 and ZrF_7 polyhedra increased as summarised in table 2. The linkage between ZrF_n units also changed. Further Molecular Dynamics simulations were also performed starting from various crystal structures, adjusted to give the correct composition and density. The simulations which agree most closely with experiment contain ZrF_8 units with a preference for corner sharing, although there is also some edge sharing associated with Ba^{2+} ions. Inoue et al. therefore conclude that the most likely structural unit in BaF_2-ZrF_4 glasses is ZrF_8. These polyhedra share corners when the concentration of BaF_2 is small, forming a three-dimensional continuous network, but as the BaF_2 content increases edge-sharing is introduced as just described.

Other Molecular Dynamics simulations have been reported for the BaF_2-ZrF_4 system [26,64,77-80] and correlation functions for vitreous $2BaF_2 \cdot 3ZrF_4$ are shown in figures 14 and 15 taken from a simulation by Hamill and Parker [78,81] and weighted respectively for X-ray (ionic form factors) and neutron diffraction but **with no resolution broadening.** The simulation involved equilibriation at 6000K followed by quenching and further equilibriation at 300K. Several points may be noted. First the different weighting factors $K_j K_k$ and $\bar{b}_j \bar{b}_k$ are immediately apparent. Notice in particular that the F-F contribution is much greater in the neutron case, as indicated in section 2.6, while that due to Ba-Ba is negligible. Even with no resolution broadening the first two peaks are not resolved and the second peak involves both Ba-F and F-F components. The first Zr-F peak is asymmetric (skewed to high r) and so a determination of the Zr(F) co-ordination number assuming a symmetric first peak will give a value which is too low. Although the composition is different, so that a direct comparison is not meaningful, the Molecular Dynamics simulation nevertheless does account qualitatively for the features in the X-ray and neutron data of Etherington et al. [66-68] (figure 12) and in particular for the relative heights of the first three peaks.

TABLE 2   Change in structural models of vitreous $2BaF_2 \cdot 3ZrF_4$ after Molecular Dynamics Simulation [69].

Initial Model		Final Model	
Structural Unit	Connectivity	$n_{Zr(F)}$	Connectivity
ZrF_6	Chains–corner sharing	7.8-8.0	Network
ZrF_7	Chains–corner and edge sharing	7.3-8.0	Network
ZrF_8	Edge sharing	7.3-8.0	Corner sharing
ZrF_8	Edge and corner sharing	7.4	Corner sharing
ZrF_8	Corner sharing	8.0	Corner sharing

The Molecular Dynamics simulations of BaF_2-ZrF_4 glasses all indicate a Zr(F) co-ordination number of ~8 (c.f. Inoue et al. above) and a Ba(F) co-ordination number of ~10-11. The F^- ions surrounding the Ba^{2+} ions are predominantly non-bridging and the Zr-F-Zr angle is close to 180° [78] for F^- ions involved in corner sharing. At the composition shown in figures 14 and 15 Hamill and Parker [79] find Zr(F) and Ba(F) co-ordination numbers of 7.84 and 9.90 respectively and in the former case 49.4% of the F^- ions are non-bridging, 42.0% are involved in corner-sharing bridges and 8.6% in edge-sharing bridges. As a result of their X-ray data [62,63] and Molecular Dynamics simulations [64], Lucas et al. [70,82] propose a structural model for vitreous $BaF_2\cdot2ZrF_4$ based on the $Zr_2F_{13}^{5-}$ unit which comprises ZrF_7 and ZrF_8 polyhedra sharing a common edge, the internal Zr-Zr distance being 3.6Å. The ZrF_7 and ZrF_8 polyhedra each share 3 corners with other ZrF_n polyhedra (r_{Zr-Zr}~4.2Å) and there are a total of 5 non-bridging F^- ions per $Zr_2F_{13}^{5-}$ unit.

The results just described illustrate the usefulness of Molecular Dynamics simulations in interpreting diffraction data for the heavy metal fluoride glasses and shows the need for a rigorous comparison between simulation and experiment for glasses of identical composition – a comparison which should be made using accurate data with high real space resolution and *must* include convolution of the simulated component correlation functions with the correct component peak functions $P_{jk}(r)$ (c.f. figure 10 for vitreous BeF_2). Frequently the radial density functions, $\rho_{jk}(r)/\rho_k^0$, from Molecular Dynamics simulations are calculated with a fairly broad histogram column width, Δr, (typically ~0.1Å). This does, however, cause extra broadening when convolving with the component peak functions $P_{jk}(r)$ and so a much narrower column width should be employed (Δr~0.01Å). The increased statistical noise thus incurred is smoothed by the convolution process. Given that the co-ordination numbers agree with those from diffraction experiments it is now important to investigate the *widths* of early peaks in the correlation function to see whether the Zr(F) and Ba(F) co-ordination polyhedra in the simulations are as well-defined (regular) as those indicated by diffraction data. Another aspect which should be carefully checked with experiment is the Zr-F-Zr angle since it is in predicting bond angles that Molecular Dynamics simulations have failed for materials such as vitreous BeF_2 and $ZnCl_2$ (c.f. sections 5.1 and 5.2).

A comprehensive study has been performed by Le Ball et al. [16,28,83-87] of three series of heavy metal fluoride glasses:-

	(i)	$2PbF_2\cdot TF_2\cdot MF_3$
	(ii)	$AF_2\cdot TF_2\cdot MF_3$ ($A^{2+} = Pb^{2+}$, Ba^{2+})
	(iii)	$6NaF\cdot5AF_2\cdot9MF_3$ ($A^{2+} = Ca^{2+}$, Sr^{2+}, Pb^{2+})

with T^{2+} = Mn^{2+}, Zn^{2+}, Cu^{2+} and M^{3+} = V^{3+}, Cr^{3+}, Fe^{3+}, Ga^{3+}. The techniques used include neutron diffraction (assuming isomorphous substitution), neutron magnetic diffraction, X-ray diffraction and EXAFS spectroscopy, together with the modified Rietveld technique discussed in section 4.2. The structure of glasses in the first series [85] is envisaged as comprising a relatively close packing of F^- and Pb^{2+} ions with no Pb-Pb contacts and the T^{2+} and M^{3+} ions in octahedral sites. Alternatively this structure may be viewed as a corner-sharing octahedral network with the Pb^{2+} ions in interstitial sites and is thus analagous to that of vitreous $ZnCl_2$, except that it is octahedral (rather than tetrahedral) holes which are filled, together with larger holes for the Pb^{2+} ions. A very similar structure is suggested for glasses in the second series [83]. The magnetic radial distribution functions for vitreous $2PbF_2\cdot MnF_2\cdot FeF_3$ and the same glass after dilution with diamagnetic ions (Mn^{2+} substituted by Zn^{2+} and Fe^{3+} by Ga^{3+}) [16] show that Mn-Mn and Fe-Fe first neighbour spins tend to align antiferromagnetically and have a separation of 3.57–3.65Å. The Mn-Fe contribution is predominant for second neighbours and is ferromagnetic. A

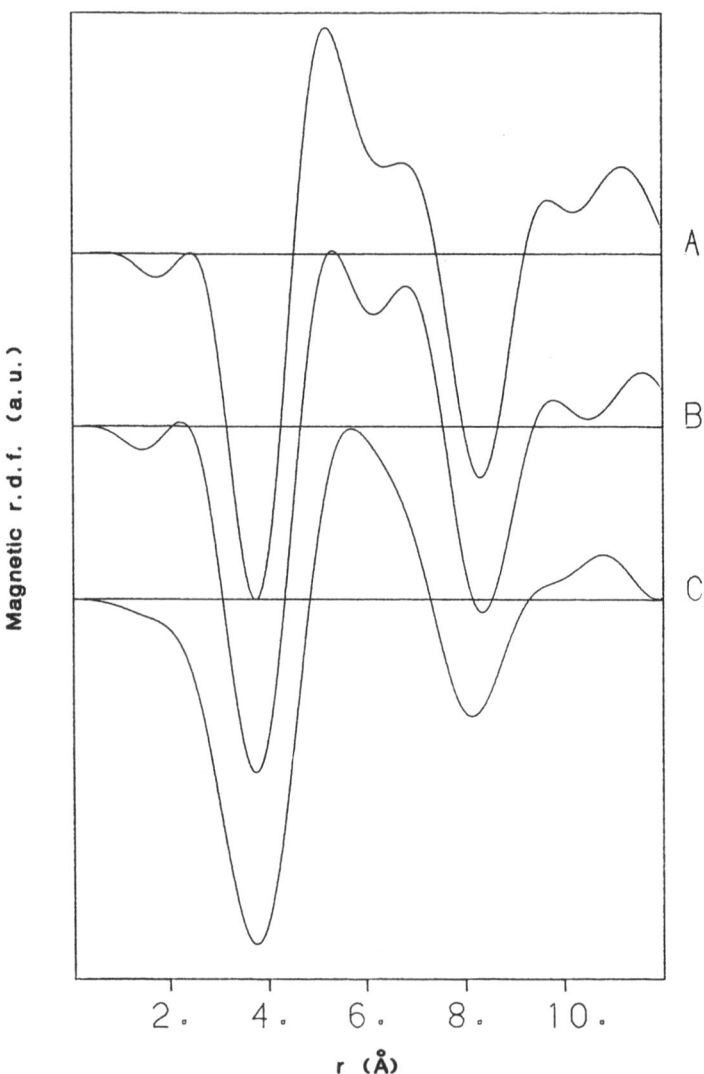

FIGURE 16 Magnetic radial distribution functions for (A) $PbF_2 \cdot MnF_2 \cdot FeF_3$, (B) $2PbF_2 \cdot MnF_2 \cdot FeF_3$ and (C) $6NaF \cdot 5SrF_2 \cdot 9FeF_3$ ($Q_{max} = 4.5Å^{-1}$) [86].

combination of the Fe-Fe and Mn-Mn spacings with EXAFS data for the Fe-F (1.93Å) and Mn-F (2.10Å) bond lengths yields Fe-F-Fe and Mn-F-Mn angles of 135° and 117° respectively. The first peak in the magnetic correlation function for glasses in the second and third series also indicates antiferromagnetic first neighbour ordering at 3.6 ± 0.1Å [86], as indicated in figure 16, which is consistent with corner-sharing TF_6 and MF_6 octahedra.

The only other reported diffraction data for heavy metal fluoride glasses is an X-ray diffraction study of vitreous $BaF_2 \cdot MnF_2 \cdot UF_4$ and vitreous $BaF_2 \cdot VF_3 \cdot UF_4$ [88]. The U^{4+} exists as UF_8 units (r_{U-F}=2.28Å) and the authors suggest a structure comprising clusters of UF_8 polyhedra linked by MF_6 (M = Mn or V) octahedra to explain the magnetic properties of these glasses.

5.4 Oxy Halide Glasses

Although strictly outside the scope of this review it is perhaps useful to briefly summarise the few diffraction studies to date on oxyhalide glasses. Matz et al. [89] have studied a range of fluorophosphate glasses of composition $(1-x)(CaF_2, AlF_3) \cdot xBa(PO_3)_2$ with x = 0.05, 0.1, 0.2, 0.5 and 1.0 together with vitreous $0.95(CaF_2, AlF_3) \cdot 0.05Sr(PO_3)_2$, the CaF_2 to AlF_3 ratio being 4:3. For x=1.0 the first (P-O) peak is at 1.55Å and yields a P(O) co-ordination number of 3.9 ± 0.1 confirming that the basic structural unit is the PO_4 tetrahedron, while analysis of the differential correlation function for x = 0.05 indicate that Al is present as AlF_6 octahedra (r_{Al-F} =1.65Å) as in crystalline Ca_2AlF_7.

Glasses can be prepared in the $PbO-PbCl_2$ and $PbO-PbF_2$ systems although in both cases silica contamination is present from the melting tube. The structure of these glasses has been investigated using a range of techniques including X-ray diffraction and EXAFS spectroscopy by Rao and co-workers [90-95] who suggest a predominance of PbO_2X_4 units (X = Cl, F) over the entire composition range and that -Pb-O-Pb-O- linkages play a crucial role in determining the limits of glass formation. A neutron diffraction study of $PbO-PbCl_2$ glasses has been reported by Wright et al. [93] which confirms previous EXAFS results [91] that whereas the first Pb-O distance is well defined, the distribution of Pb-Cl distances is much broader. This is consistent with the decrease in intermediate range order with increasing $PbCl_2$ content as evidenced by the reduction in the structure in D(r) at higher r.

6. THE LOCALLY-ORDERED RANDOM CLOSE PACKING MODEL

While a fairly clear first order structural model has emerged for the fluoroberyllate glasses and vitreous $ZnCl_2$ the picture for the heavy metal fluoride glasses is much less certain, particularly in the case of the fluorozirconates where even the identity of the basic structural unit, if any, is still the subject of some controversy. It is therefore of interest to consider the possible structure of these glasses from first principles by extending the crystallo-chemical considerations discussed in section 1.1 which are equally applicable to the amorphous state, if not more so due to the lack of the constraints imposed by periodic symmetry.

Brunner [96] has presented an unconventional view of crystalline close packing in which he argues that, given a finite volume, the close packed arrangement results from the requirement that the separation between neighbouring atoms/ions should be a maximum. In this case the spacial distribution of the centres is considered and assumptions of spherical shape and mutual contact are in fact unnecessary. The mutual repulsion argument can clearly be applied to an array of F^- ions and in the amorphous state the crystalline close packing will be replaced by random close packing. This will then be modified to a greater or lesser extent by the various cations present.

In crystallographic terms there is much in common between metallic and ionic structures since in both cases the atoms/ions can be considered as spheres. A useful starting point for a model of heavy metal fluoride glasses is thus the structural models that have been proposed for amorphous transition metal – metalloid alloys. Polk [97,98] has suggested a model for the latter in which the metalloid atoms are inserted into the larger Bernal holes (figure 2(c)-(e)) in a random close packed array of transition metal atoms. If all such holes are filled the composition is calculated to be $T_{0.791}M_{0.209}$. very close to that of the deep eutectic ($\sim T_{0.8}M_{0.2}$) which occurs in many of these systems. Optimum glass formation occurs around this eutectic composition and Polk argues that the presence of the metalloid atoms in the holes stabilises the structure (pure elemental amorphous metals are extremely unstable) and that as the fraction of holes filled is reduced the structure becomes increasingly unstable until the lower limit of glass formation is reached at $\sim T_{0.85}M_{0.15}$. The number of holes large enough to incorporate a metalloid can be increased if the atoms surrounding a given hole relax outwards slightly when the hole is filled and/or the packing density is reduced somewhat to a value nearer that for a loose random packing of the type studied by Scott [99,100]. ($\eta \sim 0.60$ – packings in which the spheres are not compacted to give the maximum possible packing density but left as initially formed). The essence of the model is that it avoids M-M contacts, which are not observed experimentally, and it is this feature which differentiates the present structure from a simple random packing of spheres of two different sizes.

The main objection to Polk's model as originally proposed is that the metalloid atoms are considered to be almost totally passive which is contrary to experience with amorphous solids where the absence of a periodic structure means that an added atom/ion can very much more easily adapt its immediate environment, to satisfy its own local bonding/radius ratio requirements, than is the case in crystalline materials. It is for this reason that the technique of doping is rarely successful in amorphous systems. As discussed by Gaskell [101,102] the metalloid co-ordination polyhedron in the corresponding crystalline T-M alloys is the trigonal prism (c.f. figure 2(c)) and hence it is likely that metalloid atoms inserted initially into holes (d) and (e) would modify their environment to (c) thus adjusting the relative numbers of each type of hole. A method of modelling T-M amorphous alloys therefore is to fill the Bernal holes as just described and then to relax the model with a suitable potential appropriate to the desired trigonal prismatic metalloid co-ordination [103]. The distribution of holes originally found by Bernal ((a) 86.2%, (b) 5.9%, (c) 3.8%, (d) 0.5% and (e) 3.7% [98]) is only appropriate for a packing unperturbed by hole filling. Note also that the hole characteristics of an unfilled packing are dramatically altered if the potential is softened from a hard sphere to a truncated Morse or Lennard-Jones potential, the larger holes being destroyed in favour of further (distorted) tetrahedra and octahedra [104].

An alternative approach has been suggested by Gaskell [101] who proposes a model for the T_4M composition based on chains of trigonal prismatic units sharing one edge of each of their triangular faces with the remaining atoms on these faces fitting into a square face of the neighbouring unit to form the half octahedron cap shown in figure 2(c). Randomness is introduced into the model in that the shared edge at each end can be chosen in one of three ways and the chains are arranged such that the transition metal atoms are approximately random close packed. The structure is then relaxed in such a manner as to conserve the transition metal co-ordination shell around each of the metalloid atoms. If however the relaxation were carried out with a less stringent conservation of topology allowing interchange of T neighbours between adjacent M atoms, while still preventing M-M contacts, it is likely that this model would approach that discussed in the previous paragraph. The two methods may thus merely be different ways of generating (viewing) the same basic structure. In this respect the situation is reminiscent of the often pointless arguments between protagonists of the extreme random network and new crystallite theories concerning the extent of the spatial fluctuation in the degree of order present in glasses.

The number of larger holes ((c)-(e)) is almost exactly the number required for modelling the T_4M alloys, and the optimum configuration around the metalloid does not correspond to the largest hole and so the perturbation to the T random close packing by the redistribution of hole types is likely to be relatively small and not lead to very significant changes in packing density or the T-T component correlation function. This situation may be likened to that for vitreous $ZnCl_2$ where the perturbation of the random close packing (of Cl^- ions) merely involves a selection of a subset of the possible topologies so that each $ZnCl_4$ tetrahedron can share four corners with its neighbours. (It is impossible to generate a perfect corner-sharing $ZnCl_4$ network starting with, for example, the Finney packing [4] – it is necessary to constrain the topology as the packing is formed).

Similar structural principles will apply for the heavy metal fluoride glasses as for the T-M alloys. The cation radius r_M is in general smaller than that for the fluoride ion (r_F) and the relatively higher cation charge will prevent M-M contacts. The proposed structural units, however, are not trigonal prisms. In the case of the fluorozirconate glasses the principal suggestions for the ZrF_n unit ($r_{Zr}/r_F = 0.65$ [105]) are the Archimedean antiprism (figure 2(d)) or the octahedron (figure 2(b)) with a somewhat higher co-ordination for Ba^{2+} ($r_{Ba}/r_F > 1$). Octahedral co-ordination is also present for the transition metal ions in the glasses studied by Le Bail et al. while the co-ordination number for Pb^{2+} ($r_{Pb}/r_F > 1$) is 9±2. The F^- ion packing density is given by

$$\eta_F = \frac{4}{3} \pi \ m_F \rho^o r_F^3 \qquad (11)$$

where m_F is the number of F^- ions in the composition unit and ρ^o is the sample density in composition units per $\mathring{A}^3$. From the densities given by Inoue et al. [69] the F^- ion packing densities for the glasses $BaF_2 \cdot 3ZrF_4$ and $2BaF_2 \cdot 3ZrF_4$, assuming $r_F = 1.33\mathring{A}$ [105], are 0.531 and 0.511 respectively. The overall compositions of these glasses ($M_{0.222}F_{0.778}$ and $M_{0.238}F_{0.762}$ respectively; M = Ba or Zr) require slightly more large holes than are available in an unperturbed random close packing and, assuming ZrF_8 Archimedean antiprismatic units together with a Ba(F) co-ordination number of ~10, all the holes to be transformed to the largest Bernal hole (of which there are only 0.5% in an unperturbed packing [98]) or even larger (for Ba^{2+}). It is therefore to be expected that η_F would be less than the value for an unperturbed packing as indeed is the case. Polyanions such as the $Zr_2F_{13}^{5-}$ unit suggested by Lucas et al. [70,82] can also be incorporated by filling two adjacent holes (followed by relaxation). A similar situation exists for the $2PbF_2 \cdot TF_2 \cdot MF_3$ glasses studied by Le Bail et al. [85] in which large holes are necessary to accommodate the Pb^{2+} ions (15.4 atom %). The transition metal ions (15.4 atom %) occupy octahedral holes (5.9% in an unperturbed packing), the overall composition being $M_{0.308}F_{0.692}$ and the F^- ion packing density is calculated to be 0.454.

As discussed in section 5.2, Molecular Dynamics simulations of halide glasses have clearly demonstrated the role of radius ratio in determining the basic structural units present and the critical radius ratios for various co-ordination polyedra have been given in table 1. When the co-ordination number is small, and the radius ratio not much above its critical value, the structural units tend to be well defined, as for example in the simulations of BeF_2 and $ZnCl_2$, and in real materials this regularity is further increased by the presence of some covalent character in the bonding. With larger co-ordination polyhedra, however, there is a greater possibility of distortion and also for a distribution of polyhedral types as demonstrated by the rich co-ordination chemistry of the crystalline fluorozirconates [71]. In studying glasses such as the fluorozirconates, therefore, an important first step for any given composition is to establish the detailed quantitative short range order. This involves not merely the average bond length and co-ordination number but the detailed distribution of M-F and F-F distances associated with the structural units present which may comprise:-

(i) a single well-defined (relatively undistorted) structural unit,

(ii) a single structural unit with varying distortions,

(iii) a distribution of structural units,

or

(iv) no identifiable structural unit.

If reasonably well-defined structural units can be identified the next stage is to determine the manner in which these are interconnected. The possibilities include:-

(i) corner sharing (1 common atom),

(ii) edge sharing (2 common atoms),

(iii) face sharing (3 common atoms),

or

(iv) a combination of any two or all three of the above.

and in case (iv) where there is a distribution of connection modes it is necessary to find the percentage of each mode. Both the structural unit(s) present and their connectivity may be a function of composition. However, only when the short range order is clearly established is it profitable to investigate the intermediate range order.

There is an obvious similarity between the chain models proposed for the heavy metal fluoride glasses and that proposed by Gaskell for the T-M alloys, although the schematic representations of these structures (e.g. figure. 13) seem much too crystalline in character. In practice the chains would be randomly coiled as in the Gaskell model and amorphous polymers. (see for example, Zallen [106] for a description of the random coil model) The question therefore arises as to the extent of chain ordering, if any, and to verify that this exists it is necessary to prove that each structural unit of the type responsible for chain formation is connected to at least two others, not merely that these units on average have approximately two neighbours. A possible structural similarity between metallic glasses and organic polymers has been discussed by Gordon [107] (see also [103]) who suggests a chain structure very similar to that postulated for the heavy metal fluoride glasses. In terms of the latter the structure comprises linear and/or branched chains with a backbone of alternating M^{n+} and F^- ions with the larger cations (e.g. Ba^{2+} or Pb^{2+}) acting as a space-filler (plasticiser) between chains.

7. CONCLUSIONS

The ideas of radius ratio and hole filling discussed in this review are not new but an attempt has been made to bring them together in the context of the heavy metal fluoride glasses and to stress the amorphographic relationships between halide glasses and other classes of amorphous solid. The study of heavy metal halide glasses is at a very early stage and it is important to distinguish between experimental fact (average bond lengths, co-ordination numbers etc.) and hypothesis in the form of the various structural models which have been proposed. In general, the structures of the heavy metal fluoride glasses, particularly the fluorozirconates where even the basic structural unit remains to be firmly established, are much less understood than those of vitreous $ZnCl_2$ and the fluoroberyllate glasses. There is consequently a great need for further diffraction studies of these systems to provide accurate data with high real space resolution, together with similar studies of the corresponding glass-forming liquids above T_g.

REFERENCES

1. Zachariasen, W.H., J.Am.Chem.Soc. 54, 3841 (1932).
2. Greaves, G.N., Fontaine, A., Lagarde, P., Raoux, D. and Gurman, S.J., Nature 293, 611 (1981).
3. Trap, H.J.L. and Stevels, J.M., Glastech.Ber. 32K, 31 (1959).
4. Finney, J.L., Proc.Roy.Soc.Lond. A319, 479 (1970).
5. Bernal, J.D., in *Liquids : Structure, Properties and Solid Interactions*, Ed. Hugel T.J. (Elsevier, Amsterdam, 1965), 25.
6. Cundy, H.M., Math.Gaz. 36, 263 (1952).
7. Wright, A.C., J.Non-Cryst. Solids 75, 15 (1985).
8. Wright, A.C. and Leadbetter, A.J., Phys.Chem.Glasses 17, 122 (1976).
9. Brawer, S.A., private communication.
10. Wright, A.C., Hannon, A.C., Sinclair, R.N., Johnson, W.L. and Atzmon, M., J.Phys.F 14, L201 (1984).
11. Hayes, T.M. and Wright, A.C., in *The Structure of Non-Crystalline Materials 1982*, Eds Gaskell, P.H., Parker, J.M. and Davis, E.A. (Taylor and Francis, London, 1983), 108.
12. Krogh-Moe, J., Acta Chem.Scand. 20, 2890 (1966).
13. Wright, A.C., Etherington, G., Desa, J.A.E. and Sinclair, R.N., J.Phys.Coll. C9, 31 (1982).
14. Wright, A.C., J.Non-Cryst. Solids 76, 187 (1985).
15. Wright, A.C., J.Non-Cryst. Solids 40, 325 (1980).
16. Le Ball, A., Jacoboni, C. and De Pape, R., J.Non-Cryst. Solids 74, 205 (1985).
17. Prins, J.A., in *Physics of Non-Crystalline Solids*, Ed.Prins, J.A. (North-Holland, Amsterdam, 1965), 39.
18. Wright, A.F., J.Non-Cryst. Solids 76, 43 (1985).
19. Wright, A.C., Adv. Struct. Res. Diffr. Meth. 5, 1 (1974).
20. Roth, M. and Zarzycki, J., J.Non-Cryst. Solids 16, 93 (1974).
21. Santini, F., Elarby-Aouizerat, A., Jal, J.F., Dupuy, J., Claudy,

114

    P., Letoffe, J.M., Bellissent-Funel, M.C., Wright, A.F., Francois, B. and Lucas, J., Mat. Sci. Forum 5, 229 (1985).

22.   Sinclair, R.N., Clare, A.G., Seddon, A.B., Parker, J.M. and Wright, A.C., unpublished work.

23.   Desa, J.A.E., Wright, A.C., Wong, J., and Sinclair, R.N., J. Non-Cryst. Solids 51, 57 (1982).

24.   Da Silva, J.R.G., Pinatti, D.G., Anderson, C.E. and Rudee, M.L., Phil. Mag. 31, 713 (1975).

25.   Coombs, P.G., De Natale, J.F., Hood, P.J., McElfresh, D.K., Wortman, R.S. and Shackelford, J.F., Phil. Mag. B51, L39 (1985).

26.   Parker, J.M., this proceedings.

27.   Leadbetter, A.J. and Wright, A.C., J. Non-Cryst. Solids 7, 23 (1972).

28.   Le Ball, A., Jacobonl, C. and De Pape, R., J. Phys. Coll. C8, 163 (1985).

29.   Rietveld, H.M., J. Appl. Cryst. 2, 65 (1969).

30.   Le Ball, A., private communication.

31.   Kaplow, R., Rowe, T.A. and Averbach, B.L., Phys. Rev. 168, 1068 (1968).

32.   Rechtin, M.D. and Averbach, B.L., Solid State Commun. 13, 491 (1973).

33.   Rechtin, M.D., Renninger, A.L. and Averbach, B.L., J. Non-Cryst. Solids 15, 74 (1974).

34.   Renninger, A.L., Rechtin, M.D. and Averbach, B.L., J. Non-Cryst. Solids 16, 1 (1974).

35.   Everest, D.A., *The Chemistry of Beryllium* (Elsevier, Amsterdam, 1964).

36.   Warren, B.E. and Hill, C.F., Z. Krist. 89, 481 (1934).

37.   Brusentsev, F.A. and Yur'ev, G.S., J. Struct. Chem. 8, 159 (1967).

38.   Batsanova, L.R., Yur'ev, G.S. and Doronina, V.P., J. Struct. Chem. 9, 63 (1968).

39.   Yur'ev, G.S. and Brusentsev, F.A., J. Struct. Chem. 9, 279 (1968).

40.   Zarzycki, J., Phys. Chem. Glasses 12, 97 (1971).

41.   Narten, A.H. and Vaslow, F., U.S.A.E.C. Rept ORNL-4706, 212 (1971).

42.   Narten, A.H., J. Chem. Phys. 56, 1905 (1972).

43.   Vaslow, F. and Narten, A.H., J. Chem. Phys. 59, 4949 (1973).

44.   Leadbetter, A.J., Wright, A.C. and Apling, A.J., in *Amorphous Materials*, Eds Douglas, R.W. and Ellis, B. (Wiley, London, 1972), 423.

45.   Leadbetter, A.J. and Wright, A.C., J. Non-Cryst. Solids 7, 156 (1972).

46.   Wright, A.C. et al., to be published.

47.   Wright, A.F., Fitch, A.N. and Wright, A.C., to be published.

48.   Rahman, A., Fowler, R.H. and Narten, A.H., J. Chem. Phys. 57, 3010 (1972).

49.   Woodcock, L.V., Angell, C.A. and Cheeseman, P., J. Chem. Phys. 65, 1565 (1976).

50.   Brawer, S.A., J. Chem. Phys. 72, 4264 (1980).

51.   Brawer, S.A. and Weber, M.J., J. Chem. Phys. 75, 3522 (1981).

52.   Hirao, K. and Soga, N., J. Non-Cryst. Solids 57, 109 (1983).

53. Weber, M.J., Cryst. Latt. Def. and Amorph. Mat. 12, 527 (1985).

54. Umesaki, N., Iwamoto, N., Ohno, H. and Furukawa, K., J. Chem. Soc. Faraday Trans. 1, 78, 2051 (1982).

55. Iwamoto, N., Umesaki, N., Ohno, H. and Furukawa, K., High Temp. Sci., in press.

56. Imaoka, M., Konagaya, Y. and Hasegawa, H., Yogyo-Kyokai-Shi 79, 97 (1971).

57. Shevchik, N.J., O.N.R. Rept HP-29 (ARPA-44) (1972) [see reference 23].

58. Wright, A.C., Etherington, G., Desa, J.A.E., Sinclair, R.N., Connell, G.A.N. and Mikkelsen, J.C. Jr., J. Non-Cryst. Solids 49, 63 (1982).

59. Biggin, S. and Enderby, J.E., J. Phys. C 14, 3129 (1981).

60. Angell, C.A. and Cheeseman, P., private communication [see reference 23].

61. Hirao, K. and Soga, N., Yogyo-Kyokai-Shi 91, 11 (1983).

62. Louër, D., Coupé, R., Lucas, J. and Léonard, A.J., 1st Int. Symp. on Halide and Other Nonoxide Glasses, Extended Abstracts, Cambridge, March 1982.

63. Coupé, R., Louër, D., Lucas, J. and Léonard, A.J., J. Am. Ceram. Soc. 66, 523 (1983).

64. Lucas, J., Angell, C.A. and Tamaddon, S., Mat. Res. Bull. 19, 945 (1984).

65. Kawamoto, Y. and Horisaka, T., J. Non-Cryst. Solids 56, 39 (1983).

66. Etherington, G., Keller, L., Lee, A., Wagner, C.N.J. and Almeida, R.M., J. Non-Cryst. Solids 69, 69 (1984).

67. Etherington, G., Wagner, C.N.J., Almeida, R.M. and Faber, J. Jr., Repts Hahn-Neitner Inst. B411, 64 (1984).

68. Almeida, R.M. and Mackenzie, J.D., J. Phys. Coll. C8, 75 (1975).

69. Inoue, H., Hasegawa, H. and Yasui, I., Phys. Chem. Glasses 26, 74 (1985).

70. Lucas, J., in *Halide Glasses: Preparation and Properties*, Eds Drexhage, M.G. and Moynihan, C.T. (Telford, New Jersey, 1986), in press.

71. Laval, J.P., Frit, B. and Lucas, J., Mat. Sci. Forum 6, 457 (1985).

72. Burbank, R.D. and Bensey, F.N. Jr., Union Carbide Nuclear Co. Rept K-1280 (1956).

73. Baldwin, C.M., Almeida, R.M. and Mackenzie, J.D., J. Non-Cryst. Solids 43, 309 (1981).

74. Almeida, R.M. and Mackenzie, J.D., J. Chem. Phys. 74, 5954 (1981).

75. Almeida, R.M. and Mackenzie, J.D., J. Chem. Phys. 78, 6502 (1983).

76. Yasui, I. and Inoue, H., J. Non-Cryst. Solids 71, 39 (1985).

77. Angell, C.A., 1st Int. Symp. on Halide and Other Nonoxide Glasses, Extended Abstracts, Cambridge, March 1982.

78. Hamill, L.T. and Parker, J.M., Mat. Sci. Forum 6, 437 (1985).

79. Hamill, L.T. and Parker, J.M., Phys. Chem. Glasses 26, 52 (1985).

80. Kawamoto, Y., Horisaka, T., Hirao, K. and Soga, N., J. Chem. Phys. 83, 2398 (1985).

81. Parker, J.M., private communication.

82. Lucas, J., Louër, D. and Angell, C.A., Mat.Sci.Forum 6, 449 (1985).
83. Le Ball, A., Jacoboni, C. and De Pape, R., J.Solid State Chem. 48, 168 (1983).
84. Le Ball, A., Jacoboni, C. and De Pape, R., J.Solid State Chem. 52, 32 (1984).
85. Le Ball, A., Jacoboni, C. and De Pape, R., J.Non-Cryst. Solids 74, 213 (1985).
86. Le Ball, A., Jacoboni, C. and De Pape, R., Mat.Sci. Forum 6, 441 (1985).
87. Le Ball, A., Ph.D. Thesis, Université du Maine, 1985.
88. Courbion, G., Guery, J.O., Le Ball, A. and Jacoboni, C., Mat.Sci.Forum 6, 739 (1985).
89. Matz, W., Bärenwald, U. and Dubiel, M., Phys.Stat.Sol. (a) 90. 107 (1985).
90. Rao, B.G. and Rao, K.J., Phys.Chem.Glasses 25, 11 (1984).
91. Rao, K.J., Wong, J. and Rao, B.G., Phys.Chem.Glasses 25, 57 (1984).
92. Rao, K.J. and Wong, J., J.Chem.Phys. 81, 4832 (1984).
93. Wright, A.C., Grimley, D.I., Sinclair, R.N., Rao, K.J. and Rao, B.G., J.Phys.Coll. C8, 305 (1985).
94. Rao, B.G., Sundar, H.G.K. and Rao, K.J., J.Chem.Soc.Faraday Trans. 1, 80, 3491 (1984).
95. Rao, K.J., Rao, B.G. and Wong, J., J.Chem.Phys., in press.
96. Brunner, G.O., Acta Cryst. A27, 388 (1971).
97. Polk, D.E., Scr. Metall. 4, 117 (1970).
98. Polk, D.E., Acta Metall. 20, 485 (1972).
99. Scott, G.D., Nature 188, 908 (1960).
100. Scott, G.D., Nature 194, 956 (1962).
101. Gaskell, P.H., J.Non-Cryst. Solids, 32, 207 (1979).
102. Gaskell, P.H., J.Phys.C 12, 4337 (1979).
103. Wright, A.C., Connell, G.A.N. and Allen, J.W., J.Non-Cryst. Solids 42, 69 (1980).
104. Finney, J.L. and Wallace, J., J.Non-Cryst. Solids 43, 165 (1981).
105. Shannon, R.D., Acta Cryst. A32, 751 (1976).
106. Zallen, R., J.Non-Cryst. Solids 75, 3 (1985).
107. Gordon, J.M., J.Phys.Chem. 83, 889 (1979).

DISCUSSION

C: J. Lucas. It is difficult to assign a unique value of ionic radius to Zr^{4+}, because of the widely varying Zr-F bond lengths and the large variety of ZrF_n polyhedral geometries.

Q: J. D. Mackenzie. In 1933, B.E. Warren published a paper on X-ray diffraction analysis of BeF_2 glass. It said that every Be atom was surrounded by 4 F atoms to form a regular tetrahedron and that these tetrahedra joined together to form 3-D random networks. What else is new, today?

A: A.C. Wright. Warren and Hill (referenced in my paper) suggested a corner-shared BeF_4 tetrahedral network structure for vitreous BeF_2, as a result of a simple fit to their reciprocal (intensity space) data. They only obtained an approximate Be-F bond length and they did not use Fourier transform techniques.
Since this first study, there have been several other investigations, as discussed in the present paper. Today, accurate quantitative diffraction data are available with high real space resolution which will provide a very strong test for models of short and intermediate range order. Both the Be-F bond length and its rms deviation are now accurately known. As with other glasses, the greatest barrier to progress in understanding structure of heavy metal halide glasses lies not with the gathering of good diffraction data, but with the development of modelling techniques whose model parameters (bond and torsion angle distributions, etc.) can easily be varied in a systematic way.

MOLECULAR DYNAMICS SIMULATIONS OF FLUOROZIRCONATE GLASS STRUCTURE

J.M. PARKER
DEPT OF CERAMICS, GLASSES AND POLYMERS, SHEFFIELD UNIVERSITY,
ELMFIELD, NORTHUMBERLAND RD., SHEFFIELD, S10 2TZ, U.K.

ABSTRACT
    The results of several different simulations of fluorozirconate glass structure using Molecular Dynamics modelling are compared. Most workers using this technique have concluded that Zr ions are on average in eightfold coordinated sites, whilst barium ions have coordination numbers between 10 and 11. ZrF_8 coordination polyhedra are principally linked by corner sharing although edge sharing also occurs, the proportion varying somewhat between models. The role of stabilising ions (Al), modifier ions (Na and Li), and transition metal ions is considered and the model is also examined on the basis of a close packed arrangement of ions. The results are related to experimental studies of structure and to the observed behaviour of the melts and glasses, particularly diffusion coefficients, glass transition temperatures, free volume and density. Future possibilities for study include an extension of the use of the models to interpret structural information obtained by other techniques and a widening of the range of physical and thermodynamic properties modelled, both to provide extra physical insight and to improve the parameters used in the simulation.

1. INTRODUCTION
    Molecular Dynamics simulation has been developed principally as a tool for understanding the physical chemistry of liquids and melts, because the many-body nature of the interactions between their components cannot be solved analytically. Simple molten salts have been extensively simulated and the results tested against experimental data. More recently several glass forming systems have been studied, including silicates (1), $ZnCl_2$ (2-4) and BeF_2 (6-15) based systems, by first modelling the melt and then quenching into the glassy state. These simulations have successfully predicted primary coordination spheres and have generally provided useful models for behaviour.
    Fluorozirconate glasses provide an interesting area for study not only because of the current interest in their application in IR optical systems but also because they are unusual among glass-forming systems in not being based on a network forming ion of low coordination number. Structural studies are therefore of scientific interest as well as providing a basis for interpreting their properties. MD simulation gives one way of setting up a model for their structure and therefore of analysing structural information.
    The measure of agreement between the model and the real structure will depend on the correctness of the potential used to define interactions and is limited by the simplifications involved in the modelling technique. For the systems studied previously there remain some unsatisfactory features of the results. For example models of $ZnCl_2$ fail to predict the short Zn-Zn distance which occurs in its melts (2). A major part of the analysis must

therefore involve refinement of the technique and the parameters used, the goal being to gain a reliable insight into the way the melts and glasses behave, and ultimately to predict the effect of new conditions of use and new compositions.

2. METHOD

MD simulations of melts or glasses require the form of the interatomic interactions between all species present to be known. The net forces acting on individual ions are then calculated and the consequent trajectories determined using Newton's laws of motion. By recalculating at intervals which are short compared with the vibrational frequency of the lightest atom present, the evolution of the melt can be followed. Once equilibrium is achieved from a random starting point, the average structure and many physical and thermodynamic properties can be determined.

2.1. Potentials used

For HMF melts the individual components are normally assumed to be present as ions carrying the theoretical charge. The difference in electronegativities for Zr and F ions is such that bonding might be expected to have a reasonable ionic component although Zr^{4+} is unlikely to exist. A Coulombic attractive potential is therefore adopted and the repulsive interaction is assumed to be of the Born-Huggins-Mayer type. Standard computer algorithms for solving the equations of motion have been adopted in the studies reported and an Ewald summation method has been used for evaluating the long range component of the electrostatic interactions. Typical model parameters are given in Table 1.

The correct choice of parameters to use in the equations is not straightforward and different workers have adopted different values. Tosi and Fumi (24) determined effective ion sizes and hardness parameters for a range of crystalline ionic solids by fitting to physical property data. These form a sensible basis for starting a simulation but small adjustments can often improve the fit of any particular model to structural data. A further difficulty is that data are not available for many of the ions of interest in Heavy Metal Fluoride glasses.

The first simulation reported was carried out by Angell, Cheeseman and Tamaddon (16) and used ion size parameters from earlier studies of molten BaF_2 and ZrF_4. Because the BaF_2 simulation had not given a perfect fit to experiment, Lucas, Angell and Tamaddon (17) subsequently refined this model with, in particular, a larger value for σ_F, consistent with the Tosi-Fumi data and which had also been used successfully to simulate BeF_2 melts and to determine the lattice energy of crystalline $NaMgF_3$ to within 1% of the experimental value. As a result their coordination of Zr by F rose from 7 to 7.7.

As part of our own studies we have examined the effect of varying the simulation parameters on the structural features of the model obtained and typical results are summarised in Figs. 1 and 2. In order to minimise the computational effort involved in these preliminary studies, the models used contained just 100 atoms.

TABLE 1. Typical parameters used in Molecular Dynamics simulations of fluorozirconate glasses.

Average model size = 100-200 atoms.
Model density taken as best estimate of experimental value.
Cell is normally chosen to be cubic with periodic boundary conditions, except where a specific structural model is used as a starting point (19).

Time step = 2×10^{-15} secs (18) to 11×10^{-15} secs (23).
Typical computer run length = 3000 time steps.
Form of the potential used:

$$\phi = \frac{z_i z_j}{r_{ij}} + b_{ij} \exp[(\sigma_i + \sigma_j - r_{ij})/\rho]$$

where ϕ = potential energy
z_i = ionic charge
r_{ij} = atomic separation
$b_{ij} = B \left\{ 1 + \frac{Z_i}{N_i} + \frac{Z_j}{N_j} \right\}$
N_i = number of outer shell electrons
ρ = hardness parameter

Inoue et al. (19) assumed the same b_{ij} values for all atom pairs in their models but used a special parameterisation to allow for their non-cubic unit cells.

Basic parameters used in the simulations:
Size parameters are in nm.

Ref.	ρ	$B \times 10^{19}$, J	σ_{Zr}	σ_{Ba}	σ_F	σ_{Li}	σ_{Na}	σ_{Al}
16	0.290	0.190	0.128	0.149	0.133			
17	0.266	0.318	0.138	0.165	0.1237			
18	0.290	0.190	0.141	0.164	0.137	0.092	0.124	0.1175
23	0.290	0.190	0.183	0.219	0.1237			

Using such information, size parameters can be adjusted empirically to fit the observed interatomic distances in glasses and/or crystalline fluoro-zirconates. Whilst it has not proved possible to define a set of size parameters for Zr, F and Ba which fit exactly all the observed distances, nevertheless the final parameters selected by this approach are in reasonable agreement with those used by Lucas, Angell and Tamaddon. The reported coordination numbers have not been fitted in this way but will depend on the size parameters chosen because the relative radius ratios define the space available for fitting in the ions (see Fig. 2).

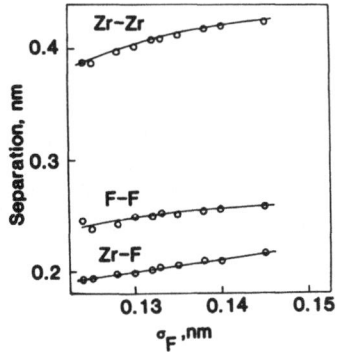

FIGURE 1. Showing the effect of varying σ_F on mean Zr-Zr, Zr-F and F-F distances obtained in MD simulations of ZrF$_4$.

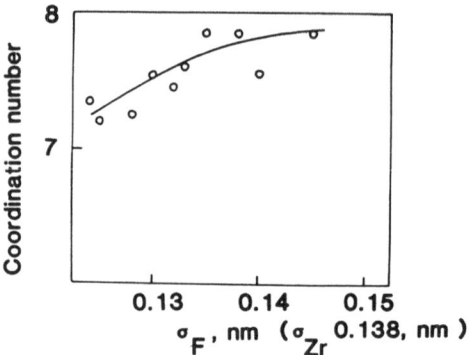

FIGURE 2. The change in Zr coordination number as the size of the fluorine ion is varied is shown.

Another significant model parameter is the cell density. The effect on the main features of the calculated structure of varying its density is shown in Fig. 3. As the model density rises so the interatomic distances fall and coordination numbers rise as might be expected. The effect is relatively small however. Further if melt densities are overestimated, this together with the fact that a relatively small box size is simulated (in some models as small as 100 atoms) with periodic boundary conditions, may impose artificially restrictive constraints on the atomic positions and the final relaxed structure will fail to represent reality.

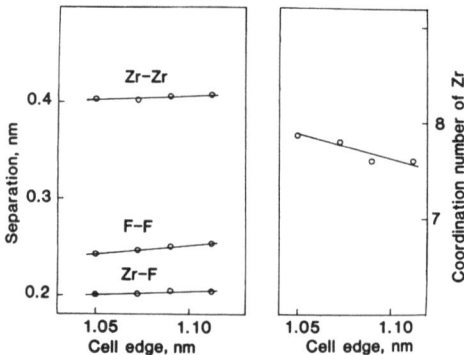

FIGURE 3. The effect on atomic separations and coordination numbers of varying the model density.

Most workers have used experimental densities based on room temperature measurements. These are not always available however, for example where the glasses generated in the computer are not within the experimentally accessible glass-forming region. Similarly high temperature measurements of melt densities have rarely been made. In these cases extrapolated values have had to be used. Changing melt density during a computer experiment as when quenching to room temperature is computationally expensive and the melt density simulated at high temperatures has for some studies been taken as the room temperature value.

For the melts an effective pressure at a particular density can be calculated via the Virial equation. Our results for the variations of pressure with density of the model chosen are plotted in Fig. 4. Zero pressure does not correspond exactly with the ideal model box size, but the pressures are acceptable at the experimental density, as has been reported by other workers (17).

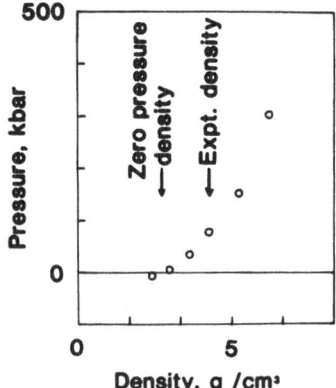

FIGURE 4. Calculated melt pressures as a function of density.

The initial arrangements and velocities of ions in MD models are usually selected randomly, velocities being scaled so that the overall kinetic energy corresponds to the required temperature. The models are then allowed to come to equilibrium at high temperatures when relaxation rates will be very fast. Kawamoto et al. (23) for example selected 1673K whereas Yasui (19) used a somewhat different approach in their modelling study. Starting from various idealised model structures 1) based on different coordination numbers and polyhedra linkage schemes and 2) related to the structures of known compounds in the BaF_2-ZrF_4 system, they used MD simulations at T_g to relax their structural models to the lowest energy configurations. This they followed by an equilibration run at room temperature. Such an approach means that the final structure is partly controlled by the starting point, T_g being too low to allow more than a local relaxation. Different models could therefore be constructed, equilibrated and compared with structural information.

Cooling of the model structure to a realistic working temperature has in some studies been achieved by instantaneously rescaling velocities to the new value required and then allowing a short period for equilibrium to be achieved before commencing any property or structure averaging. Alternatively this process may be carried out over a number of time steps. We have typically used 250. Nevertheless structural disorder will remain in the model because of the high effective quenching rate. Therefore coordination polyhedra will not be as well defined as in the real materials either in terms of the angular relationships between ions or in their separations. Whilst the Zr-F nearest neighbour cut-off is clearcut, this is less so for Ba-F etc. As a result there is a certain arbitrariness of choice in defining which are nearest neighbour atoms; slight differences in the criteria adopted inevitably lead to difficulties in making precise comparisons between models and with experiments and the assumptions made are not always reported.

3. RESULTS

In spite of variations in the way MD simulations have been carried out, the studies documented on HMF glasses (16-23) show a considerable measure of agreement. Their main conclusions are summarised in Tables 2,6 and 7.

3.1. Zr-F coordination

Coordination numbers and Zr-F separation distances reported for different simulations are summarised in Table 2. The Zr-F separations are normally quoted as the peak positions on the pair distribution functions. Mean positions are often slightly longer because the Zr-F pair distribution function peaks are asymmetric and do not return to zero at longer distances. A typical Zr-F pair distribution function is shown in Fig. 5.

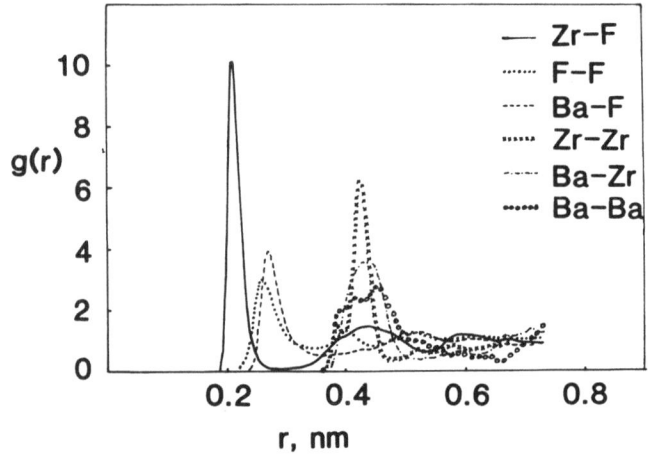

FIGURE 5. Pair distribution functions obtained by MD simulation of $20BaF_2.30ZrF_4$ composition glass.

Most model studies give coordination numbers just below 8 for simulations at room temperature. This figure is expected by analogy with the crystalline forms of ZrF_4, $BaF_2.ZrF_4$ and $BaF_2.2ZrF_4$. That the glasses and related crystalline species have similar structures is confirmed by the small change in density which occurs on producing a vitreous phase (see Table 4). The size and shape of coordination polyhedra in the crystalline forms and the way they are linked together are summarised in Table 3.

Most x-ray diffraction studies have given results consistent with MD models (see Table 5). For example Coupé et al. (39) quoted coordination numbers and Zr-F distances of 7.1 and 8.4 and 0.209 to 0.211 nm respectively (the precise values varying with composition, the high figures being for higher levels of ZrF_4), while Kawamoto et al. (31) quoted 8 and 0.220 nm for a glass much richer in BaF_2 than other workers, obtained by roller quenching. Etherington et al. studied both Zr and Hf containing glasses to aid separation of the pair distribution functions and quoted a Zr-F coordination number of 7.6 with a Zr-F distance of 0.208 nm (32). Almeida (35) has reanalysed the Etherington data however with a readjusted density correction and found a somewhat lower coordination number of 6.5.

TABLE 2. MD simulation results for Zr primary coordination.

Ref.	Composition	Zr-F,nm	Zr C.N.	Comments
16	$36BaF_2.64ZrF_4$		7	4 bridging F/Zr. Zr-F-Zr~180°
17	$4BaF_2.7ZrF_4$		7.7	1/4 of Zr polyhedra linkages are edge sharing.
18	ZrF_4	0.209	8.0	All Fs bridging. 1/18 of Zr polyhedra in edge sharing linkages.
18	$2BaF_2.3ZrF_4$	0.210	7.8	Half of the Fs bridging 1/9 of Zr polyhedra linkages are edge sharing.
19	$2BaF_2.3ZrF_4$		7.3 to 8	Edge and corner sharing. Edge linking encouraged by increasing BaF_2 content.
21	$3BaF_2.7ZrF_4$	0.209	8.0	2/3 Fs are bridging. 1/12 of polyhedra linkages are edge sharing.
23	$BaF_2.ZrF_4$	0.212	<8	Coordination polyhedron for 8 C.N. is dodecahedron. 60% linkages are edge sharing, 40% corner sharing. A few 7 coordinated Zrs exist, coordination polyhedron being monocapped trigonal prism.

TABLE 3. Zr-F coordination polyhedra in crystalline compounds in the BaF_2-ZrF_4 system.

Ref.	Composition	N	Zr-Zr nm	Structure
25	βZrF_4	8	0.416	Corner sharing square Archimidean antiprisms.
26	αZrF_4	8	0.415 0.356	Corner sharing and edge linked triangular dodecahedra.
27	$\beta BaF_2.ZrF_4$	8	0.382	Edge sharing triangular dodecahedra.
28	$\alpha BaF_2.ZrF_4$	7	0.382	Edge sharing monocapped trigonal prism.
29	$\beta BaF_2.2ZrF_4$	7		Edge and corner sharing.

Inoue, Hasegawa and Yasui (19) started from various idealised model structures based on different coordination numbers and polyhedra linkage schemes and related to the structures of known compounds in the BaF_2-ZrF_4 system. They concluded that the structural units in the glass were fluorozirconate polyhedra in 8 fold coordination. These mostly shared corners but some shared edges, the proportion varying with Ba content. The edge shared linkage of the polyhedra was surrounded by two or more Ba atoms and this was explained by reference to the structures of ZrF_4 and $BaF_2.ZrF_4$.

They also compared their models with radial distribution functions determined using x-rays and concluded that the latter could only be explained on the basis of the model described.

Coupé et al. have given an extensive list of the structure of a range of fluorozirconates. These data show a clear shift in the mean Zr-F distance according to the coordination number (0.200 nm to 0.211 nm for a change in C.N. from 6 to 8). The mean distances obtained in experimental studies are consistent with the higher coordination numbers deduced by integrating the area under the first peak in the x-ray radial distribution functions.

TABLE 4. A comparison of the densities of crystalline barium fluorozirconates and glasses of similar composition. The data for crystals are x-ray calculated values. Measured values are slightly smaller.

Crystalline cmpd. densities			Glass densities		
Ref.	Composition	Density $kg\ m^{-3} \times 10^{-3}$	Ref.	Composition	Density $kg\ m^{-3} \times 10^{-3}$
27	$\beta BaF_2.ZrF_4$	4.73	23	$BaF_2.ZrF_4$	4.6
33	$\alpha BaF_2.ZrF_4$	5.04	30	$4BaF_2.6ZrF_4$	4.57
33	$\beta BaF_2.2ZrF_4$	4.35	19	$4BaF_2.6ZrF_4$	4.587
33	$\alpha BaF_2.2ZrF_4$	4.35	32	$36BaF_2.64ZrF_4$	4.64
			19	$BaF_2.3ZrF_4$	4.326

Ota and Soga (34) have also compared molar volumes of a range of fluorozirconate glasses with the molar volumes of their separate components and comment that the very small differences suggest a close similarity in structures.

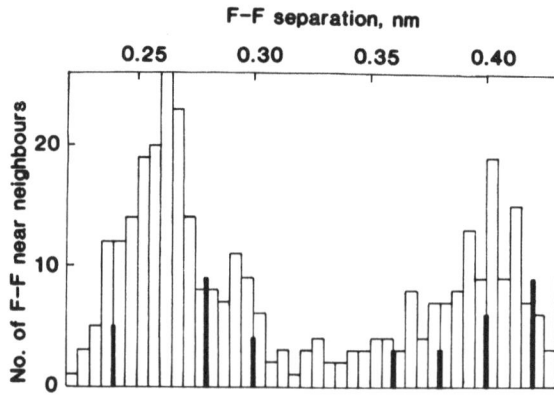

FIGURE 6. Giving histograms for the distribution of intrapolyhedral F-F distances for an MD simulation and $\beta BaZrF_6$. (crystal data as thick bars).

TABLE 5. Summarising the principal results of diffraction studies.

Ref.	Composition	C.N.	Zr-F nm	Structure
31	$BaF_2.ZrF_4$	8	0.220	Edge shared chains.
32	$36BaF_2.64ZrF_4$	7.6±0.5	0.208	
30	$BaF_2.2ZrF_4$ to $BaF_2.3ZrF_4$	7.1 to 8.4	0.210 to 0.211	
19	$2BaF_2.3ZrF_4$ $BaF_2.3ZrF_4$	7-8		Corner and edge sharing.

In barium fluorozirconate glasses some fluorine ions are shared between two Zrs while others are coordinated with one Zr only but with at least one Ba neighbour instead. Etherington et al. (32) commented on the widths of the Zr-F peaks observed in rdfs which they attributed to different Zr-F separations according to whether the F is bridging or non-bridging. MD studies allow these two values to be separately determined and our results are 0.2060±0.0101 nm for the non-bridged distance and 0.2175 ± 0.0099 nm for the bridged ions. Because the cut-off chosen for the Zr-F bonds was relatively large the quoted standard deviations for these results are probably biased by a few exceptionally long Zr-F bonds included in the calculation. We have therefore also measured the widths of our peaks both for the model at room temperature and by averaging the coordinates over along period of time to eliminate thermal broadening. Our half height peak width including thermal effects was 0.023 nm and with thermal effects removed was 0.020 nm. These figures correspond to rms deviations of 0.0098 and 0.0085 nm respectively and compare with values of 0.01 nm reported by Etherington et al. for ZB glasses and 0.014 nm for HB glasses using x-ray diffraction measurements. A detailed comparison is unreliable however because the latter data suffer from broadening associated with termination effects and the MD results have not been corrected for this.

The cation to anion ratio for Zr and F is 0.69, near the critical value for a transition between 6 and 8 fold coordination. Among the range of fluorozirconate structures 6,7 and 8 fold coordination occurs. Thus while there is considerable experimental evidence for a coordination number between 7 and 8, consistent with MD simulations, the possibility of a smaller value must be examined. Indeed to find that the melt structure differed significantly from that of the relevant crystalline compounds would make it easier to explain the occurrence of glass formation in this system. An alternative model has in fact been proposed by Almeida and Mackenzie (36,37). That was based on Raman data which was interpreted by comparison with crystalline forms of known structure, and analysis of the vibrational modes of different polyhedral units. From this work barium difluorozirconate glasses were thought to be based on sixfold coordination of zirconium ions, with the octahedra sharing corners to produce chains. Barium metafluorozirconate glasses were believed to contain 7 fold coordinated Zr ions. Etherington et al. rationalised their diffraction evidence in terms of the Almeida and Mackenzie model by proposing that octahedra were linked into chains but with Fs on adjacent chains being not more than 0.01 nm away and so contributing the observed coordination number of >7. Analysis of Raman spectra however is not straightforward and an alternative interpretation based on 8 fold coordinated Zr has been proposed by Kawamoto and co-workers (38-40).

The effect of the BaF_2/ZrF_4 ratio on the Zr-F bond length and coordination number is unclear. X-ray diffraction studies (30) have suggested a decreasing C.N. with rising BaF_2 content and this is support-ed by our MD simulations (18) although in the latter case the effect is much smaller. In general we have found that as the proportion of non-bridged Fs rises the Zr-F distance slowly decreases and the mean Zr co-ordination number falls as might be expected from simple electrostatic considerations. On the other hand Raman spectroscopic results have suggest-ed the reverse (36).

3.2. F-F distances

A typical F-F correlation function obtained in an MD simulation is shown in Fig. 5. The first two maxima in this distribution are at 0.256 nm and 0.401 nm and the area under the first peak using a rather large cut-off distance of 0.35 nm is 9. Models generated in computer simulations allow the F-F correlations within the Zr polyhedra to be separated from those from adjacent units. Fig. 6 shows such data calculated from a simul-ation of a $2BaF_2.3ZrF_4$ glass plotted as a histogram. The peaks are better defined than in the pair distribution function given in Fig. 5 because interpolyhedra contributions have been excluded. Equivalent results for $BaF_2.ZrF_4$ in its crystalline state (27) are also presented and the simil-arity is clear, confirming Kawamoto's conclusion, based on Raman spectro-scopy and a comparison with crystalline species, that the structural units in the simulations are triangular dodecahedra (23). Hoard and Silverton(41) have commented that the triangular dodecahedron and the square antiprism are very similar energetically as ways of providing 8 fold coordination. These authors and Burns, Ellison and Levy (42) also discuss the typical edge length values for a hard sphere approximation. Topological and en-ergetic constraints prevent all the short F-F edge distances having the same value and this accounts partly for the breadth of the F-F peak ob-served in pair correlation functions (see Figs. 5 and 6). Broadening is further exaggerated because some polyhedra share edges as opposed to corn-ers (see Section 3.4). The resulting F-F distances along these edges are considerably reduced. The average length of this edge in one of our simu-lations was 0.226 nm with the corresponding Zr-F distance being 0.223 nm, i.e. the F-F distance is shortened and the Zr-F distance lengthened comp-ared with the norm. Equivalent results are found for crystalline compounds. For αBZ Laval, Papiernik and Frit (28) give a shared edge length of 0.236 nm, for β BZ the value is variously given as 0.223 nm and 0.232 nm (27,33) and in $7NaF.6ZrF_4$ the result is 0.238 nm (42). These authors also comment on the lengthening of the Zr-F bonds for the shared edges in these compounds.

The ratio of observed Zr-F distances to F-F distance provides additional insight into the nature of the coordination polyhedra. Typical values found in model studies and expected for various idealised polyhed-ral shapes are given in Table 6.

Because the F-F peak in the correlation function overlaps with the Ba-F peak and because of the low scattering power of F ions for x-rays it is difficult to derive an experimental value for the F-F distance. A comparison of the ratio of the peak positions of the first two peaks (32) gives a value of 1.308 for fluorozirconate glasses and 1.341 for flurohafnates. However Almeida (35) has reported a combined x-ray and neutron diffraction study allowing him to separate out the two overlapping components to some extent. These rdfs suggest a lower value for the F-F distance than the position of the double peak and give a value for the

TABLE 6. Principal dimensions of ZrF$_n$ polyhedra.

Ref.	Composition	C.N.	Zr-F nm	F-F nm	r(F-F)/r(Zr-F)
MD results					
3	ZrF$_4$	8	0.209	0.251	1.201
3	2BaF$_2$.3ZrF$_4$	7.8	0.210	0.255	1.214
Crystal structure results					
27	βBaF$_2$.ZrF$_4$	8	0.213	0.257	1.265
28	αBaF$_2$.ZrF$_4$	7	0.208	0.265	1.271
42	7NaF.6ZrF$_4$	8	0.211	0.264	1.247
43	Li$_2$ZrF$_6$	6	0.202	0.285	1.414
44	CuZrF$_6$.4H$_2$O	6	0.200	0.279	1.400
45	Rb$_5$Zr$_4$F$_{21}$	6	1.99	0.281	1.413
		7	0.206	0.266	1.289
		8	0.211	0.264	1.248

Theoretical ratios:

Polyhedron	r(BB)/r(AB)
Octahedron	1.414
Simple cube	1.155
Square antiprism[*]	1.216

[*]This figure has been calculated assuming all the short edges of the antiprism are the same length. Burns et al (42) however point out that the optimum shape is when the square edge lengths are 1.05 times the lengths of the triangular edges.

The theoretical figure for coordination polyhedra other than those given is not known but is believed to be close to that of the Archimidean antiprism, since energetically they are very similar.

r(F-F)/r(Zr-F) ratio of 1.23 consistent with 8 fold but not octahedral coordination.

Few workers have reported F coordination results because of the difficulties of measurement. Etherington et al. have reported a total coordination number for the combined F-F and Ba-F peak of 16. In our own MD studies each fluorine is typically surrounded by 9 others, 6.6 of which are intrapolyhedral distances, the rest interpolyhedral for a glass of composition 2OBaF$_2$.3OZrF$_4$. Making due allowance for the fluorines which are non-bridging and therefore linked to only one polyhedron, the average number of intrapolyhedral short distances (0.26 nm) is 4.3 and the average number of long distances (0.40 nm) is 2.3. These are consistent with the results expected for a triangular dodecahedron where the equivalent figures would be 4.5 and 2.5. The observed results are slightly lower because of the small proportion of 7 coordinated units present. The overall F-F coordination number, while consistent with typical results for crystalline compounds and with the observed high packing density, is at variance with the result of Álmeida from diffraction studies, that it is only 4.8.

3.3. Ba-F distances

Mean Ba-F distances and Ba coordination numbers are reported in Table 7 and typical pair correlation functions are given in Fig. 5.

TABLE 7. Giving principal results on Ba coordination derived in different MD simulations.

Ref.	Composition	Ba-F nm	Ba C.N.
23	$BaZrF_6$	0.275	10
21	$3BaF_2.7ZrF_4$	0.264	10.3
22	$BaF_2.2ZrF_4$	0.270	8 + 3
21	$4BaF_2.6ZrF_4$	0.265	10.8
18	$4BaF_2.6ZrF_4$	0.263	9.9

Coupé et al (30) have reported typical Ba coordination results for a range of crystalline species. Average coordination numbers vary between 8 and 12 with the mean Ba-F distances ranging from 0.265 to 0.291 nm. A detailed inspection of x-ray diffraction results (30) for barium fluorozirconate glasses and comparison with crystalline structures indicates this distance is often split into two components one involving a short Ba-F distance and another a longer bond. Lucas et al. (22) show that the Ba primary coordination sphere for their MD simulation is ill defined and suggest that it can be interpreted as consisting of 7-8 near neighbours with further atoms at a range of distances.

A possible way of rationalising this result is in terms of the nature of the F neighbours, and whether they are linked to one or two Zrs. The proportion of bridging vs. non-bridging F ions in the structure is of course defined by the overall Zr/F ratio (R) and the Zr coordination (N), the average number of bridging Fs per Zr being 2(N-R). What is not fixed by the simple requirements of stoichiometry is the number of bridging and non-bridging ions present around each Ba ion. For one of our recent models of a $2BaF_2.3ZrF_4$ composition glass the mean number of fluorine near neighbours of Ba was 9.2, although the range was from 8 to 12. Of the near neighbour ions, 1.4 on average were also coordinated by 2 Zrs, 1.2 were only coordinated by Ba ions and the remainder were linked to non-bridging Fs, the highest coordination numbers being associated with the Bas having the greatest number of bridging Fs as near neighbours. The average Ba-F distances for bridged Fs was 0.299 nm, for non-bridged Fs 0.281 nm and for isolated Fs 0.269 nm. Incidentally just 4% of the Fs were 'free' i.e. not bonded to Zr. Whether this is a realistic feature of the model is not yet clear.

Almeida (46) has examined HMF glasses using XPS and has been unable to find any shifts for the F energy levels which might allow non-bridged and bridged ions to be distinguished.

As found by Coupé et al. (30) MD simulations show a decrease in Ba C.N. (10.8 to 10.3) as the BaF_2 concentration falls (from 40 to 30 mol %). Since this is apparently accompanied by an increase in the proportion of bridging Fs in the primary coordination shell of barium (22% to 30%) the result is surprising.

3.4. Zr-Zr distances

One way of describing fluorozirconate glass structures is from a polymeric point of view i.e. the way the polyhedral units link.

Most simulations suggest that the ZrF_8 polyhedra are linked by corner and edge sharing, but in varying proportions as indicated in Tables 2 and 8. We have found that this ratio varies with model parameters chosen which is no doubt one reason for the discrepancies between different papers.

There is no evidence to suggest from our studies however (21) and those of Lucas et al. (17) that their proportion varies with the Ba content of the melt as proposed by Inoue et al. (19). Coupé et al. (30) have reported that x-ray evidence confirms the existence of these edge sharing units and they are certainly a common feature of crystalline fluorozirconates (see Table 3). However the first phases to crystallise from HMF glasses are highly disordered compared with the stable forms and the latter should therefore be used as a guide to structure with caution.

TABLE 8.   Zr-Zr near neighbour distances.

Reference	Short Zr-Zr nm	Long Zr-Zr nm	Ratio of short to long distances present
17 (Expt. x-rays)	0.360	0.413	1/4
22 (MD)	0.385	0.430	1/4
21 (MD)	0.384	0.421	1/9
23 (MD)	0.381	0.418	3/2
23 (Crystals)	0.370	0.416	

The length of the long Zr-Zr bond is almost exactly twice that of the Zr-F bond in all reported simulations confirming a Zr-F-Zr bond angle of $180°$.

On the basis of an analysis of the linkages between ZrF_n polyhedra Lucas et al. (22) have proposed a detailed model for fluorozirconate glasses based on ZrF_8 and ZrF_7 polyhedra sharing an edge to give Zr_2F_{13} units which then share corners with similar units nearby. Our MD simulations display considerably less ordering of units however e.g. for a $2BaF_2.3ZrF_4$ there were on average 4.7 Zr-Zr polyhedra linkages but with a range from 2 to 7, and with an occasional polyhedron sharing two edges.

3.5. Zr-Ba and Ba-Ba distances

Most MD simulations show a broad distribution of Zr-Zr and Ba-Ba distances between 0.38 and 0.47 nm approximately (see Fig. 5). These correspond only poorly with the peaks observed by Coupé et al. (30) at 0.46 and 0.50 nm in an experimental correlation function but are consistent with the broader, less sharply defined features in this distance range reported by other workers (23,32).

3.6. Effect of adding modifier ions on glass structure

As part of our studies we have examined the role of Na and Li atoms in melts (21). Principal results are summarised in Table 9.

An examination of the F ions near the alkali ions shows some to be bridging and others non-bridging with respect to Zr as expected for a typical glass network modifying ion. Because of the relatively low charge on the alkali metal cation the non-bridged F ions move closer to the Zr and there is a small decrease in the Zr coordination number as the alkali ion concentration rises. The coordination numbers obtained for Na and Li, while similar to those observed in crystalline fluorozirconates, are not consistent with the suggestions of Lecoq and Poulain (47) who proposed on the basis of observations of discontinuities in physical properties with composition that Li was in tetrahedral coordination at low concentrations, and octahedral coordination as its concentration rose, the discontinuity being at approximately 20 mol % addition. Differences are observed between the coordination spheres around Li at low and high concentrations but the same trends are repeated for Na.

TABLE 9. Coordination data for fluorozirconate glasses containing alkali metal ions.

Composition		$2LiF.40ZrF_4$	$22LiF.33ZrF_4$	$2NaF.40ZrF_4$	$22NaF.33ZrF_4$
r(Zr-F)	nm	0.210	0.209	0.210	0.210
r(X-F)	nm	0.216	0.197	0.240	0.229
C.N. Zr		8.0	7.8	8.0	7.8
C.N. X		7.2	5.6	8.8	7.2
% F near X					
Non-bridging		23	69	29	66
Bridging		77	31	71	34

3.7. TM ion environments

The role of Mn and Fe ions in the glass structures has been consider-ed (21) in the latter case for both oxidation states. They appear to sit in octahedrally coordinated sites in agreement with the interpretation of optical spectra (48), ESR data (49) and Mössbauer results (50). Any site stabilisation energy associated with Ligand field effects and Jahn-Teller distortions have not been included in the calculations.

3.8. The role of stabilising ions

The role of Al ions in the structure has been examined in detail (21). They sit in octahedrally coordinated sites surrounded entirely by fluorines coordinated only to one Zr, i.e. non-bridging Fs. In view of the high charge on the ion this result is not surprising but is entirely consistent with the picture of this ion as a network forming ion. The different structural role of this ion compared with Zr may explain why the stabilis-ing ions so easily crystallise from the melt at low concentrations (cf. phase separation in borosilicates).

3.9. Role of oxygen ions

We have examined the sites occupied by O^{2-} ions in the fluorozirconate glass structure and found that it occupies bridging positions between two Zrs. This result is as expected given its double charge but is also consistent with the interpretation of a Raman spectrum from an oxide doped glass reported by Almeida and Mackenzie (51).

4. SIMULATIONS OF MELT PROPERTIES
4.1. Diffusion studies

Apart from testing the fit to structural data, the ability of the model to explain and predict the behaviour of melts and glasses can be ex-amined. One area which has been extensively studied is diffusion. By fol-lowing the motion of individual ions in simulations of melts, self diffusion coefficients can be determined. MD studies have shown fluoride glass melts to be anion conductors (16,21,23), in agreement with NMR exper-iments (52) and measured electrical conductivities (23). A mechanism for F ion diffusion which differentiates the role of bridged and non-bridged ions has also been suggested by Angell et al. (16), the latter diffusing more quickly, although a continuous interchange between these two species is expected. By extrapolating predicted D values to zero a T_g can be estimated (21,23). A more accurate figure is obtained by plotting calculated pressures (or cell volume if allowed to vary) or total energies against temperature. The results for HMF glasses are near 360°C (23) for $BaZrF_6$ and range between 230°C and 430°C for ZrF_4 according to the method of calculation used (21), in good agreement with experiment. Diffusion coefficients also fall rapidly with ion charge explaining the stabilising action of e.g. Al and La ions. Thus at 1250K the only ions we find to be diffusing at an appreciable rate are Na, Li and F.

Our determinations of diffusion coefficient for F and Zr ions are plotted against temperature in Fig. 7. From the temperature dependence of the results activation energies can be determined (40 kJ/mol.). The results are probably unrealistically low and should be compared with values of 600 kJ/mol. or higher for viscous flow near T_g. On the other hand the activation energy for F- diffusion is reported as 80 kJ/mol. (52) and at higher temperatures in real melts the activation energy for viscous flow is considerably lower.

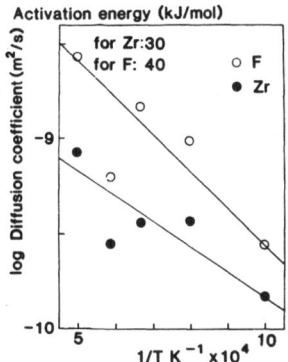

FIGURE 7. An Arrhenius plot of self diffusion coefficients for Zr and F ions calculated in MD simulations.

4.2. Melt and glass density

Another important test of the model is whether it predicts the correct density (see Section 2.1). In terms of the volume occupied by ions, HMF glasses are considerably more dense than oxide glasses. 64% of the available space is occupied in typical barium fluorozirconate glasses, a figure identical to that for random close packing of identical sized spheres. This suggests a model for the structure based on close packing as was also suggested by Ota and Soga (34), who measured acoustic velocities and concluded that the structure of HMF glasses was quite different to that of conventional glasses. An interesting implication of these results is that H_2 diffusion will not be a problem in fluoride glass fibres (21).

5. FUTURE STUDIES
5.1. Improvements to modelling techniques

The measure of agreement between model and experiment inevitably depends in part on how well the initial potential has been chosen. In oxide based systems considerable strides are now being made in improving the form of the potential by including 3 body interactions and polarisation effects. Interestingly we have found that short range approximations to the standard Born-Huggins-Mayer interionic interactions give substantially the same results but with a considerable saving of computational effort.

In view of the apparent disagreement between model studies and the Raman data of Almeida and Mackenzie (36) and also of Toth et al. (53) who found that an LiF (NaF)-ZrF$_4$ melt at 650°C consisted primarily of isolated ZrF$_6$ units, one sensible modification of the interatomic potentials would be to define three body potentials for the F-Zr-F interaction which would constrain the Zr coordination polyhedron to be octahedral and to examine the effect on the model results. Incidentally Kawamoto et al. (23) have shown how analysis of the vibrational frequency spectrum can be carried out as part of the MD simulation and use this to confirm the identification of one particular vibrational mode in the Raman spectrum of fluorozirconate glasses. Simulation of AF$_3$ glasses, where A is a TM ion, would also be of interest since these are largely believed to be based on octahedral coordination. The effect on total lattice energies of different coordination models should be calculated and the results compared with those for crystalline forms in order to assess relative stabilities and to allow estimation of enthalpies of transformation.

Other investigation methods worth applying to fluoride glasses include NMR spectroscopy, particularly MAS NMR, and EXAFS, with MD simulation structures providing a useful starting point for interpretation of results. Weber (15) has also described how computer simulations can be useful for setting up models to explain optical absorption and fluorescence spectra of RE and TM ions in glass, with laser-induced fluorescence line narrowing techniques providing a useful experimental probe of different site geometries (11).

There is considerable opportunity for calculation of thermodynamic data, elastic properties, etc. using simulations which can then be compared with experiment (3,4). Elastic coefficients are of interest with respect to intrinsic Brillouin scattering (54). Other workers have studied crystal/liquid interfacial energies (55), surface states and radiation damage effects using MD modelling (56,57).

Another important area of study is the use of the models to determine changes in the energy of an assemblage of ions when the oxidation state of one of the components changes. This has been done using Monte Carlo techniques for crystalline solids such as spinels. Applied to glass forming melts containing transition metal ions it has the potential of providing valuable information on redox equilibria and compositional effects and some experimental information is already available (48,50).

REFERENCES

1. Soules, T.F., J. Non-Cryst. Solids 49, 29 (1982).
2. Desa, J.A.E., Wright, A.C., Wong, J. and Sinclair, R.N., J. Non-Cryst. Solids 51, 57 (1982).
3. Hirao, K. and Soga, N., J. Ceram. Soc. Japan 91, 11 (1983).
4. Hirao, K. and Soga, N., J. Ceram. Soc. Japan 92, 112 (1984).
5. Rahman, A., Fowler, R.H. and Narten, A.H., J. Chem. Phys. 57, 3010 (1972).
6. Brawer, S.A. and Weber, M.J., Phys. Rev. Lett. 45, 460 (1980).
7. Brawer, S.A. and Weber, M.J., J. Non-Cryst. Solids 38 & 39, 9 (1980).
8. Brawer, S.A., Phys. Rev. Lett. 46, 778 (1981).
9. Brawer, S.A. and Weber, M.J., J. Chem. Phys. 75, 3522 (1981).
10. Weber, M.J. and Brawer, S.A., J. Non-Cryst. Solids 52, 321 (1982).
11. Weber, M.J. and Brawer, S.A., J. Physique C9, 43, 291 (1982).

12. Brawer, S.A., J. Chem. Phys. 79, 4539 (1983).
13. Hirao, K. and Soga, N., J. Non-Cryst. Solids 57, 109 (1983).
14. Dell, W.J. et al., Phys. Rev. B 31, 2624 (1985).
15. Weber, M.J., J. Non-Cryst. Solids 73, 351 (1985).
16. Angell, C.A., Cheeseman, P.A- and Tamaddon, S., J. Physique Coll. C9, 43, 381 (1982).
17. Lucas, J., Angell, C.A. and Tamaddon, S., Mater. Res. Bull. 19, 945 (1984).
18. Hamill, L.T. and Parker, J.M., Phys. Chem. Glasses 26, 52 (1985).
19. Inoue, H., Hasegawa, H. and Yasui, I., Phys. Chem. Glasses 26, 74 (1985).
20. Kawamoto, Y., Mater. Sci. Forum 6, 417 (1985).
21. Hamill, L.T. and Parker, J.M., Mater. Sci. Forum 6, 437 (1985).
22. Lucas, J., Louer, D. and Angell, C.A., Mater. Sci. Forum 6, 449 (1985).
23. Kawamoto, Y., Horisaka, T., Hirao, R. and Soga, N., J. Chem. Phys. 83, 2398 (1985).
24. Fumi, F.G. and Tosi, M.P., J. Phys. Chem. Solids 25, 31 (1964) and 25, 45 (1964).
25. Wyckoff, R.W.G., Crystal Structures, Wiley and Sons, New York, Vol.2, 127 (1964).
26. Papiernik, R., Mercurio, D. and Frit, B., Acta Cryst. B38, 2347 (1982)
27. Mehlhorn, B. and Hoppe, R., Z. Anorg. Allgemein. Chem. 425, 180 (1976).
28. Laval, J.-P., Papiernik, R. and Frit, B., Acta Cryst. B34, 1070 (1978).
29. Laval, J.-P., Frit, B. and Lucas, J., Mater. Sci. Forum 6, 457 (1985).
30. Coupé, R., Louer, D., Lucas, J. and Leonard, J. Amer. Ceram. Soc. 66, 523 (1983).
31. Kawamoto, Y. and Horisaki, T., J. Non-Cryst. Solids 56, 39 (1983).
32. Etherington, G., Keller, L., Lee, A., Wagner, C.N.J. and Almeida, R.M. J. Non-Cryst. Solids 69, (1984).
33. Laval, J.-P., Frit, B. and Gaudreau, B., Rev. de Chim. Minerale 16, 509 (1979).
34. Ota, R. and Soga, N., J. Non-Cryst. Solids 56, 105 (1983).
35. Almeida, R.M., Mater. Sci. Forum, 6, 427 (1985).
36. Almeida, R.M. and Mackenzie, J.D., J. Chem. Phys. 74, 5954 (1981).
37. Almeida, R.M. and Mackenzie, J.D., Phys. Chem. Glasses 26, 189 (1985).
38. Kawamoto, Y. and Sakaguchi, F., Bull. Chem. Soc. Japan 56, 2138 (1983).
39. Kawamoto, Y., Phys. Chem. Glasses 25, 88 (1984).
40. Kawamoto, Y., Phys. Chem. Glasses 26, 190 (1985).
41. Hoard, J.L. and Silverman, J.V., Inorg. Chem. 2, 235 (1963).
42. Burns, J.H., Ellison, R.D. and Levy, H.A., Acta Cryst. B24, 230 (1968).
43. Brunton, G., Acta Cryst. B29, 2294 (1973).
44. Fischer, J. and Weiss, R., Acta Cryst. B29, 2955 (1973).
45. Brunton, G., Acta Cryst. B27, 1944 (1971).
46. Almeida, R.M., Lau, J. and Mackenzie, J.D., Mater. Sci. Forum 6, 465 (1985).
47. Lecoq, A. and Poulain, M., J. Non-Cryst. Solids 34, 101 (1979).
48. France, P.W., Carter, S.F. and Parker, J.M., Phys. Chem.Glasses 27, 32 (1986).
49. Bogomolova, L.D., Grechko, E.G., Krasil'Nikova, N.A. and Sakharov, V.V., J. Non-Cryst. Solids 69, 299 (1985).
50. Kawamoto, Y., Horisaki, T., Hirao, K. and Soga, N., Chem. Lett. (8), 1441 (1984).
51. Almeida, R.M. and Mackenzie, J.D., J. Non-Cryst. Solids 56, 63 (1983).
52. Ravaine, D., Perera, W.G. and Minier, M., J. Physique Colloq. C9 43, 407 (1982).

53. Toth, L.M., Quist, A.S. and Boyd, G.E., J. Phys. Chem. 77, 1384 (1973).
54. Schroeder, J., Fox-Bilmont, M., Pazol, B.G., Tsoukala, V., Drexhage, M.G. and El-Bayoumi, O.H., Opt. Eng. 24, 697 (1985).
55. Heyes, D.M. and Clarke, J.H.R., J. Chem. Soc. Farad. Trans. (2) 77, 1089 (1981).
56. Garofalini, S.H., J. Chem. Phys. 78, 2069 (1983).
57. Doan, N.V., Phil. Mag. A 49, 683 (1984).

DISCUSSION

C: C. Moynihan. It is impossible at the moment to compare your
molecular dynamics activation energy values with experimental ones. The reason is that the fluorozirconate melt vis
cosities and diffusivities are very non-Arrhenian, so that
the "activation energies" are very temperature-dependent.

Q: C. Jacoboni. Have you had a chance to gather all the M.D.
partial results in order to build a neutron or X-ray interference function that can be compared with an experimental
one?

A: J. Parker. We have not yet done this. However, A.C. Wright
has taken our results and he has combined them into total
correlation functions, although he has not used experimental broadening corrections, because the experimental and
simulated compositions were different. These correlation
functions (published elsewhere in the current proceedings)
show satisfactory agreement when one bears in mind the
mentioned limitations.

C: R. Almeida. You cannot infer the coordination number solely
based on the Zr-F distance. Sixfold coordinated Li_2ZrF_6 has
only F_{nb} atoms, whereas the fluorozirconate glasses have
both F_b and F_{nb} species.

R: J. Parker. While I agree that mean Zr-F distances depend on
the degree of bridging as well as the coordination number,
the evidence from a wide range of fluorozirconate structures,
as previously listed by Coupé et al., does suggest that a
close link exists between the Zr-F distance and the C.N..
Furthermore, lower C.N. values imply a higher proposition
of the shorter non-bridging bonds, so that the two factors
affecting the bond length are closely related.

VISCOSITY BEHAVIOR OF HALIDE GLASSES AND MELTS

J.D. MACKENZIE, H. NASU AND J. SANGHERA

Department of Materials Science and Engineering
School of Engineering and Applied Science
University of California
Los Angeles, CA 90024, USA

1. INTRODUCTION

Viscosity is one of the most important properties of glass-forming systems. At temperatures above the melting point or the liquidus, viscosity and the activation energy for viscous flow provide indirect information on the structure of melts. They also furnish information on the ease of glass-formation and, obviously, the ease of crystallization. At temperatures between the melting point or liquidus and the glass transition temperature T_g, knowledge of viscosity and viscoelasticity is critical to the fabrication of the molten glass into various geometries such as plates and fibers. At temperatures below T_g, the deformation of glass under stress is of vital importance to the use of glass objects at elevated temperatures. Viscosity is of course governed by the structure of the melt and the glass which in turn is influenced by chemical composition. The variations of viscosity of glass-forming systems with temperature and with chemical composition have been well-studied for oxides (1). This is not so for halides. The objectives of this report are to review the status of our knowledge and understanding of halide glasses and melts. Arbitrarily, this review is divided into three sections separated by the three temperature zones as described above.

2. VISCOSITY OF MOLTEN HALIDES AT ABOVE T_m AND T_L

The viscosity of molten fluorozirconates at temperatures above the liquidus T_L has been systematically studied (2). In almost all systems, the viscosity at T_L is less than 1 poise. In this temperature range, the simple Arrhenius equation

$$\eta = \eta_0 \exp(E/RT) \tag{1}$$

where E is the so-called activation energy and η_0 is a constant is not applicable. The plots of log η against $1/T$ are non-linear (2,3). Perhaps it is justified to interpret this as a decreasing activation energy with increasing temperature. The Fulcher equation

$$\eta = A \exp(B/T-T_0) \tag{2}$$

where A, B and T_0 are temperature independent constants, is, however, applicable.

A comparison of the viscosity of various liquids is shown in Table 1. Some preliminary results on molten bromides are included. The viscosities of the molten halides are clearly much less than those of the glass-forming oxides. Their values are, however, higher than those

Table 1

Comparison of halide melts, other glass-forming
liquids and molten salts at their liquidus
temperatures (2).

Substance	Structural type	t_m (°C)	Viscosity (P)	Activation energy (kcal/mol)
SiO_2	polymeric	1710	10^7	180
BeF_2	polymeric	540	$>10^6$	>100
B_2O_3	polymeric	450	10^5	40
P_2O_5	polymeric	572	$>10^6$	41.5
$ZnCl_2$	polymeric	318	50	40
Li_2SiO_3	chains/rings	1200	4	24
$NaPO_3$	chains/rings	615	17	16.5
$LiBeF_3$	chains/rings	365	5.6	14
NaCl	ionic	800	0.015	9.3
KNO_3	ionic	337	0.03	4.3
$BaZr_2F_{10}$	----	590	0.6	16.1
$30BaF_2 10LaF_3$ $60ZrF_4$	----	540	3	25
$ZnBr_2$	----	394	2	28
$55ZnBr_2-45KI$	----	187	1	18
$55ZnBr_2-45KCl$	----	190	1	12

Note: Last three results are unpublished work from UCLA

for the molten salts. The activation energies are in the same trend.
It has been suggested that the viscosities and activation energies of
the molten halides are indicative of the presence of small rings and
short chains similar to the structures of molten Li_2SiO_3 and $NaPO_3$ (2).
    The relatively low viscosities of the molten halides at the
liquidus temperature would imply difficulties for glass-formation. For
instance, the viscosity of BeF_2 at the melting point is 10^6 poises
whereas that of a fluorozirconate is 1 poise. The differences between
these values and the viscosity at T_g (10^{13} poises) are immense and very
rapid quenching is required to prevent crystallization.

3. VISCOSITY BETWEEN T_M AND T_G
    For most glass-forming liquids, viscosity increases rapidly with
decreasing temperatures below T_M or T_g. Although Eq. (2) appears to be
very compatible with the experimental results at $T > T_m$, extrapolation
to higher viscosities was not satisfactory. For instance, T_o values by

extrapolation are higher than T_G obtained from DTA (2). Theoretically, T_O cannot be higher than T_G(4). Moynihan et al (5) have also shown that Eq. (2) does not satisfactorily fit viscosity data for fluorozirconates (5). However, a new free volume model which gives the expression (6)

$$\log\eta = A + 2B/(T - T_O + [(T - T_O)^2 + 4CT]^{1/2})$$ (3)

where A,B,C and T_O are constants [different from the constants of Eq. (2)] was shown to fit the experimental data better than the Fulcher equation (5). Perhaps equations of this type can be of use for the interpolation of viscosities in temperature regimes where crystallization would prevent the actual measurement of viscosity. The extraction of structural information from them is not possible at present.

Figure 1 is a plot of the log viscosity against the reduced temperature, T_G/T for a variety of liquids. The ZBLA glass shows a very sharp rise of viscosity from T_M to T_G, similar to the behavior of a glassy nitrate and O-terphenyl, an organic glass. The other glasses are very different. From a technological point of view, the very sharp rise of viscosity with decreasing temperature is an obviously serious problem for the fabrication of glass fibers. Scientifically, this sharp variation of viscosity is an interesting problem. The two obvious causes are (a) a changing structure, such as rapid increase of the degree of polymerization and (b) a rapid decrease of free volume for viscous flow. Cause (a) is unlikely. Halides undergo large volumetric increase on melting, as much as 25% (7). Thereafter the melts have high thermal expansion coefficients. Typically, the expansion coefficients of molten salts are around 4×10^{-4}/°C. Silicate melts have much lower expansion coefficients, typically 4×10^{-5}/°C. If these values are not changed drastically when the temperature of the undercooled melts reaches T_G, then the specific volume of the halide glass will be much smaller than that of the melt at T_M. If molten fluorozirconates behave like molten salts, then the drastic decrease of the specific volume with decreasing temperature is likely to contribute significantly to the steep rise of the viscosity curve in Fig. 1.

A typical undercooled molten silicate would give a maximum crystallization rate of 1 mm per hour at some 30° below the liquidus (8). The viscosity is approximately $10^{4.5}$ poises (1). The rate is usually described by

$$U = K.\Delta T/\eta$$ (4)

where K is an empirical constant.
Various attempts have been made to calculate U. One model leads to the expression.

$$U = \Delta H_f (\Delta T)/T_M 3\pi \lambda^2 \eta N$$ (5)

where ΔH_f is the latent heat of fusion, λ is some jump distance and N is Avogadro's number (9). Such models are not highly accurate in the calculation of U. However, if nucleation is unavoidable and ΔH_f is approximately equal for salts and silicates (3-10 kcals/g.mole), U can

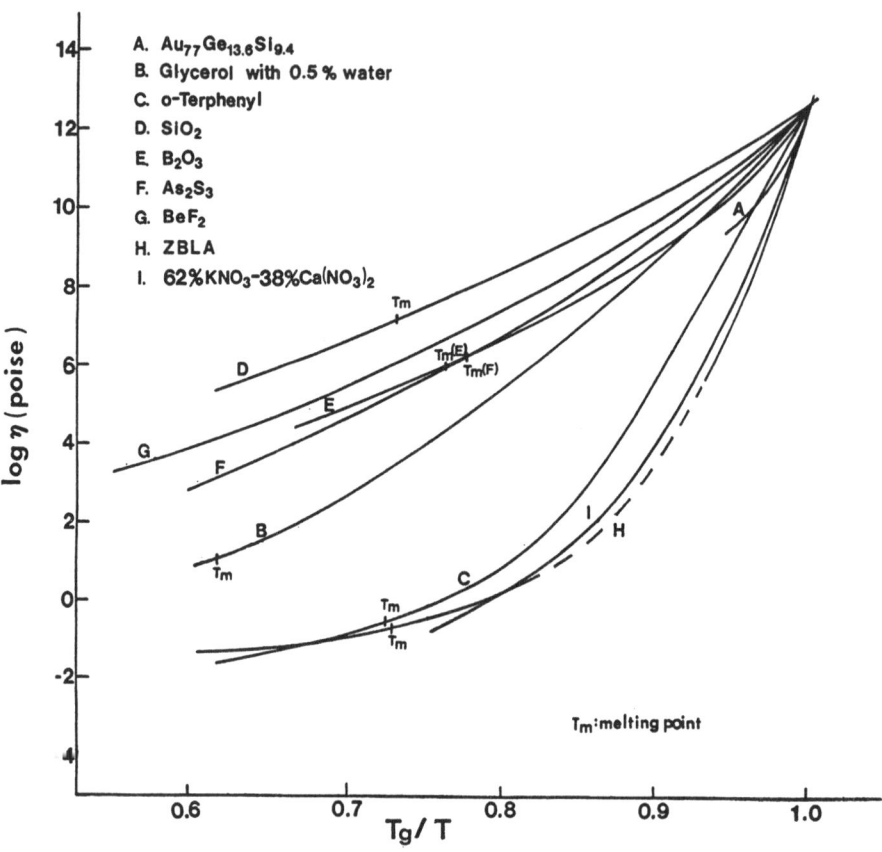

Fig. 1   Log viscosity of various glass-forming liquids as a function of reduced temperature.

be estimated by considering the viscosity alone. Since the viscosity at some $10°$ to $20°$ below T_M is in the order of 1 to 10 poises, U can be 10^2 to 10^3 higher for the fluorozirconates. The crystallization rate can thus be as high as 10^5 mm per hour. This again would imply the difficulty of drawing fibers from the melt.

4. VISCOSITY BEHAVIOR BELOW T_G

At temperatures below T_G but not too far below ($< 50°C$, say), oxide glasses are well-known to exhibit viscoelasticity (8). At temperatures far below T_G, a glass is of course elastic. In a little known publication, Charles and Fisher had shown that when a silicate glass fiber with a T_G of 515°C was subject to a fiber stress of 250,000 p.s.i. for 2 hours at 150°C, the fiber was **permanently** bent at room temperature (10). After heating at 150°C for 2 hours without an applied stress, the bent fiber almost resumed its original shape. Such experiments have not been carried out for halide glass fibers. This is a good example of the anelastic effects obtainable at moderately low temperatures. Because T_G for halide glasses is much lower than for oxide glasses, it seemed desirable to examine whether such unusual behavior is observable for fluorozirconates.

The density of an annealed bulk glass is known to be greater than that of a glass fiber of the same composition because of the rapid quenching. Some results are shown in Table 2. The larger specific volume of a glass fiber is probably the main reason why Charles and Fisher were able to obtain **permanent** bending of their fiber at 150°C when T_G was at 515°C. Secondly, when even bulk glass is under very high shear stresses at relatively low temperatures, **permanent** densification can occur because of the unique random network structure of oxide glasses (15, 16). No such experiments have been made for halide glasses. Some preliminary experiments have been carried out recently in the author's laboratory with ZBLAN glass fibers pulled from a rod preform (Fig.2). A 100 μm fiber was bent at 100°C for two days in an inert atmosphere. After two days, a **permanent** deformation had occurred, similar to the work of Charles and Fisher. The T_G of the ZBLAN glass was 268°C. Fiber stress σ was calculated from (17):

Table 2

Comparison of densities of bulk glass and glass
fibers.

Materials	Bulk Density	Fiber Density	% Change in Density	Reference
$Na_2O-Li_2O-P_2O_5$	2.4238 g/cc	2.4106 g/cc	0.55	11
$CaO-BaO-P_2O_5$	3.1336	3.1025	0.99	11
E-Glass	2.5732	2.5500	0.9	12
ZBLAN	4.9084	4.5556	7.2	13
Alumino-Boro-Silicate	---	---	1.3	14

144

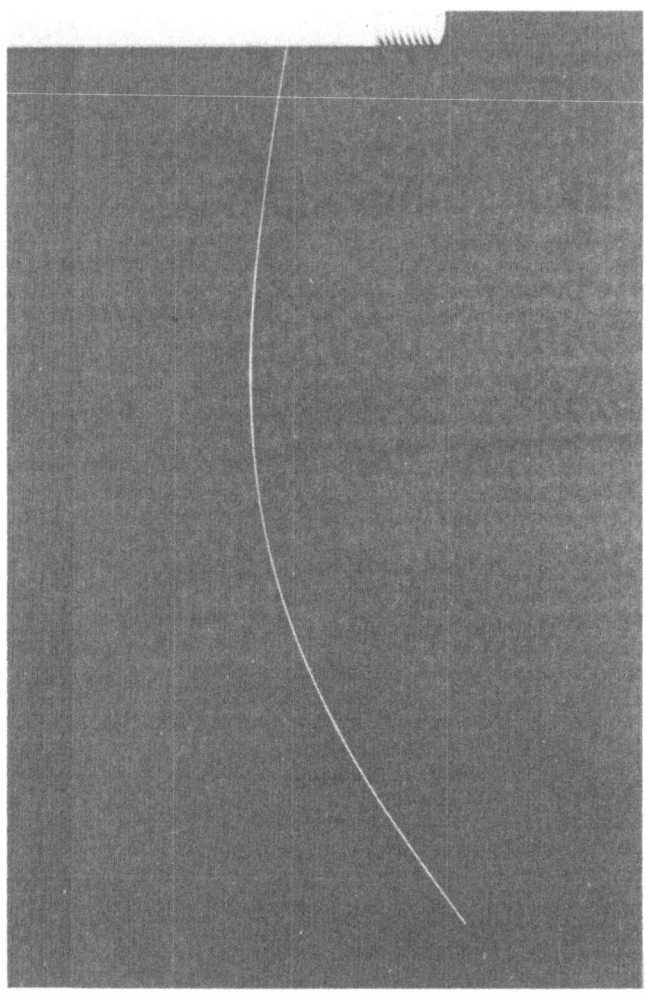

Fig. 2   ZBLAN fiber, 100 micron diameter, after 2 days at 100°C and
          a fiber stress of 10,000 p.s.i (T_g = 268°C).

$$\sigma = Er/\rho \qquad (6)$$

where E is the elastic modulus, r the fiber radius and ρ the radius of curvature of the bent fiber. Eq. (6) was obtained from

$$\sigma = Mr/I \qquad (7)$$

where M is the bending moment and I is the moment of inertia and

$$\rho = EI/M \qquad (8)$$

It seems therefore that even for a fairly low fiber stress of about 10,000 p.s.i., permanent deformation can occur for a fluorozirconate glass fiber. Perhaps this is due to the fairly large specific volume of the fiber as compared to the bulk glass as shown in Table 2. For oxide glasses, bending can induce ionic transport so that the chemical compositions of the glass in tension and that in compression can change with time. Because of the relatively low T_G of fluorozirconates, the behavior of fibers under stress at $T < T_G$ is a topic worthy of further research.

5. CONCLUSIONS

The viscosity of most halide melts are relatively low at temperatures above the liquidus. Viscosity rises very rapidly between T_M and T_G. These properties constitute difficult problems for the direct drawing of fibers from the melt as well as the pulling of fibers from rods. The careful suppression of nucleation during rod-pulling is critical. At temperatures well below T_G, fluorozirconate fibers can undergo permanent deformation under relatively low stress. This phenomenon is not well understood at present.

ACKNOWLEDGEMENTS

The authors are grateful to the Directorate of Chemical and Atmospheric Sciences, Air Force Office of Scientific Research, for the support of this work.

146

REFERENCES

1. Rawson H: Properties and Applications of Glass. New York: Elsevier Publishing Co., 1980.
2. Hu H and Mackenzie JD: J. Non-Cryst. Solids **54**, 241, 1983.
3. Mackenzie JD: this Proceedings.
4. Ferry JD: Viscoelastic Properties of Polymers. New York: Wiley, 1961.
5. Moynihan CT et al: Materials Science Forum **6**, 655 (1985).
6. Cohen MH and Grest GS: J. Non-Cryst. Solids **61** and **62**, 749 (1984).
7. Janz GJ: Molten Salts Handbook. New York: Academic Press, 1967.
8. Stanworth JE: Physical Properties of Glass. Oxford: Clarendon Press, 1950.
9. Doremus RH: Glass Science. New York: Wiley, 1973.
10. Charles RJ and Fisher JC: Chap 19, Frechette VD(ed): in Non-Crystalline Solids. New York: Wiley, 1960.
11. Stockhorst H and Bruckner R: J. Non-Cryst. Solids, to be published.
12. Stockhorst H and Bruckner R: J. Non-Cryst. Solids, **49**, 471, 1982.
13. Mackenzie JD: unpublished results.
14. Bateson S: J. Soc. Glass Tech. **37**, 303, 1953.
15. Mackenzie JD: J. Am. Ceram. Soc. **47**, 76, 1964.
16. Mackenzie JD: J. Am. Ceram. Soc. **46**, 470, 1963.
17. Laurson PG and Cox WJ: Mechanics of Materials. New York: Wiley, 1947.

DISCUSSION

C: J. Lucas. I agree with your views that one needs much infor mation about molten halides and that the relationship bet- ween viscosity and the modification of local structure comes from bond breaking. This leads to a marked change in the degree of polymerization.

Q: C. Pantano. Do you think that the permanent deformation of bent fibers might be promoted by enhanced hydration on the surfaces under tension? I believe that Charles et al. have attributed their similar observations in silicate fibers to hydration effects.

A: J. D. Mackenzie. Our experiments were done under very dry conditions and I do not think that hydration was a factor.

C: R. Almeida. It Appears that we very much need density mea- surements for molten glass-forming halides.

C: C. Moynihan. ΔC_p, the heat capacity difference between supercooled liquid and glass, is very large for fluorozir- conates, compared to other systems. This probably means a correspondingly large expansion coefficient difference $\Delta \alpha_T$. Hence, both volume and structure are expected to chan- ge rapidly above T_g, giving rise to the rapid drop in vis- cosity above T_g. Likewise, it suggests a large density dif- ference between fibers and bulk glasses. Structural relaxa- tion has been observed for these glasses well below T_g, so it was not surprising that you observed a rapid shear stress relaxation and permanent deformation of fibers well below T_g.

Q: G. Frischat. In order to correctly measure the viscosity near T_g, one has to make sure that the melt or glass struc- ture is in a metastable equilibrium. Does the data you showed obey this requirement?

A: J.D. Mackenkie. That data were not mine, but it were taken by others and has been published. The effect you mentioned is certainly important because, near T_g, a change in tem- perature of $10^{\circ}C$ may cause a 10^4 change in viscosity.

Q: G.Frischat. If $10^{\circ}C$ may change the viscosity by up to 4 orders of magnitude, how can you be certain that the Cohen- -Crest viscosity equation gives the best fit to the expe- rimental data?

A: J.D. Mackenzie. We are not certain that it is the best fit, but it seems to fit well the available data. Don't forget that, since there are four adjustable parameters, one could fit almost anything.

THE THERMAL PROPERTIES OF FLUORIDE GLASS

ALLAN J. BRUCE

AT&T BELL LABORATORIES
600 MOUNTAIN AVE.
MURRAY HILL, New Jersey, U.S.A., 07974

ABSTRACT

The thermal properties of fluoride glasses, including their characteristic transformation temperatures, thermal expansion, specific heat capacity and thermal conductivity, are reviewed in this paper. The significance of these parameters both in terms of the glass structure and their technological importance is discussed, with a particular emphasis on the data for ZrF_4- and HfF_4-based glasses. It has also been attempted to derive simple predictive relationships for the thermal properties, in order to fill existing gaps in the available data, either by extrapolation or by analogy to other glass systems. Finally, the possible magnitude and effect of thermally induced stress in fluoride glass optical waveguides has been considered.

1. INTRODUCTION

Broadly defined, the thermal properties (T.P's) of a material, include all physical properties which are measured during controlled temperature programs. To review such a large area, comprehensively, for any group of materials, is a daunting task and it is certainly beyond the scope of this paper on fluoride glasses. The present discussion, will, therefore, be limited to a number of the most relevant T.P's including; the characteristic transformation temperatures, thermal expansion, specific heat capacity and thermal conductivity.

The T.P's of a glass are important from two standpoints. Firstly, they dictate a number of the limiting conditions which are encountered in the glass preparation and subsequent use, and secondly, they can provide useful information on the stucture and bonding of these materials. The latter, is largely due to the relationships which exist between a number of T.P's and the lattice vibrations, or phonons, in a glass. In addition, many workers have proposed, on the basis of an equipartition of energy, that these phonons can be resolved into individual atomic, or molecular, vibrations and that a number of T.P parameters (A), including expansion coefficient, specific heat and thermal conductivity, can therefore be described by simple additive relations [1,2] i.e;

$$A = \sum_i x_i a_i \qquad (1)$$

where x_i is the molar fraction of constituent i in the glass and a_i is an empirical constant describing the contribution of this component to the net value of parameter A. Such relationships are found to be adequate for predicting the T.P's of many silicate glass compositions [1] and they are both useful to the engineer and time saving for the experimenter. The establishment of similar additive relations for the T.P's of fluoride glasses is clearly a worthwhile objective. However, although, these, and other, relationships will be considered below, the accurate prediction of the T.P's for such materials should be regarded as a long-term goal, the attainment of which is primarily dependent on the availability of a substantially larger volume of verified experimental data.

Specific reference will only be made in this text to the fluoro-zirconate (ZrF_4) and fluorohafnate (HfF_4) glasses which are listed in table 1. These are, generally, those for which a number of different T.P. data are available. They also provide a representative cross section of the glass compositions which have been reported for these families. Other families of fluoride glasses including the fluoro-berrylates have, generally, been less well studied. However, much of the present discussion is also relevant for all of these materials [3].

Table 1

Glass Acronym	Composition (MOL %)											
	ZrF_4	BaF_2	ThF_4	LaF_3	AlF_3	NaF	LiF	SrF_2	CaF_2	PbF_2	GdF_3	HfF_4
ZB (1)	50	50										
ZB (11)	65	35										
ZT	57		43									
ZS	70							30				
ZP	50									50		
ZBL	62	33		5								
ZBT	58	33	9									
ZBN	50	25				25						
ZBG	63	33									4	
ZBS	70	15						15				
ZTL	60		30	10								
ZTA	50		43		7							
ZTS	65		15					20				
ZBTN	50	20	7.5			22.5						
ZBTL	60	25	8	7								
ZBTA	65	10	21		4							
ZBLN	58	15		6		21						
ZBLLi	58	15		6			21					
ZBLA	58	33		5	4							
ZBAN	52	26			4	18						
ZBALi	52	24			4		20					
ZBAG	61.8	32.3			2.0						3.9	
ZBLAN	55.8	14.4		5.8	3.8	20.2						
ZBLALi	53	19		5	3		20					
ZBTLN	47.5	12.5	8	7		25						
ZBTLANLiCaP	53	22	7.7	4.2	0.6	3.4	6.9		0.4	1.8		
HB (1)		50										50
HB (11)		35										65
HT			40									60
HBL		33		5								62
HBT		33	7									60
HBLA		33		5	4							58

2. CHARACTERISTIC TRANSFORMATION TEMPERATURES

Each glass forming system has at least three characteristic temperatures, corresponding to both first and second order phase transformations. These are: the glass transition temperature (T_g) the equilibrium liquidus temperature (T_l) and the vaporization temperature (T_v). From a technological standpoint, these temperatures respectively, dictate: the upper limit for using glass components, the lower limit for melt equilibration and chemical processing and the upper limit for the accurate compositional control of the glass forming melt. In the case of fluoride glasses, T_g is one of the most commonly reported parameters, T_l is generally ill-defined and T_v is virtually unknown, except for individual glass components [4,5].

The T_g and T_l values have usually been measured using DTA or DSC techniques, it is therefore appropriate to illustrate their assignment by referring to a typical DSC trace for a ZBLA glass as shown in figure 1(a). The complexity of this trace is typical for most fluoride glasses and it exhibits the following features in order of increasing temperature; a glass transition endotherm, crystallization exotherms, successive exo- and endo-thermic solid-solid phase transformations and finally a series of endothermic melting peaks. The T_g values are assigned as shown in figure 1(b) as the intercept of the extrapolated glass heat capacity curve and the leading edge of the endotherm.

When melting data have been reported for fluoride glasses the attention has usually been focussed on the initial stages of the process. The most commonly reported parameters are therefore the fusion (T_f) and melting (T_m) temperatures. These correspond to the extropolated onset and peak temperatures of the first, or major, melting endotherms, respectively, as shown in figure 1(c). The point which has often been overlooked, however, is that T_l, which corresponds to the peak temperature of the final melting endotherm and which correlates with the transition to a single liquid phase, often lies at significantly higher temperatures. The case of ZBLA is quite common for many fluoride glasses, where there are several, well separated melting endotherms. The higher temperature endotherms are often weak and can easily be overlooked, either if the scans are prematurely terminated or if the sensitivity of the equipment is in doubt. The present assignment of T_l for ZBLA is consistent with the empirical predictions discussed below.

Characteristic temperatures for most of the glasses listed earlier are compiled in table 2. The greatest difficulty encountered in this task was to provide a consistency in the assignment of the respective T_f, T_m and T_l values, which were obtained from the literature. As a general rule, T_l values have only been assigned if the actual DTA/DSC traces were available for consultation. If this was not the case, the peak temperatures for the reported melting exotherms have been assigned as T_m. The ratios of the T_m/T_g and T_l/T_g values are also given in table 2. The constancy of these parameters is remarkable. Their statistical averages, to within one standard deviation, are 1.41 ± 0.03 and 1.56 ± 0.08 respectively. Similar relationships, ($T_l/T_g \approx 1.5$) have been observed in glass forming systems for many years [6,7]. Although, their orgin is still not fully understood, they provide a convenient method for predicting the T_l values of new compositions and also the minimum temperature ranges which should be covered when studying the melting behavior of glass forming systems.

The range of T_g values reported for the ZrF_4/HfF_4 glasses is quite extensive (~ 450-$800K$) and, so far, the widest for any family of heavy metal fluoride glasses [3]. The lower T_g values usually pertain to those glasses with high alkali and/or alkali earth content components and the higher to those containing highly polarizing cations (e.g. Th^{4+}).

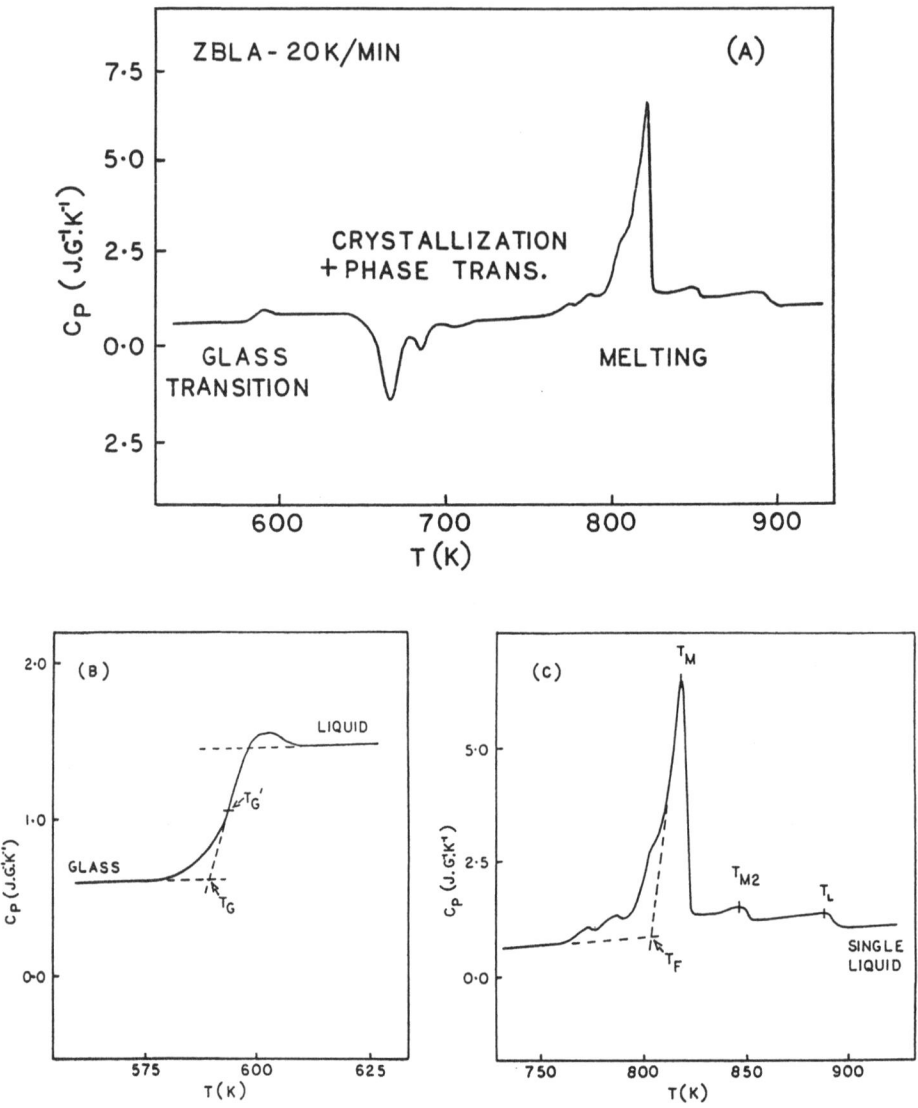

Figure 1

DSC trace for a ZBLA glass (A) with assignments
of the glass transition (B) and melting
endotherm (C) temperatures.

Table 2

Glass Acronym	Characteristic Temperatures (K)						
	T_g	T_F	T_m	T_1	T_m/T_g	T_1/T_g	REF
ZB(1)	608	848	860	922	1.41	1.52	8,9
ZB(11)	570	787	798	860	1.41	1.51	9
ZT	763	–	–	1188	–	1.54	10
ZS	581	–	–	869	–	1.49	11
ZP	453	–	–	823	–	1.82	9
ZBL	580	810	820	883	1.41	1.52	8,12
ZBT	593	823	833	901	1.40	1.52	13
ZBN	513	–	–	843	–	1.64	14
ZBS	577	–	815	–	1.41	–	11
ZBG	583	839	848	903	1.45	1.55	15
ZTL	723	–	–	1083	–	1.50	16
ZTA	788	–	1098	–	1.39	–	10
ZTS	601	–	867	–	1.44	–	11
ZBTN	543	–	803	863	1.48	1.59	14
ZBTL	583	–	813	–	1.39	–	17
ZBTA	582	–	853	–	1.47	–	18
ZBLN	535	728	735	821	1.37	1.53	8
ZBLLi	515	740	750	815	1.46	1.58	8
ZBLA	588	805	820	883	1.39	1.50	8
ZBAN	520	–	751	–	1.44	–	18
ZBALi	511	–	–	835	–	1.63	18
ZBAG	579	–	834	–	1.44	–	19
ZBLAN	543	–	745	831	1.37	1.53	8
ZBLALi	530	–	–	820	–	1.55	8
ZBTLN	563	–	752	–	1.34	–	17
ZBTLANLiCaP	598	–	823	–	1.38	–	14
HB(1)	574	797	808	–	1.41	–	8,20
HB(11)	580	815	825	880	1.42	1.52	8
HT	757	–	–	–	–	–	21
HBL	605	805	832	920	1.38	1.52	8
HBT	593	–	853	–	1.44	–	22
HBLA	580	–	835	–	1.44	–	8

3. THERMAL EXPANSION

The expansion of a material with increasing temperature is often pictured in terms of a collection of anharmonic oscillators (see Figure 2), in which the attainment of higher energy vibrational states (a→b) results in a net increase in the interatomic spacing. Further, if, as in the case of fluoride glasses, the material expands rapidly with temperature, this is thought to imply that   a significant proportion of vibrational transitions are between states near the top of the potential energy well (a'→b'). In other words, the binding energy in these oscillators is relatively weak.

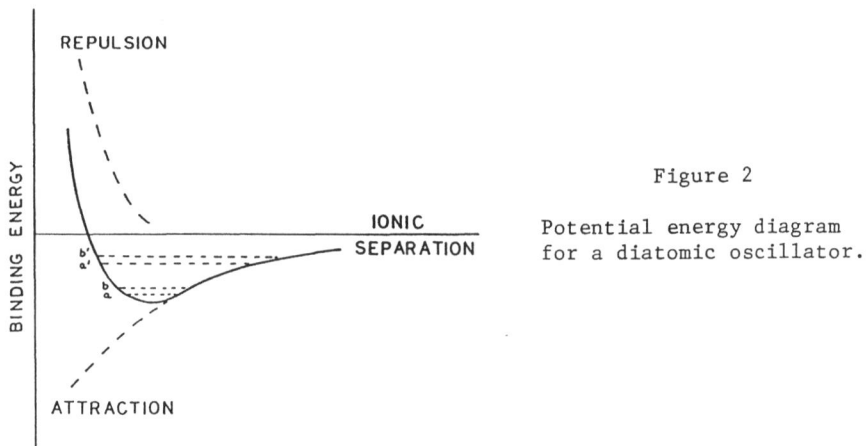

Figure 2

Potential energy diagram for a diatomic oscillator.

Technologically, the thermal expansion of a glass forming system is very important. It, effectively, determines the volumetric compatibility of a glass with other materials, including other glasses, throughout the temperature regimes appropriate for melting, processing, quenching and ultimately using glass components. Volumetric incompatibilities, particularly during the quenching stages, can generate both internal and external stresses in, or on, the resulting glass. These may produce significant modifications in the physical properties of the glass and in extreme cases lead to the catastrophic failure of glass components in use. To avoid, or even to usefully employ, such phenomena, one requires a good understanding of the thermal expansion behavior, particularly for high expansion materials such as fluoride glasses. The specific case of the stresses which are generated in core/clad optical fibers will be considered in a later section.

Thermal expansion data, from dilatometric or TMA studies, have been reported for a number of fluoride glasses, in the temperature range between ambient and T_g. In this range, as for most glasses, the degree of expansion is almost linearly dependent on temperature. The expansion behavior is therefore adequately described by a mean thermal expansion coefficient ($\bar{\alpha}$), i.e.;

$$\bar{\alpha} = \frac{1}{l_o} \frac{\Delta l}{\Delta T} \qquad (2)$$

where l_o is the unidimensional sample length at ambient temperature and Δl is the observed change in length over a temperature difference ΔT. The units of $\bar{\alpha}$ are those of reciprocal temperature (K^{-1}).

The $\bar{\alpha}$ values which have been reported for the glasses listed previously are compiled in table 3. The average value for these glasses, to within one standard deviation, is $17 \pm 4 \cdot 10^{-6}$ K^{-1}. This value is, generally, much higher than those for silicate glasses, being approximately forty times that for fused silica ($\sim 0.5 \times 10^{-6}$ K^{-1}), and it is only approached for some of the higher alkali silicate compositions (e.g. $13.5 \cdot 10^{-6}$ K^{-1} for $Na_2O \cdot (SiO_2)_2$, [1]).

Table 3

Glass Acronym	$\bar{\alpha}.10^6$ (K^{-1})	$\bar{\alpha}.T_g^2$ (K)	Ref
ZB(11)	18.6	6.0	24
ZT	9.0	5.2	25
ZBL	17.0	5.7	20
ZBT	18.0	6.3	26
ZBG	17.5	6.0	15
ZBS	16.7	5.6	11
ZTL	25.0	5.8	25
ZBTL	16.8	5.7	17
ZBLA	27.7	5.9	17
ZBAG	16.9	5.7	28
ZBLAN	27.5	8.1	29
ZBLALi	14.1	3.7	27
ZBTLN	18.0	5.7	17
ZBTLANLiCaP	16.5	5.9	14
HBL	18.2	6.2	30
HBLA	17.3	5.8	31

A few systematic studies on the compositional dependence of $\bar{\alpha}$ have been performed (see Figure 3) [15,17]. However, in the absence of more studies like these, it is not yet possible to derive meaningful additive relations for the $\bar{\alpha}$ values of fluoride glasses. An alternative method of predicting $\bar{\alpha}$ values can be based on the work of Van Uitert [23], who noted that the product of $\bar{\alpha} \cdot T_g^2$ is approximately constant for many glasses. Using this formalism for the $\bar{\alpha}$ data given in table 3, we find that the average relation

$$\bar{\alpha} = 5.8/T_g^2 \qquad (3)$$

holds to $\pm 2 \times 10^{-6}$K^{-1}, within one standard deviation. This parameterization has the virtue that the low $\bar{\alpha}$ values for the ZT and ZTL glass compositions now fall into line. The data for the ZBLAN and ZBLALi compositions, however, are still somewhat anomalous, and until a sufficiency of data has been obtained to fully test such relations they should be applied with caution.

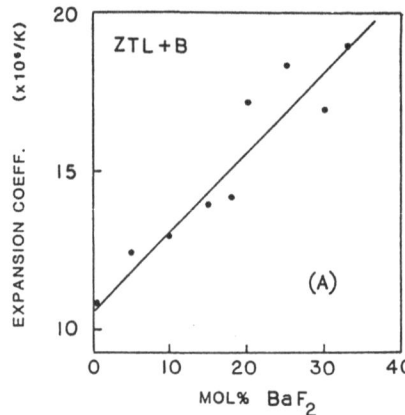

Figure 3

Thermal expansion coefficient
($\bar{\alpha}$) as a function of composition in $ZrF_4(0.6)-LaF_3(0.07)-$
$ThF_4(33-x) -- BaF_2(x)$
glasses [17].

4. HEAT CAPACITY

The isostatic specific heat capacity (C_p, $J \cdot g^{-1} \cdot K^{-1}$) of a material is a measure of the energy required to raise the temperature of a unit mass by one degree, at some specified temperature. In terms of an oscillator model, this simply reflects the change in the average vibrational energy, for the given material, as a function of temperature. As a further refinement, the experimental data for solids, may also be fit to a Debye heat capacity function [32] and average frequencies (v_D) for their fundamental vibrations determined.

In the case of fluoride glasses, although all DSC/DTA studies are potential sources of data, only a few sets of precisely measured C_p values have been reported [33-36]. These, however, appear to be sufficient to provide a comprehensive picture of the C_p behavior with temperature, at least for the ZrF_4/HfF_4 glasses. At present, ZBLA is the only composition for which precise C_p data are available at both high (> 300K) and low (< 100K) temperatures (see figure 4).

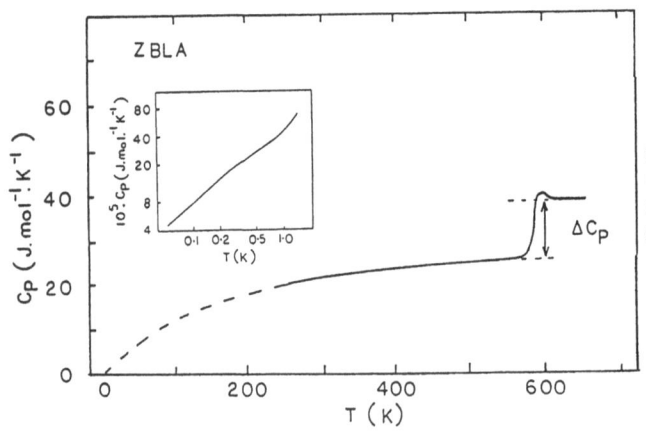

Figure 4

Specific heat capacity
(C_p) as a function of
temperature for a ZBLA
glass [33,35].

The low temperature data for ZBLA conform to the general pattern previously observed for other, non-halide, glasses, where the C_p values approach zero as an almost linear function of temperature [35]. In the higher temperature regime it is convenient to consider the data on a per mole of atoms basis, i.e., in terms of $\bar{C}_p$ ($J \cdot mol^{-1} \cdot K^{-1}$) values, where:

$$\bar{C}_p = C_p \cdot \bar{M} \tag{4}$$

and $\bar{M}$ is the average atomic weight of the glass. Using this rationale, it can be seen that the $\bar{C}_p$ values for ZBLA near T_g, approach, but do no exceed, the Dulong-Petit limit of 3R ($24.9 \ J.mol^{-1}K^{-1}$) for the vibrational heat capacity of solids [2]. At 300K the $\bar{C}_p$ value ($21.1 \ J.mol^{-1}K^{-1}$) is close to 90% of this limit.

Table 4

Glass Acronym	$\bar{M}$ ($g \cdot mol^{-1}$)	$\bar{C}_p(J \cdot mol^{-1}K^{-1})$		$\Delta\bar{C}_p$ ($J \cdot mol^{-1} \cdot K^{-1}$)	V_D (cm^{-1})	REF
		300K	T_g-35K			
ZB(11)	39.55	22.7*	24.9	14.1	–	36
ZBL	39.94	21.5	25.4	14.5	375	33
ZBT	42.07	21.5	–	–	–	34
ZBLA	39.53	21.1	24.8	13.9	403	33
ZBLN	35.88	22.2*	24.2	13.7	–	36
ZBLLi	35.04	22.4*	24.2	14.1		36
ZBLAN	35.25	22.3*	24.2	14.2		36
ZBLALi	35.73	23.4*	24.5	14.5	–	36
HB(11)	52.74	22.5*	25.1	15.4	–	36
HBL	52.55	21.7	25.5	13.8	354	33
HBT	53.48	22.9	–	–	–	34
HBLA	51.44	21.3	25.1	14.3	382	33

*extrapolated values

From the data compiled in table 4 for other ZrF_4/HfF_4 glasses it is observed that the $\bar{C}_p$ values for these glasses, at the corresponding reference temperatures, are idential to those for ZBLA (within experimental error). In addition, the $\Delta\bar{C}_p$ values which correspond to the difference, at T_g, between the extrapolated $\bar{C}_p$ curves for a glass and its supercooled liquid (see figure 4), are also approximately constant ($14.3 \pm 0.5 \ J.mol^{-1}K^{-1}$). The later values are considerably larger than those observed for network oxide glasses ($5 - 8 \ J.mol^{-1}K^{-1}$) and this suggests that more substantial changes occur in the structure of the fluoride systems when their glass transitions are travesed [33].

Based on the above, it would seem that C_p values for ZrF_4/HfF_4 glasses may be predicted fairly accurately between 300K and T_g, simply with a fore knowledge of its average atomic weight. Further, since the $\bar{C}_p$ values for the undercooled liquid are not expected to increase significantly with temperature, the thermal energy per gram (ΔQ) which is liberated on quenching a glass forming ZrF_4/HfF_4 melt, from T_1 say, is approximated by;

$$\Delta Q(J \cdot g^{-1}) \approx \frac{39.2(T_1-T_s)}{\overline{M}} \tag{5}$$

For ZBLA this yields a ΔQ value of approximately 300 J·g⁻¹.

5. THERMAL CONDUCTIVITY

The thermal conductivity of a material (κ in $W \cdot m^{-1} \cdot K^{-1}$) is defined as the rate of heat flow between opposite faces of a unit cube, under a driving force of a one Kelvin temperature difference. In a glass at, or below, ambient temperature this transfer of thermal energy is facilitated by the propogation of phonons in the materal [2] and κ is often simply expressed as;

$$\kappa = (\rho C_v \cdot v \cdot l)/3 \tag{6}$$

where, ρ is the density, C_v is the, constant volume, heat capacity, v is the speed of sound in the material and l is the mean free path for the phonons. At substantially higher temperatures other mechanisms of heat transfer, including radiation and convection (in melts), can contribute significantly to, and will eventually dominate the effective thermal conductivity [37]. Radiative contributions to K, become apparent at temperatures which correlate quite well with those at which the glasses are expected to start emitting radiation of shorter wavelength than their I.R. edges (see Figure 5).

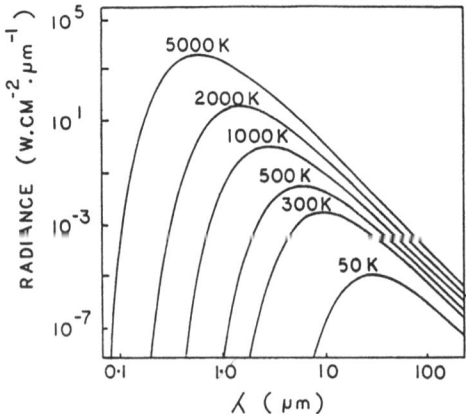

Figure 5

Black body radiation as a function of wavelength (λ) and temperature.

Thermal conductivity data have only been reported for the ZrF_4 compositions ZBLAN and ZBLALi [38]. These data pertain to a low temperature regime (<100K) and they generally follow the trend observed for other non-halide glasses, approaching zero near 0K and exhibiting small plateau regions near 10K. The values of v($\sim$ 240 m.s^{-1}) and l($\sim 1.5 \times 10^{-10}$ m) obtained at these low temperatures can however be used in equation 5 together with ρ($\sim$ 4.5 g.cm^{-3}) and C_v ($\sim C_p$ at T > 300K) data to estimate κ at temperatures between ambient and T_g. By this method, κ values of approximately 0.4 W$\cdot$m$^{-1}\cdot$K^{-1} are predicted for both glasses. These are typically three to four times lower than the values observed for most silicate glasses over similar temperature ranges.

Intuitively, the removal of thermal energy from glass-forming fluoride melts is likely to be a major limiting factor in the fabrication of large glass components. In this regard we can only surmize at present that (1) The radiative contributions to the thermal conductivity will be significant at somewhat lower tempertures than in silicate glasses (due to the longer wavelength transmission of fluoride glasses) and (11) the convective transfer of heat in the low viscosity fluoride melts may be substantial.

6. THERMAL STRESSES IN OPTICAL FIBERS

If there is a mismatch in the thermal expansion of the core and clad glass compositions in an optical fiber, between ambient and the fiber drawing temperature, this can result in the generation of stresses in the fiber. The effect of these stresses on the physical properties of the fibers is an important design consideration for such components [39]. Pimarily, the stresses derive from differing $\bar{\alpha}$ values in the ambient to T_g range. However, significantly higher stresses, than those predicted by this mechanism, can be generated if the cladding glass cools through its glass transition before the core glass [40]. Fortunately, the latter contributions can usually be eliminated either by using glasses where T_g (clad) is slightly lower than T_g (core) and/or by carefully annealing the fibers.

The distribution of stresses in optical fibers, is essentially the same as that determined for other cylindrical composite materials, such as glass to metal seals [41]. At ambient temperature (T_o), the axial component of the stress (S_z) is given by [40]

$$S_z = \frac{\Delta\bar{\alpha}(T_g - T_o) \cdot E}{(r_{clad}/r_{core})^2 \cdot (1 - \mu)} \qquad (7)$$

where $\Delta\bar{\alpha}$ is ($\bar{\alpha}_{core} - \bar{\alpha}_{clad}$), E is Youngs Modulus, r is the radius of each respective component, and μ is Poissons ratio. Using, typical values for fluoride glasses [42] (E $\sim$ 60 GPa, $\mu \sim$ 0.3 and ($T_g - T_o$) $\sim$ 300K) we obtain

$$S_z(T \cdot Pa) \approx 25.7 \, (\Delta\bar{\alpha} \cdot r_{core}^2)/r_{clad}^2 \qquad (8)$$

The axial stresses for fluoride glasses of different core/clad dimensions can therefore be estimated from their $\Delta\bar{\alpha}$ values. Radial (S_R) and tangential (S_O) stresses in the fibers can also be predicted from similar formalisms.

The effects of the thermal stress on the physical properties of fluoride glass fibers are as yet undetermined. However, by analogy to silica fibers, we can expect a beneficial strengthening to occur if the net stresses are slightly compressure ($\Delta\bar{\alpha} \geq 0$). This factor has almost certainly been considered by Takiwa et al. [29] in their selection of potential core/clad compositions.

7. SUMMARY

The thermal properties of fluoride glasses appear to conform to the trends previously observed for other, non-halide, glass composition. However, from the magnitude of these properties and their dependence on temperature, we can infer that the extended bonding in such glasses is generally much weaker than in the covalent-network oxide glasses and that they undergo more extensive structural reorganization at T_g. Also, as outlined above, it is possible to estimate many of the thermal property values for new compositions from a minimum amount of data. This is seen as a significant time-saving development in the study of fluoride glasses.

Additional data on the thermal transport behavior of fluoride glasses at high temperatures ($> T_g$) would be useful for determining the maximum achievable cooling rates for fluoride glass forming melts. These, in competition to the crystallization kinetics, are likely to set the upper size limits for high quality fluoride glass components.

8. ACKNOWLEGEMENTS

The author is most grateful to Dr's L. G. Van Uitert and C. R. Kurkjian at Bell Laboratories for useful discussions and comments on the melting point relations for glasses and thermal stresses in fibers, respectively.

9. REFERENCES

1. Morey, G. W.: The properties of Glass, Reinhold Publishing (1954).

2. Holloway, D. G.: The Physical Properties of Glass, Springer-Verlag Publishing (1973).

3. Bruce, A. J.: In Halide Glasses: Preparation, Structure and Properties; Drexhage, M. G. and Moynihan, C. T. (Eds), Telford Press (in preparation).

4. Margrave, J. L. (ed): Characterization Of High Temperature Vapors, Wiley Publishing (1967).

5. Mitachi, S., Terunuma, Y., Ohishi, Y. and Takahashi, S.: J. Lightwave Tech. LT-2 587-592 (1984).

6. Kauzmann, W.: Chem. Rev. 43, 219 (1948).

7. Sakka, S., Mackenzie, J. D.: J. Non. Cryst. Solids 6, 145-62, (1971).

8. Bruce, A. J., Moynihan, C. T.: Unpublished data.

9. Kawamoto, Y., Sakaguchi, F.: Bull. Chem. Soc. JPN. 56, 2138-41, (1983).

10. Matecki, M., Poulain, M., Poulain, M.: Mat. Res. Bull. 17, 1035-43, (1982).

11. Saad, M., Poulain, M.: Mat. Sci. Forum 5, 105-12, (1985).

12. Lecoq, A., Poulain, M.: J. Non. Cryst. Solids 34, 101-10, (1979).

13. Poulain, M., Poulain, M., Matecki, M.: Mat. Res. Bull 17, 1275-81, (1982).

14. Poulain, M., Lucas, J.: Verres Refract. 32, 505-13, (1978).

15. Takahashi, S., Shibita, S., Kanamori, T., Mitachi, S., Manabe, T.: Adv. in Ceramics 2, 74-83, (1981).

16. Matecki, M., Poulain, M., Poulain, M. Lucas, J.: Mat. Res. Bull 13, 1039-46, (1979).

17. Poulain, M., Grosdemouge, M. A.: Verres Refract. 36, 853-8, (1982).

18. Lecoq, A., Poulain, M.: Verres Refract. 34, 333-42, (1980).

19. Kanamori, T., Takahashi, S.: Jpn. J. Appl. Phys. 24, L758-60, (1985).

20. Almeida, R. M., Mackenzie, J. D.: J. Chem. Phys. 78, 6502-11, (1983).

21. Almeida, R. M., Mackenzie, J. D.: J. Non. Cryst. Solids *68*, 203-17, (1984).

22. Robinson, M., Drexhage, M. G.: Mat. Res. Bull. *18*, 1101-12, (1983).

23. Van Uitert, L. G.: J. Appl. Phys. *50*, 8052-61, (1979).

24. Almeida, R. M., Mackenzie, J. D.: J. Chem. Phys. *74*, 5954-61, (1981).

25. Matecki, M., Poulain, M., Poulain, M.: Mat. Res. Bull. *17*, 1035-43, (1982).

26. Poulain, M., Poulain, M., Matecki, M.: Studies in Inorganic Chemistry *3*, 461-4 (1983).

27. Tran, D. C., Ginther, R. J., Sigel, G. H.: Mat. Res. Bull. *17*, 1177-84, (1982).

28. Mitachi, S., Shibata, S., Manabe, T.: Jpn. J. Appl. Phys. *20*, L337-8 (1981).

29. Takiwa, H., Mimura, Y., Shinbori, O., Nakai, T.: J. Lightwave Tech. *LT3*, 569-73 (1985).

30. Poulain, M.: J. Non. Cryst. Solids *56*, 1-14, (1983).

31. Drexhage, M. G., El-Bayoumi, O. H., Moynihan, C. T., Bruce, A. J. Chung, K. H., Gavin, D. C., Loretz, T. J.: J. Am. Ceram. Soc. *65*, C168-71 (1982).

32. Kittel, C.: Introduction to Solid State Physics, pp. 125-152, Wiley Publishers (1976).

33. Gavin, D. G., Chung, K.-H., Bruce, A. J., Moynihan, C. T., Drexhage, M. G., El-Bayoumi, O. H.: J. Am. Ceram. Soc. *65*, c182-3, (1982).

34. Robinson, M., Pastor, R. C., Turk, R. R., Devor, D. P., Braunstein, M., Braunstein, R.: Proc. SPIE, *266*, 78-83, (1981).

35. Lasjaunias, J. C., Grosdemouge, M. A.: J. Non. Cryst. Solids, *54*, 183-6 (1983).

36. Bruce, A. J., Gavin, D. L., Moynihan, C. T., Drexhage, M. G.: J. Am. Ceram. Soc. (in press).

37. Blazek (Ch): Review of Thermal Conductivity Data in Glass, Intnl. Comm. on Glass (1983).

38. McCarthy, K. A., Sample, H. H., Dharmaratna, W. G. D.: J. Non Cryst. Solids, *64*, 445-8 (1985).

39. Miller, S. E., Chynoweth, A. G.: Optical Fiber Telecommunications, Academic Press, (1979).

40. Krohn, D. A.: J. Am. Ceram. Soc. *53*, 505-7 (1970).

41. Varshneya, A. K.: In Treatise on Materials Science and Technology, Vol. 22, pp. 241-306 Academic Press (1982).

42. Drexhage, M. G.: In Treatise on Materials Science and Technology, Vol. 26, Academic Press (1985).

DISCUSSION

C: J. D. Mackenzie. The occurence of large ΔC_p, as compared to oxide glasses, is perhaps an important area for theoretical studies which may tell us something about the structural differences between fluorozirconates and silicates.

R: A.J. Bruce. The ΔC_p values observed for heavy metal fluoride glasses are approximately twice those for network oxide glasses. This is consistent with substantial changes taking place in the former materials at T_g, with probably an almost complete breakdown of extended bonding taking place. This is certainly an area which deserves further study. It is also noteworthy that smaller ΔC_p values are measured for chalcogenide glasses.

Q: D. Ulrich. The trends that were shown in transport properties and glass transition for HMF glasses appear to be similar to recent observations in some of the newer high temperature extended rigid chain polymers which are derived from liquid crystalline polymeric solutions. These trends in properties may provide some insight into the polymeric chain-like structures which have been suggested by Almeida and Mackenzie for fluorozirconate glasses, as well as the balance in these glasses between polymeric and 3-D network structures.

A: A. J. Bruce. The mentioned trends should certainly reflect the underlying structural changes in the glasses. However, it is not clear whether it will be possible to distinguish between the different structural models on this basis; this will require further study.

STRUCTURAL RELAXATION IN FLUORIDE GLASSES

C.T. MOYNIHAN, S.M. OPALKA, R. MOSSADEGH,
S.N. CRICHTON and A.J. BRUCE

Center for Glass Science and Technology, Materials Engineering
Department, Rensselaer Polytechnic Institute, Troy, NY 12180-
-3590, USA

ABSTRACT
    Structural relaxation studies during annealing of a series
of ZrF_4-based glasses below the glass transition temperature
have been carried out. Indications are that no property
changes due to structural relaxation are likely to occur at
ambient temperature over periods of tens of years. Some of the
lower Tg glasses, however, did exhibit detectable structural
relaxation on annealing at temperatures as low as 100°C over
roughly a one year time period.

1. INTRODUCTION
    Heavy metal fluoride glasses (HMFG) are a fairly new class
of materials which are currently receiving much attention
because of potential uses as optical materials in the mid-IR.
(1,2). Depending on composition, their IR transparency extends
out to 6-9 µm, much further than silicate glasses, whose IR
transparency stops at around 2-3 µm. The most widely characterized of the HMFG are those whose main component is ZrF_4.
ZrF_4-based HMFG compositions, which are typical for these
materials and which were characterized in this study, are
listed in Table I, along with their glass transition temperatures Tg measured by differential scanning calorimetry (DSC)
at a 10 K/min heating rate.

TABLE I. HMFG compositions (mol%) and glass transition
          temperatures.

Glass	ZrF_4	BaF_2	LaF_3	AlF_3	NaF	LiF	PbF_2	Tg(K)
ZBL	62	33	5	–	--	--	--	574
ZBLA	58	33	5	4	--	--	--	584
ZBLAN	56	14	6	4	20	--	--	536
ZBLALi	51	21	5	3	--	20	--	523
ZBLALiPb	50	17	5	3	--	20	4	513

    Tg's for these glasses are fairly low – 240 to 311°C – with
the alkali fluoride containing glasses having the lower Tg's.
There thus exists the possibility that at ambient temperatures
structural relaxation or physical aging of these glasses might
cause their properties to drift over long periods of time.

(See Refs. 3-10 for some recent articles on structural relaxation in glasses.) Slight changes in, for instance, refractive index over a period of years could lead to unacceptable changes in optical properties of extremely long fiber optic waveguides made from these materials. Such considerations prompted our studies of sub-Tg structural relaxation in HMFG.

2. SUB-Tg ANNEALING EXPERIMENTS

Our experimental approach has been described in detail in a report of some early results (9) and is illustrated schematically in Fig. 1. Briefly, samples of the glasses were rate cooled on the DSC at 10 or 100 K/min from above to well below the glass transition region. The samples were then annealed isothermally a temperature T_A well below Tg for times t_A ranging up to over one year. Following the isothermal anneal, the specific heats Cp were measured on the DSC at 10 K/min while reheating through the glass transition region. During this reheat the glass regains any enthalpy lost due to the earlier sub-Tg anneal, so that the Cp curves measured during reheating can be used to monitor the progress of enthalpy changes due to structural relaxation during sub-Tg annealing.

Typical DSC results are shown in Figs 2 and 3. Figure 2 is characteristic of anneals carried out at temperatures a moderate distance (10-60 K) below Tg. The effect of annealing is to increase the magnitude of the Cp maximum in the main glass transition region and shift it to higher temperature. Figure 3 is characteristic of anneals carried out a fairly large distance (100-140 K) below Tg. Here the effect of annealing shows up as an increase in Cp or even a small, extra Cp peak in the sub-Tg region just below the main glass transition region. Both of these results are in accord with predictions of the standard Tool-Narayanaswamy kinetic model for structural relaxation (3-10), which assumes that the relaxation process is non-linear and is controlled by a monotonic, single-peaked spectrum of relaxation times. The appearance of two peaks in the Cp reheating curve does not require a double-peaked spectrum of relaxation times and does not imply that there are two distinct relaxation mechanisms in the glass (6,7).

For purposes of analysis it is convenient to express the changes in enthalpy due to structural relaxation in terms of the time dependence of the fictive temperature T_f. T_f may be thought of as a measure of the average configurational enthalpy of the glass and is defined such that $T = T_f$ for a glass in the equilibrium state at temperature T (see Fig. 1). T_f is calculated from the experimental Cp data obtained during reheating via the following expression (3):

$$T_f = T^* + \int_{T^*}^{T \ll Tg} [(Cp-Cpg)/(Cpe-Cpg)] dT \tag{1}$$

where T^* is a temperature above the glass transition region and Cpg and Cpe are respectively the glass and equilibrium liquid heat capacities. This calculation is illustrated in Fig. 4 for a ZBLA glass sample. T_f in this case is the fictive temperature reached by the glass immediately after cooling at

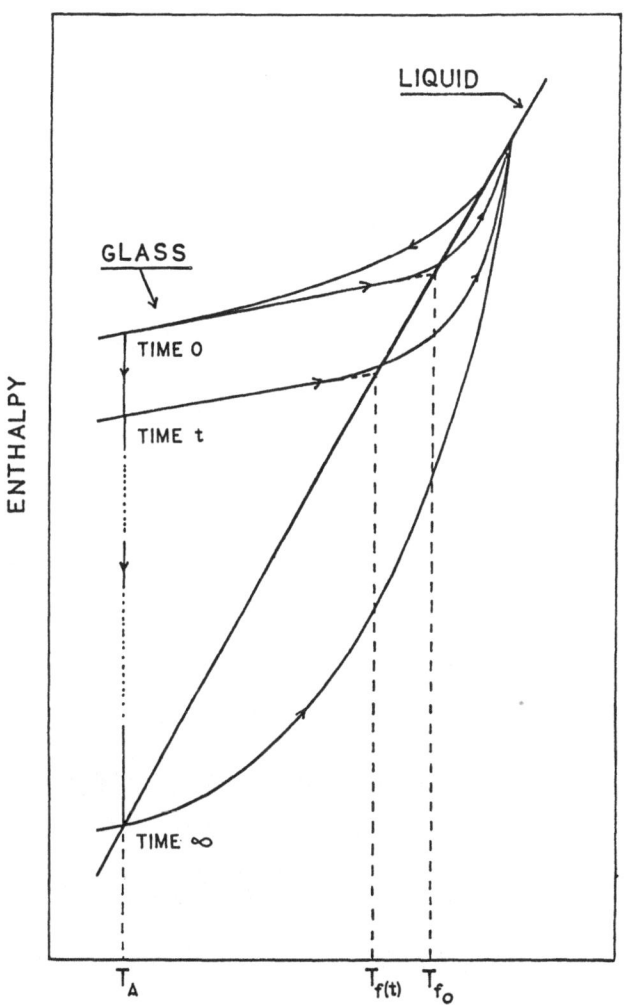

Figure 1. Schematic representation of the variation of
enthalpy and fictive temperature during rate
cooling through the glass transition region,
sub-Tg annealing and rate heating through the
glass transition region.

166

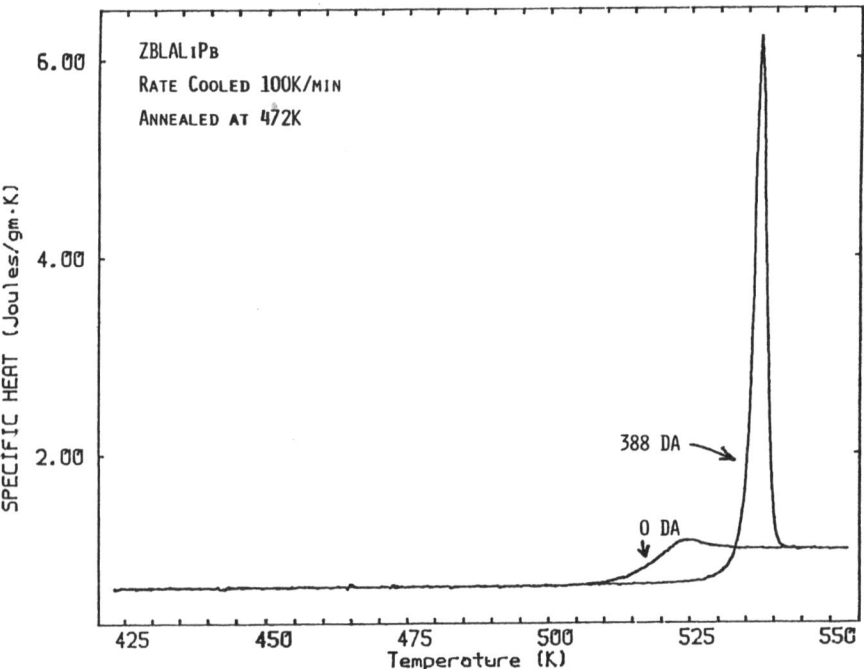

Figure 2. Specific heat measured at 10 K/min heating
rate for ZBLALiPb glass following rate cool
and sub-Tg anneal for times shown in figure.

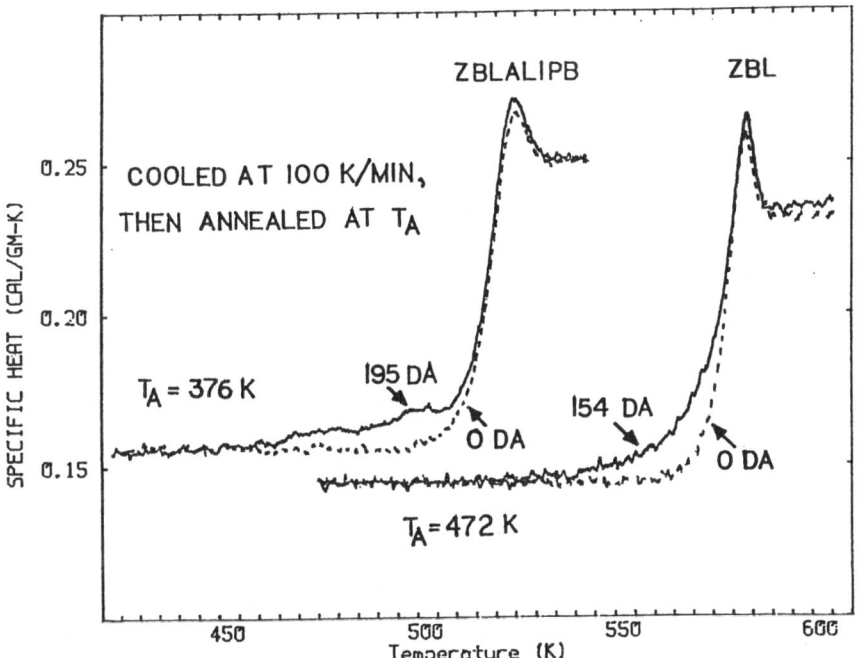

Figure 3. Specific heat measured at 10 K/min heating rate
for ZBL and ZBLALiPb glasses following rate cool
and sub-Tg anneal for times shown in figure.

168

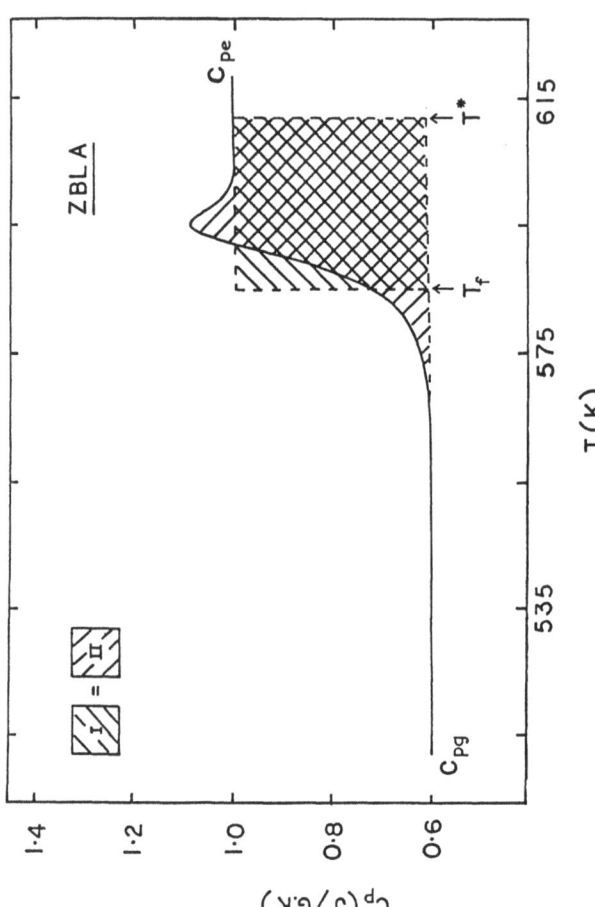

Figure 4. Illustration of fictive temperature calculation for ZBLA glass heat capacity data measured at a 10 K/min heating rate immediately following a 100 K/min rate cool through the transition region.

100 K/min through the transition region and would correspond
to T_{fo} in Fig. 1.

3. RESULTS OF SUB-Tg ANNEALING EXPERIMENTS

Figure 5 shows typical results for ZBLA glass for the evolution of T_f with time during sub-Tg annealing at several temperatures T_A. Note from Fig. 4 that the initial fictive temperature at time t=0 obtained immediately after rate cooling
through the glass transition region is approximately equal to
Tg. For the sub-Tg anneal at the highest temperature the sample comes to equilibrium (T_f = T_A = 575 K) within one day. For
annealing at the next two lower temperatures (T_A = 472 and
524 K) relaxation is slower but appreciable over a one year
time period. However, the samples are still very far from equilibrium after one year. At the lowest temperature (T_A = 376 K)
some 210 K below Tg no detectable relaxation has occurred in
nearly two years. Similar results are shown in Fig. 6 for
ZBLALiPb glass. Note that with this glass a small but detectable change in T_f does occur on annealing for several months
or longer at 376 K, as is also indicated directly in Fig. 3
by the effect of annealing on the Cp curves. Figure 7 compares
the rates of sub-Tg relaxation following a 100 K/min cool of
all five glasses of Table I at a common temperature T_A = 472 K.
As expected, the glasses with the lowest Tg relax most rapidly.

4. ANALYSIS OF SUB-Tg RELAXATION CURVES

The Tool-Narayanaswamy model (3-10) for structural relaxation during isothermal annealing at temperature T_A gives for
the time dependence of fictive temperature:

$$T_f(t) = T_A + \sum_{i=1}^{n} g_i [T_{fi}(0) - T_A] \exp[-\int_0^t dt/\tau_i(t)] \qquad (2)$$

where the relaxation process is assumed to involve n relaxing
order parameters, the contribution of each of which is
governed by a weighting coeffixient g_i and a relaxation time
τ_i. A fictive temperature $T_{fi}(t)$ is associated with each of
the order parameters, and the experimental fictive temperature
$T_f(t)$ is the weighted average of these:

$$T_f(t) = \sum_{i=1}^{n} g_i T_{fi}(t) \qquad (3)$$

$T_{fi}(0)$ in Eq. (2) is the initial fictive temperature (i.e.,
immediately after rate cooling) associated with the ith order
parameter. The relaxation is non-linear, so that the relaxation times depend both on actual temperature T and average
fictive temperature $T_f(t)$:

$$\tau_i(t) = A_i \exp[x\Delta H^*/RT + (1-x)\Delta H^*/RT_f(t)] \qquad (4)$$

where A_i is a pre-exponential constant, ΔH^* the activation
energy for structural relaxation, x($0 \leq x \leq 1$) the non-linearity parameter and R the ideal gas constant.

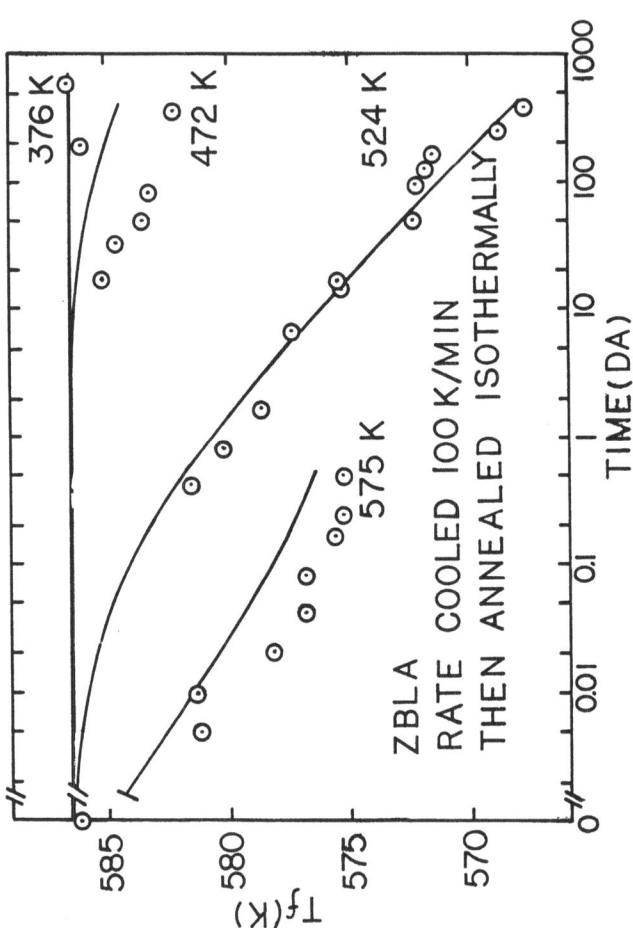

Figure 5. Evolution of fictive temperature of ZBLA glass during sub-Tg annealing following a rate cool at 100 K/min. Solid lines are calculated from parameters in Table III as described in text.

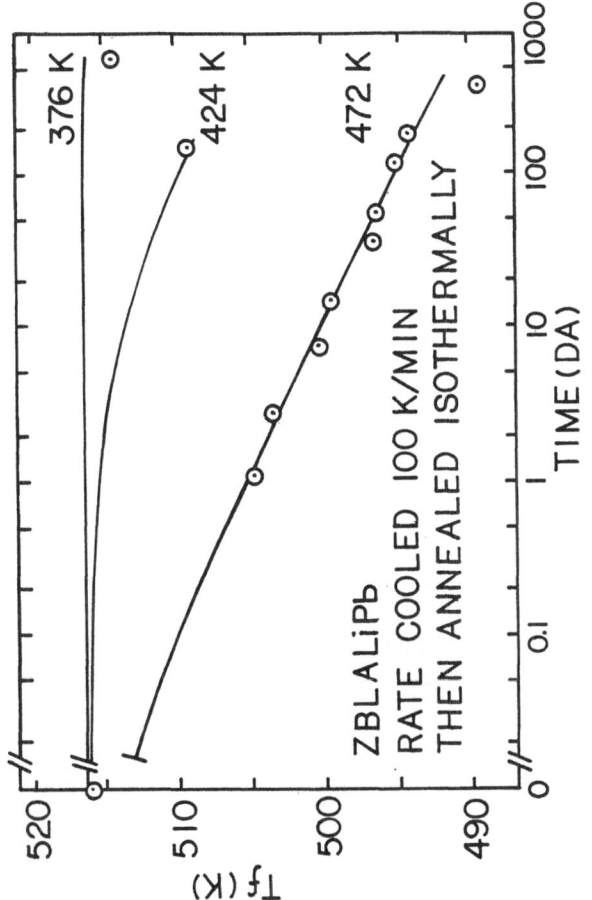

Figure 6. Evolution of fictive temperature of ZBLALiPb glass during sub-Tg annealing following a rate cool at 100 K/min. Solid lines are calculated from parameters in Table III as described in text.

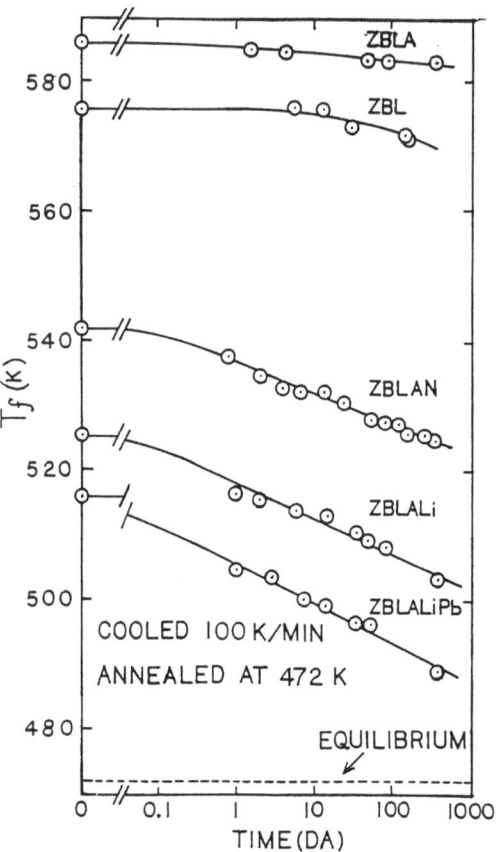

<u>Figure 7</u>. Evolution of fictive temperature of five glasses
during sub-Tg annealing at 472 K following a
rate cool. Solid lines are smooth curves drawn
through the data points.

The activation energy ΔH^* can be determined from the dependence of the fictive temperature T_{fo} obtained immediately after rate cooling the glass on the cooling rate q (3,11):

$$d\ell n|q|/d(1/T_{fo}) = -\Delta H^*/R \qquad (5)$$

This is illustrated in Fig. 8 for ZBLA glass. In Table II are listed the structural relaxation activation energies obtained in this fashion for the glass samples of the present study. ΔH^* is generally found to be the same or very nearly the same as the activation energy ΔH^*_η for shear viscosity in the glass transition region (3). He have also listed in Table II values of ΔH^*_η obtained from beam bending measurements in our laboratory for two of the glasses. The agreement between ΔH^* and ΔH^*_η is within or nearly within experimental error. An apparent large discrepancy between ΔH^* and ΔH^*_η for ZBLA glass reported by Shelby et al. (12) has been found probably   to be due to differences between their viscosity samples and our DSC samples (13). It has been suggested (11,14) that the decrease in ΔH^* or ΔH^*_η that occurs when alkali fluoride is incorporated into ZrF_4-based glasses is largely responsible for the improved stability of the alkali fluoride-containing compositions against devitrification.

ΔH^* for the HMFG is extremely large, so that at equilibrium ($T_f = T$) the relaxation times would change by an order of magnitude roughly every 5 K. That extensive relaxation can be observed in these glasses very far below Tg thus indicates that the relaxation is highly non-linear, i.e., that the relaxation times depend very strongly on T_f and that x in Eq. (4) is relatively small. (Hodge (7) has noted and rationalized a correlation between large ΔH^* and small x). This high degree of non-linearity is apparent in the data of Figs. 5, 6 and 7. For example, during the ZBLA glass anneal at 524 K shown in Fig. 5 the relaxation is initially fairly rapid, and T_f drops about 6 K in the first day. Once this has occurred, however, the relaxation times via Eq. (4) are greatly increased, and a subsequent 6 K drop in T_f takes much longer --about 30 days.

A fit of the data in Figs. 5 and 6 along with analogous data for relaxation of ZBLA and ZBLALiPb glasses initially cooled at 10 K/min to Eq. (2) was carried out. The (g_i, τ_i) pairs in Eq. (2) were chosen to correspond to the well known KWW or stretched exponential relaxation function $\exp[-(t/\tau_o)\beta]$ (8,15). Hence a fit to Eq. (2) requires the use of four adjustable parameters - β, A_o, ΔH^* and x. The last three of these determine the value of KWW relaxation time τ_o for non-linear relaxation (4-6):

$$\tau_o(t) = A_o \exp[x\Delta H^*/RT + (1-x)\Delta H^*/RT_f(t)] \qquad (6)$$

The values of A_o and ΔH^* are fixed by the slope and intercept of the best fit line through the $\ell n|q|$ vs. $1/T_{fo}$ plot (4,5) (see Fig. 8), so that to fit data such as that in Figs. 5 and 6 one need only iterate on two parameters, x and β.

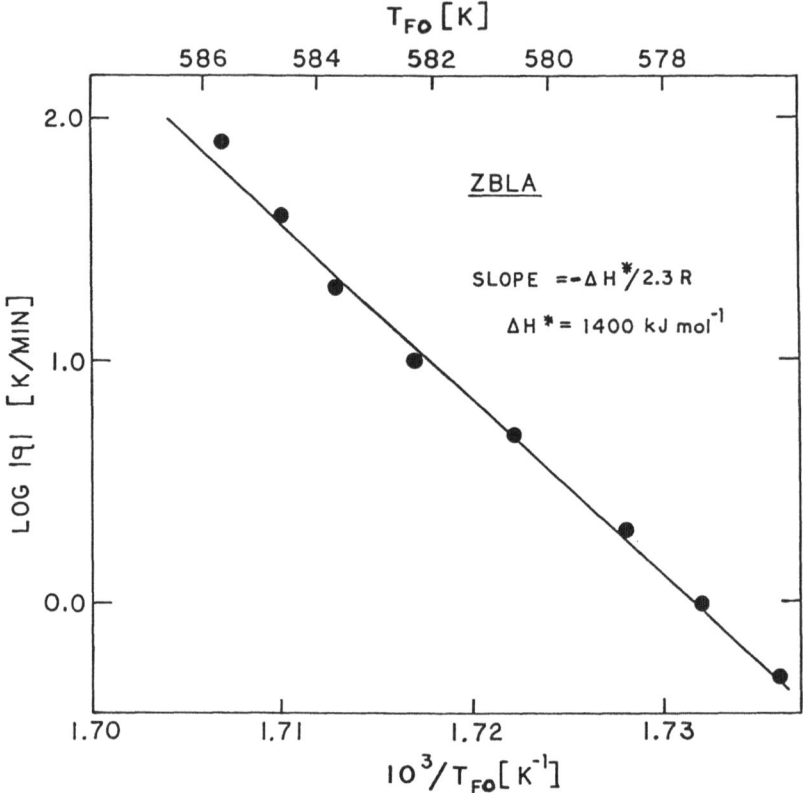

Figure 8. Logarithm of cooling rate |q| vs. reciprocal
of fictive temperature T_{fo} for ZBLA glass.

TABLE II. Activation energies ΔH^* for structural relaxation obtained from Eq. (5) and activation energies ΔH^*_η for viscous flow in the glass transition region.

Glass	ΔH^* (kJ/mol)	ΔH^*_η (kJ/mol)
ZBL	1430	–
ZBLA	1400	1150
ZBLAN	930	870
ZBLALi	1100	–
ZBLALiPb	1030	–

The fit to the sub-Tg annealing T_f vs. t data for ZBLA and ZBLALiPb glasses was carried out using procedures similar to those employed previously (4-6,8). The structural relaxation parameters are listed in Table III, and the solid curves in Figs. 5 and 6 were calculated from these parameters via Eq. (2). (That β and x are the same for the ZBLA and ZBLALiPb glasses may be fortuitous). Hodge (7) has compared the relaxation parameters A_o, ΔH^*, β and x for a variety of polymeric, organic and inorganic glasses and showed that there appear to be correlations among them, e.g., a high ΔH^* is generally associated with low values of $\ln A_o$, β and x. Interestingly, the values of the relaxation parameters for the HMFG glasses in Table III fall fairly closely on Hodge's correlation lines.

The best fits to the data using the Tool-Narayanaswamy model are in general not within experimental error, as is particularly clear from Fig. 5 in which a wide range of sub-Tg annealing temperatures is covered. This is not surprising. Scherer (16), for instance, has recently analyzed sub-Tg annealing data for a soda-lime silicate glass and found that structural relaxation rates for rapidly cooled specimens at temperatures far below Tg were considerably faster than were predicted from fits to data at higher temperatures closer to Tg. Improvements in the semi-empirical Tool-Narayanaswamy model to allow accurate modelling and prediction of rates of structural relaxation over a wide range of times and temperature is a cur-

TABLE III. Best fit structural relaxation kinetic parameters for T_f vs. t curves obtained during sub-Tg annealing of ZBLA and ZBLALiPb glasses following rate cools at 10 and 100 K/min.

Glass		ΔH^* (kJ/mol)	β	x
ZBLA	4.1×10^{-124}	1400	0.54	0.28
ZBLALiPb	2.8×10^{-103}	1030	0.54	0.28

rently active research area both at our laboratory and elsewhere.

5. CONCLUSIONS

In this study we have investigated the change in enthalpy H of HMFG during sub-Tg annealing. Of interest for fiber optic applications are changes in index of refraction. Refractive index n, which generally correlates linearly with density, does not and is not expected to relax at exactly the same rate as enthalpy (4,5,8). However, judging from results on oxide glasses (4,5,8), differences in the rates of relaxation of H and n appear to be relatively small - no more than a factor of 2 in the relaxation times. Hence one may conclude that any substantial drifts in enthalpy due to structural relaxation will be accompanied by substantial drifts in refractive index.

Lacking at the moment a good kinetic model for extrapolating rates of structural relaxation to low temperatures and long times, we can only make qualitative predictions based on the actual data of rates of drift of properties at use temperatures envisioned for HMFG. Inspection of the data for annealing at 376 K (= 103°C) indicates only minor structural relaxation over times of the order of a year for the lowest Tg, fastest relaxing glass - ZBLALiPb (see Figs. 3 and 6). Indications are thus that negligible long term structural relaxation and change of properties at ambient temperature are expected for any of the glasses of the present study. Given that more rapidly quenched glasses, e.g., optical fibers, relax more quickly than slowly cooled glasses (9,16), however, some of the low Tg HMFG might show some perceptible long term property drifts if use temperatures range up to the 100°C region. Likewise, IR transmitting glasses with substantially lower Tg's ($\leq$ about 200°C) than those considered here are very apt to suffer from long term ambient temperature structural relaxation.

ACKNOWLEDGEMENTS

The studies reported here were supported by Contract No. F19628-83-C-0016 from the Air Force Electronic Systems Division, USAF and Grant No. DMR85-10617 from the National Science Foundation. The authors are grateful to Dr. G.W. Scherer for sharing with them his versatile computer program for analysis of structural relaxation.

REFERENCES

1. Drexhage, M.G., in "Treatise on Materials Science and Technology", Vol. 26, M. Tomozawa and R.H. Doremus, Eds., Academic Press, New York, 1985, pp. 151-243.
2. Tran, D.C., Sigel, G.H. and Bendow, B., J. Lightwave Tech. LT-2, 566 (1984).
3. Moynihan, C.T., Easteal, A.J., DeBolt, M.A. and Tucker, J., J. Am. Ceram. Soc. 59, 12 (1976).
4. Debolt, M.A., Easteal, A.J., Macedo, P.B. and Moynihan, C.T., J. Am. Ceram. Soc. 59, 16 (1976).

5. Moynihan, C.T. et al., Ann. NY Acad. Sci. 279, 15 (1976).
6. Hodge, I.M. and Berens, A.R., Macromolecules 15, 762 (1982).
7. Hodge, I.M., Macromolecules 16, 898 (1983).
8. Scherer, G.W., J. Am. Ceram. Soc. 67, 504 (1984).
9. Moynihan, C.T., Bruce, A.J., Gavin, D.L., Loehr,S.R., Opalka, S.M. and Drexhage, M.G., Polym. Engin. and Sci. 24, 1117 (1984).
10. Rekhson, S.M., J. Non-Cryst. Solids 73, 151 (1985).
11. Moynihan, C.T., Gavin, D.L., Chung, K.-H., Bruce, A.J., Drexhage, M.G. and El Bayoumi, O.H., Glastechn. Ber. 56K, 862 (1983).
12. Shelby, J.E., Pantano, C.G., and Tesar, A.A., J. Am. Ceram. Soc. 67, C-164 (1984).
13. Shelby, J.E., private communication.
14. Bansal, N.P., Bruce, A.J., Doremus, R.H. and Moynihan, C.T., J. Non-Cryst. Solids 70, 379 (1985).
15. Moynihan, C.T., Boesch, L.P. and Laberge, N.L., Phys. Chem. Glasses 14, 122 (1973).
16. Scherer, G.W., unpublished manuscript.

DISCUSSION

Q: M. Robinson. Can a long term anneal of weeks or months give rise to compositional changes due to hydrolysis?

A: C. Moynihan. The samples were sealed in pyrex tubes which had been evacuated and back-filled with dry N_2. After measuring C_p following a long term anneal, we immediately cooled the sample in the DSC and did a second scan, heating through T_g. This gives us the initial fictive temperature after cooling, T_{f_0}. T_{f_0} did not drift with annealing time, suggesting that no changes occurred in the composition of the samples during long term sub-T_g annealing.

Q: P.W. France. Could you please comment on the structural relaxation effects related to fiber drawing? E.g. (1) annealing of preform and (2) fiber drawing.

A: C. Moynihan. Preforms are annealed to prevent fracture on cooling to room temperature. On reheating till the drawing temperature, all the thermal history is vanished. To prevent structural relaxation after fiber drawing, perhaps one could give the fibers a brief pass through a heated zone (on-line) slightly above T_g. This could lower T_f substantially.

Q: R. Almeida. Even well bellow T_g, there is still a certain amount of viscous flow. Although T_g is a convenient T value to use because it marks discontinuities in α_T and C_p, would there be a more fundamental, lower temperature value which could be used in order to define a liquid-solid transition?

A: C. Moynihan. The shift of T_g with the cooling or heating rate is well understood. It suffices to measure T_g at an experimentally convenient rate, e.g. 10 K min^{-1}.

Q: P. Klein. Do you sometimes anneal the glasses at fixed fractions of the absolute glass transition temperature? This might reveal common behavior that does not show up at arbitrarily selected annealing temperatures.

A: C. Moynihan. The behavior on annealing depends on several factors such as prior cooling rate or equilibration temperature, activation energy for structural relaxation, etc., so that annealing at fixed fractions of T_g would not give more information than that obtainable with an arbitrary selection of temperatures. The actual number of annealing temperatures we used was somewhat limited by the number of annealing furnaces we had, since these were tied up for about two years.

CRYSTALLIZATION BEHAVIOR OF FLUOROZIRCONATE GLASSES

C.T. Moynihan
Rensselaer Polytechnic Institute, Troy, New York 12180-3590,
USA

ABSTRACT
One of the most formidable barriers to development of fluorozirconate glasses as optical materials has been their strong
tendency to devitrify either on initial cooling from the melt
or on reheating above the glass transition temperature. It is
the purpose of the present paper to review and assess crystallization studies that have been carried out on ZrF_4-based glasses.

1. QUALITATIVE DSC/DTA CHARACTERIZATION OF CRYSTALLIZATION

Gross devitrification of glasses on cooling or reheating is
manifested by exotherms on DSC or DTA traces, as illustrated
in Fig. 1 for a 57 ZrF_4-36 BaF_2-3 LaF_3-4 AlF_3 (ZBLA) glass (1).
Lack of appearance of a crystallization exotherm starting at
T_{XC} on cooling (e.g., Ref. 2) or an increase in the difference
(T_{XH}-Tg) between the onset temperature of crystallization on
reheating and the glass transition temperature (e.g., Ref. 3)
is often taken as a crude measure of resistance to devitrification.

2. CRYSTALLINE PHASES APPEARING ON DEVITRIFICATION

A number of studies using X-ray diffraction and other means
have been made of the crystalline phases appearing in large
amount on devitrification of ZrF_4-based glasses. For glasses
containing BaF_2 (e.g., ZBLA) metastable barium fluorozirconates, β-$BaZr_2F_{10}$ and/or β-$BaZrF_6$, are usually the first to appear in quantity on reheating above Tg (4-10). At longer times
or on heating to higher temperatures other phases may appear,
such as α-$BaZrF_6$ (sometimes formed from β-$BaZrF_6$ via a solid-
-solid transformation (4,6,10)), α-$BaZr_2F_{10}$ (8) and sodium or
sodium barium fluorozirconates in NaF-containing glasses (8).
There is now accumulating a sizeable body of information
on the very first crystals to appear on cooling or reheating
of ZrF_4-based melts. These crystalline phases are present only
in minor amounts, may largely be responsible for excess light
scattering in the glasses or fibers, may serve as nucleation
sites for the major crystal phases described above and are
even more metastable (or further off from the parent melt composition) than the major phases. Minor phases of this sort include ZrF_4 (11,12), LaF_3 (11-13), AlF_3 (11,14), ZrO_2 (12) and
crystals enriched in fluorides such as GdF_3, LaF_3 and ThF_4
present only in small amounts in the bulk melt (13,15).

180

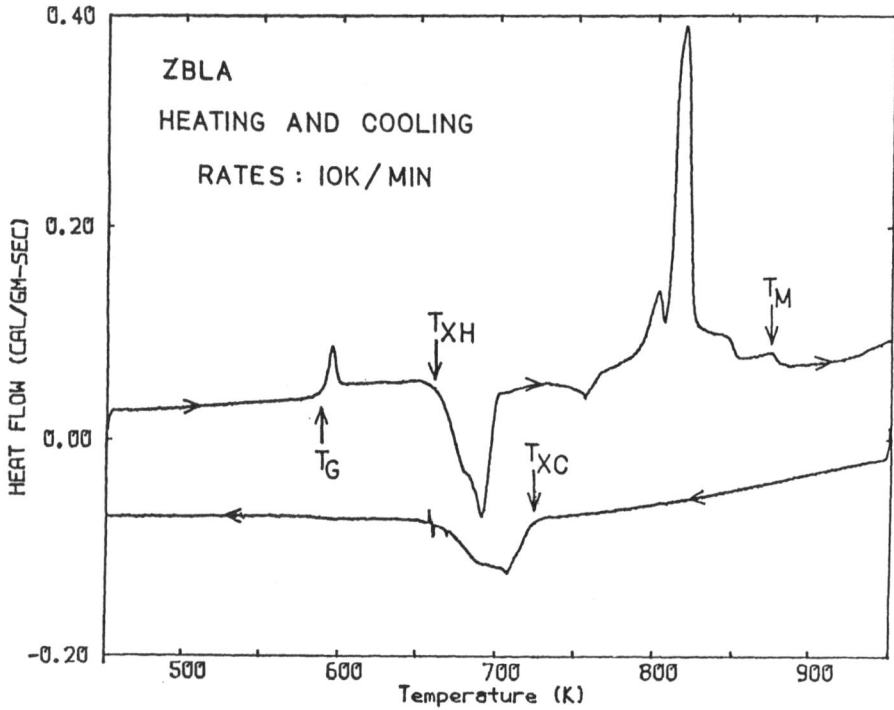

Figure 1. DSC traces of ZBLA glass.

3. CRYSTALLIZATION KINETICS

A number of studies of the kinetics of crystallization of ZrF_4-based glasses on reheating above Tg have been carried out using DSC or DTA (3, 16-21), where the fraction x of glass crystallized during isothermal or rate heating as a function of time is deduced from the growth of the crystallization exotherm (cf. Fig. 1). Analysis of the data is generally carried out via the Avrami equation

$$x = 1 - \exp[-(kt)^n] \qquad (1)$$

for isothermal crystallization. Non-isothermal crystallization during rate heating is analyzed via a suitable modification of Eq. (1). n is a constant related to the geometry of crystal growth and the nucleation frequency (e.g., n= 3 for 3-dimensional crystal growth from a constant number of nuclei), and k is a rate constant whose temperature dependence is generally assumed to be given by an Arrhenius expression:

$$k = \nu \exp(-E/RT) \qquad (2)$$

where ν is a frequency factor and E an activation energy. DSC and DTA studies of this sort for ZrF_4-based glasses generally given n values of about 3 and activation energies E in the range 200-700 kJ/mol. The activation energies are of the magnitude of those expected for viscous flow in the temperature range where crystallization occurs, and glass compositions which are found to be most resistant to devitrification are those with the lowest E values.

Fewer DSC/DTA studies exist of the kinetics of crystallization of ZrF_4-based glasses on cooling the melt from above the liquidus temperature. Kanamori and Takahashi (22) studied the variation of T_{XC} (cf. Fig. a) with cooling rate for a variety of compositions and deduced values in the range 20-400 K/min for the critical or minimum cooling rate required to prevent crystallization. Esnault-Grosdemouge et al. (3) and Busse et al. (23) determined the rate of isothermal crystallization vs. temperature for various melt compositions quenched from above the liquidus and constructed TTT curves from their data. Critical cooling rates to avoid crystallization depended on composition and ranged from 8 to 120 K/min.

A rather elegant series of studies related to crystallization kinetics were carried out by Tokiwa et al. (14,24). They monitored light scattering, which is sensitive to very small volume fractions of crystals, as a function of temperature for glasses which were reheated rapidly to temperatures in the melt viscosity range (10^3-10^7 P), held for a fixed time (15 min) and then cooled rapidly back to room temperature. The 15 min heating period was intended to approximate the amount of time a melt might have to spend at temperatures in the 10^3-10^5P viscosity range during fiber drawing from a preform. (For all ZrF_4-based glasses investigated so far the 10^3-10^5P temperature is below the liquidus (14,24,25).) Hence only compositions which could survive this heat treatment at the 10^5P temperature or above without increase in light scattering were suitable for fiber optic production by drawing from the preform. More recently (26) these workers have carried out similar studies on melts quenched directly to fiber drawing temperatures from above the liquidus to assess stability against crystallization during fiber drawing directly from the crucible.

4. BULK vs. SURFACE CRYSTALLIZATION, NUCLEATION

Crystallization of ZrF_4-based glasses starting both at the glass surface and in the bulk has been observed in a large number of the studies cited above. A study of crystallization by Bansal and Doremus (27) at polished surfaces in air indicated limited nucleation and growth at a relatively small number of surface sites. Busse et al. (23) found that in DSC crystallization studies most of the crystallization started at the interface of the glass with the metal DSC pan, although at low temperature in one composition bulk crystallization was also in evidence. MacFarlane et al. (18) in a DSC study found that finely powdered samples crystallized more rapidly than bulk samples. However, they did not attribute this to an increase in surface nuclei with surface area, but rather to internal nuclei produced by grinding the powder.

Overall, the weight of experimental evidence indicates that the initial small amounts of crystallization at low temperatures and short times which give rise to excess light scattering in ZrF_4-based glasses occur in the bulk glass. The question remains as to whether these crystals are nucleated heterogeneously or homogeneously. A comparison of the onset temperatures T_{XH} (see Fig. 1) of crystallization measured by DSC of a bulk 75 ZrF_4-25 BaF_2 glass and a sample of the same glass finely dispersed in a polymer indicated a homogeneous nucleation mechanism (28). On the other hand, Lu and Bradley (29) have observed small Pt specks in the center of isolated crystals in glasses, suggesting that a small Pt particle (from the melt crucible) nucleated the crystal. Likewise there is evidence (29,30) that undissolved oxide crystals cause excess scattering in the ZrF_4-based glasses; these could possibly also act as heterogeneous nuclei.

5. CONCLUSION

On the basis of crystallization studies carried out so far the fluorozirconate glass compositions most resistant to devitrification seem to be those referred to generically as ZBLAN (typical composition (mol%): 56 ZrF_4-14 BaF_2-6 LaF_3-4 AlF_3-20 NaF). The ZrF_4 may be partly replaced by HfF_4 with no effect on the tendency to devitrification. Part of the reason for the unusual stability of ZBLAN lies in its viscosity temperature dependence (17,20,31). Addition of NaF to ZrF_4-based melts lowers the viscosity/temperature dependence, so that the melts stay viscous and the crystallization rates small over a longer temperature range above Tg compared to melts without NaF. This decrease in viscosity/temperature dependence translates immediately into the experimentally observed decrease in crystallization activation energy on addition of NaF to ZrF_4-based compositions. Given that β-BaZrF$_6$ and β-BaZr$_2$F$_{10}$ are the first major phases to crystallize from a ZBLAN melt (8,29), dilution of the melt with NaF may also raise interdiffusion barriers to crystallization.

REFERENCES

1. Moynihan C.T., Mossadegh R., Gupta P.K., and Drexhage M.G. Mater. Sci. Forum 6, 655 (1985).
2. Mimura Y., Tokiwa H., and Shinbori O.. Electron. Lett. 20, 100 (1984).
3. Esnault-Grosdemouge M.A., Matecki M., and Poulain M.. Mater. Sci. Forum 5, 241 (1985).
4. Kawamoto Y., and Sakaguchi F..Bull. Chem. Soc. Jpn. 56, 2138 (1983).
5. Neilson G.F., Smith G.L., and Weinberg M.C., Mat. Res. Bull. 19, 279 (1984).
6. Bansal N.P., Doremus R.H., Bruce A.J., and Moynihan C.T.. Mat. Res. Bull. 19, 577 (1984).
7. Bansal N.P., and Doremus R.H.. Mater. Sci. 20, 2794 (1985).
8. Parker J.M., Clare A.G., and Seddon A.B., Mater. Sci. Forum 5, 257 (1985).
9. Neilson G.F., Smith G.L., and Weinberg M.C.. J. Am. Ceram. Soc. 68, 629 (1985).
10. Miniscalco W.J., Andrews L.J., Hall B.T., and Guenther D.E. Mater. Sci. Forum 5, 279 (1985).
11. Lu G., Fisher C.F., Burk M.J., and Tran D.C.. Bull. Am. Ceram. SOc., 63, 1416 (1984).
12. Carter S.F., France P.W., Moore M.W., and Williams J.R.. Mater. Sci. Forum 5, 397 (1985).
13. Wilson S.J., and Pitt N.J.. Mater. Sci. Forum 5, 275 (1985).
14. Tokiwa H., Mimura Y., Shinbori D., and Nakai T.. J. Lightwave Tech. LT-3, 569 (1985).
15. Boehm L., Raizman A., and Sapir Y.. Mater. Sci. Forum 5, 205 (1985).
16. Bansal N.P., Doremus R.H., Bruce A.J., and Moynihan C.T.. J. Am. Ceram. Soc. 66, 233 (1983).
17. Bansal N.P., Bruce A.J., Doremus, R.H., and Moynihan C.T.. Proc. SPIE 484, 51 (1984).
18. MacFarlane D.R., Matecki M., and Poulain M.. J. Non-Cryst. Solids 64, 351 (1984).
19. Bansal N.P., Bruce A.J., Doremus R.H., and Moynihan C.T.. J. Non-Cryst. Solids 70, 379 (1985).
20. Bansal N.P., Doremus R.H., Moynihan C.T., and Bruce A.J.. Mater. Sci. Forum 5, 211 (1985).
21. Matusita K., Miura K., and Komatsu T.. Thermochem. Acta 88, 283 (1985).
22. Kanamori T., and Takahashi S., paper presented at Colloquium on Fluoride Glass for Optical Fibers, NTT Corp., Ibaraki, Japan, August, 1985.
23. Busse L.E., Lu G., Tran D.C., and Sigel G.H.. Mater. Sci. Forum 5, 219 (1985).
24. Tokiwa H., Mimura Y., Shinbori D., and Nakai T.. J. Lightwave Tech. LT-3, 574 (1985).

25. Moynihan C.T., Mossadegh R., Gupta P.K., and Drexhage M.G.. Mater. Sci. Forum 6, 655 (1985).
26. Tokiwa H., Mimura Y., Nakai Y., and Shinbori D.. Electron Lett. 21, 1131 (1985).
27. Bansal N.P., and Doremus R.H.. J. Am. Ceram. Soc. 66, C-132 (1983).
28. MacFarlane D.R., and Fragoulis M.. Mater. Sci. Forum 5, 237 (1985).
29. Lu G., and Bradley J.. Mater. Sci. Forum 6, 551 (1985).
30. Nakai T., Mimura Y., Tokiwa H., and Shinbori D.. J. Lightwave Tech. LT-3, 565 (1985).
31. Moynihan C.T., Gavin D.L., Chung K.-H., and Bruce A.J., Drexhage M.G., and El Bayoumi O.H.. Glastechn. Ber. 56K, 862 (1983).

DISCUSSION

Q: G. Frischat. What do we know so far about the influence of all the trace impurities, e.g. Fe, C, etc., on nucleation and crystallization?

A: C. Moynihan. Very little is known. Carbon or metallic impurities can serve as nucleation sites. But an impurity such as FeF_3 is very soluble in the melt and probably does not induce nucleation.

Q: M. Robinson. Does one see any evidence for the presence of oxide in the X-ray diffraction data reported in this paper?

A: C. Moynihan. No. The amount of oxide in heavy metal fluoride glasses should be no more than one or two mol%. This would be very hard to pick up by powder diffraction.

C: J. Lucas. I would just like to confirm the author's observations on the metastable phases which first appear during crystallization. We have observed the same for Zr-free glasses (G. Fonteneau, in Rennes), where quite peculiar crystal structures are formed, leading to good ceramic materials.

Q: A.C. Wright. What is the minimum crystallite size (not volume fraction) needed for one to detect a polymorphic transition by DSC?

A: C. Moynihan. It is probably in the order of 10 nm. This is large enough to be considered a macroscopic crystal which can exhibit well defined transitions.

Q: D. Ulrich. Has the nucleation and crystallization of controlled microstructures been investigated? The glass-ceramic approach to transparent ceramics would lead to important optical materials for defense applications. In fact, the use of nucleating agents may lead to the nucleation and crystallization of very small crystallites, of a size less than the wavelength of visible light. And there is need for IR-transmitting ceramics with increased hardness and toughness relatively to the glasses.
Also, J. Lucas pointed that he has observed the crystallization of metastable phases that have not previously been obtained. This gives the solid state chemist a method for obtaining crystals for structural studies.

A: C. Moynihan. A few years ago we actually produced, by accident, some fully crystallized heavy metal fluoride glasses where the crystallite size was small enough for them to retain both visible and infrared transparency. These BaF_2-MnF_2-ThF_4 glasses were quite soft compared to other HMF glasses, but, on crystallizing, their microhardness went up by 20-30%.

Q: R. Almeida. Could you single out a best fluorozirconate glass-forming composition?

A: C. Moynihan. The ZBLAN glasses are the best so far. However, further improvements may be possible, particularly if one dilutes the minor components which first crystallize metastably from the glasses. For example, one can mix ZrF_4 with HfF_4, LaF_3 with GdF_3, or LaF_3 and AlF_3 with InF_3 and possibly improve the resistance to devitrification.

PHYSICAL, CHEMICAL PROPERTIES AND CRYSTALLIZATION TENDENCY OF THE NEW FLUOROALUMINATE GLASSES

T.Izumitani, T.Yamashita, M.Tokida, K.Miura and H.Tajima
HOYA Research Laboratories
572, Miyazawacho, Akishima-shi, Tokyo
JAPAN

1. INTRODUCTION

Weak points of fluorozirconate glasses are found in poor chemical durability, low mechanical strength and high tendency for devitrification. In order to improve these properties, the fluoroaluminate glasses were investigated.

Fig.1 shows the glass formation range of P_2O_5-AlF_3-alkali earth fluoride system. Fluorophosphate glass[1], FCD-10, containing a small amount of P_2O_5 is stable against devitrification and produced as an optical glass in large quantity and in a large size such as 50 cm diameter and 8 cm thickness as laser disk. As this glass contains very small amount of P_2O_5, only 13.4 cation %, FCD-10 is substantially more a fluoride glass than phosphate glass.

In place of phosphoric oxide, fluorozirconate or beryllium fluoride was introduced as a network former into fluoroaluminate glass. The fluoroaluminate glasses showed improved properties in chemical durability, mechanical properties and light scattering.

2. EXPERIMENTAL

2.1. Glass preparation

The commercially available 99 - 99.9% purity fluorides were used as raw materials. 50 to 100g of glasses were melted from premixed batches of fluorides and an excess of $NH_4F \cdot HF$ in a gold crucible at 900°C for 2 hours under Ar gas atmosphere. The obtained glass was a disk of about 40mm x 8 - 15mm thickness.

2.2 Fluorozirco-aluminate glass

Phosphoric oxide in P_2O_5-AlF_3-Alkali earth fluoride system was replaced by ZrF_4-BaF_2 in order to remove oxide.

Table 1 shows some examples of fluorozirco-aluminate glass compositions.

1) DTA

Fig.2 and 3 show the DTA curves of glass powder(325 - 400 mesh) and the bulk glass. The heating rate was 10°C/min. DTA curve 1 for fluorozirco-aluminate glass powder shows two exothermic peaks while the curve for the bulk glass shows only one exothermic peak. From the comparison of the DTA curves for glass powder and bulk glass, the exothermic peak on lower temperature side for the powdered glass is attributed to the surface devitrification. Therefore, it can be concluded that the fluorozirco-aluminate bulk glass does not cause any surface

188

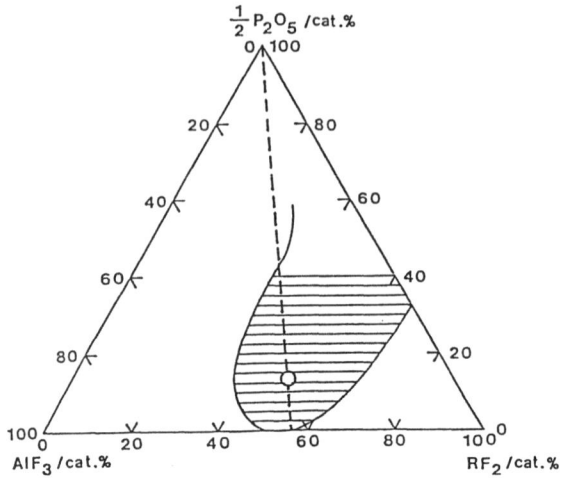

$\frac{1}{2}$ P$_2$O$_5$ /cat.%

AlF$_3$ /cat.%

RF$_2$ /cat.%

Fig.1 Glass forming region of P$_2$O$_5$-AlF$_3$-RF$_2$ system.

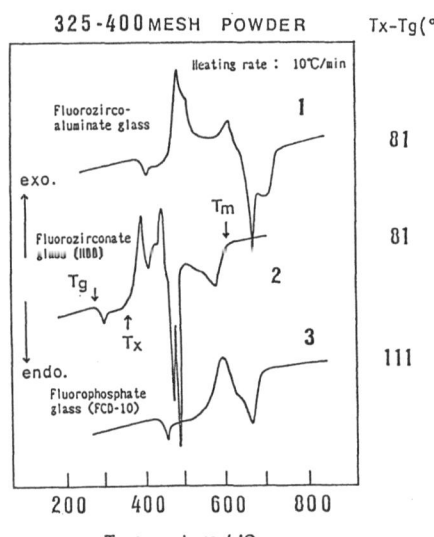

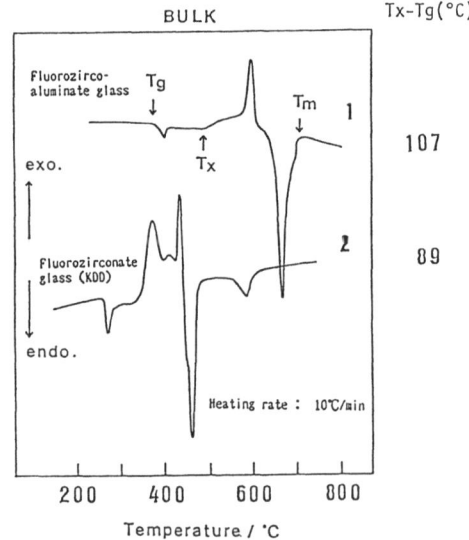

Fig.2 DTA curves of the fluoride glasses and fluorophosphate glass. (Powder)

Fig.3 DTA curves of the fluoride glasses. (Bulk)

devitrification.

DTA curve 2 of fluorozirconate glass[2] shows two exothermic peaks for both the powder glass and bulk glass. Accordingly, the fluorozirconate glass shows the surface devitrification
even for the bulk glass.

DTA curve 3 for the fluorophosphate glass does not show
any surface devitrification even for powdered glass. Therefore
the crystallization tendency decreases in the order of fluorozirconate > fluorozirco-aluminate > fluorophosphate glass.

This tendency agrees with Tx-Tg tendency, Tx-Tg values are
81°C, 81°C and 111°C for fluorozirco-aluminate, fluorozirconate
and fluorophosphate glasses in powdered form.

Tx-Tg values are 107°C and 89°C for the bulk state of
fluorozirco-aluminate glass and fluorozirconate glass, respectively. Therefore, fluorozirco-aluminate glass is more stable
than fluorozirconate glass.

Table 1 Fluorozirco-aluminate glass compositions.[3]

(mol%)

No.	AlF_3	ZrF_4	YF_3	LaF_3	MgF_2	CaF_2	SrF_2	BaF_2	NaF
1	34.1	7.1	3.7	–	4.0	23.1	15.0	10.3	2.7
2	31.7	10.7	3.8	–	3.7	21.3	13.8	11.1	4.0
3	30.2	10.2	8.3	–	3.5	20.3	13.2	10.6	3.8
4	34.1	7.1	–	3.7	4.0	23.1	15.0	10.3	2.7
5	33.7	13.4	2.6	0.6	4.0	23.1	15.0	7.6	–

2) Viscosity

Fig.4 shows the viscosity-temperature curves for the
fluorozirconate, fluorozirco-aluminate and fluorophosphate
glasses. The fluorozirco-aluminate glass shows higher viscosity and lower viscosity change against temperature than fluorozirconate glass.

3) Light scattering

Fig.5 shows the Rayleigh ratio($R_{90}°$) against wavelength.
The fluorozirco-aluminate glass show lower light scattering
intensity than the fluorozirconate glass.

4) Transmittance

Fig.6 shows the transmittance curves in infrared region. The
fluorozirco-aluminate glass shows a little bit shorter transmittance limit wavelength than the fluorozirconate glass.

5) Refractive index and Abbe's values

Fig.7 shows nd-νd relation for the fluorozirconate,fluoro-
phosphate and fluorozirco-aluminate glasses. The fluorozirco-
aluminate glass also show the lowest non-linear refractive index
n_2 and the highest UV abnormal partial dispersion P_{ig}.

6)Chemical durability

Fig.8A shows the weight loss due to dissolution of glass
plate($12 \times 12 \times 2mm^3$) in water at 50°C with time. The fluorozirco-
aluminate glass shows much improved chemical durability.

Fig.8B shows the weight loss against pH value. The fluorozirco-aluminate glass is more durable against acid solution and

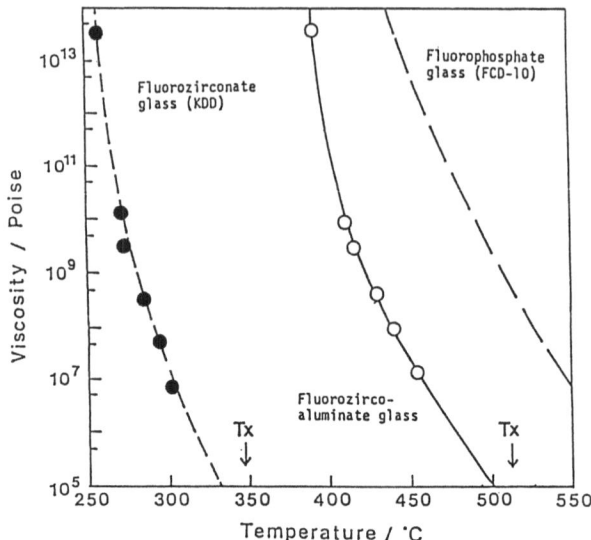

Fig.4 Viscosity - temperature curves.

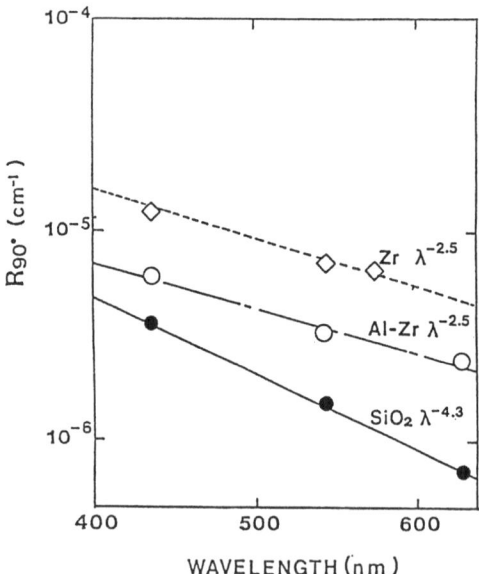

Fig.5 Rayleigh ratio ($R_{90°}$) against wavelength.

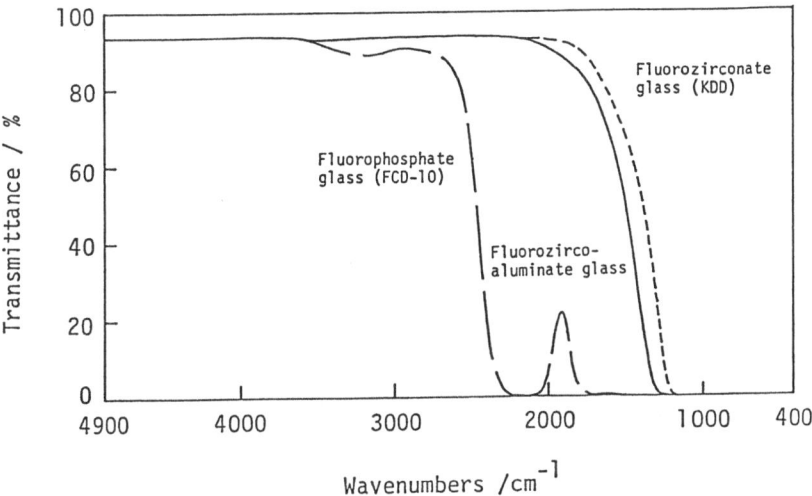

Fig.6 IR transmission spectra of fluoride glasses and fluorophosphate glass.

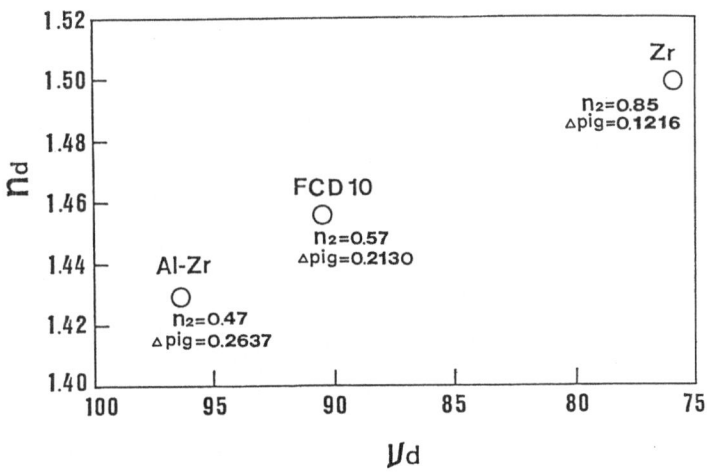

Fig.7 nd-νd diagram of the fluoride glasses and fluorophosphate glass.

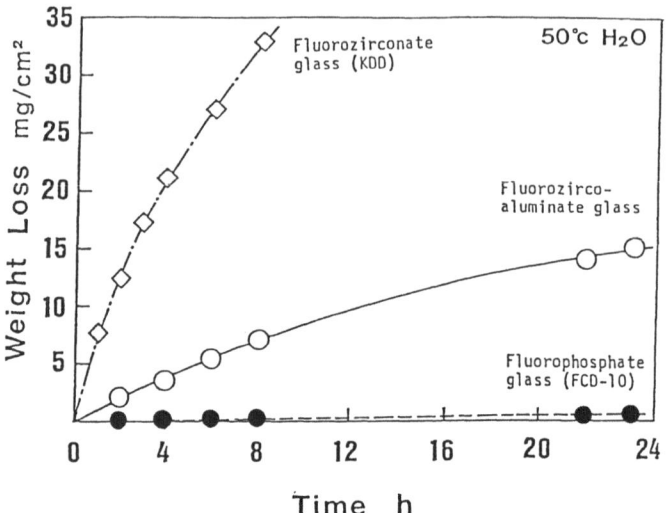

Fig.8A Chemical durability of the fluoride glasses and
fluorophosphate glass.

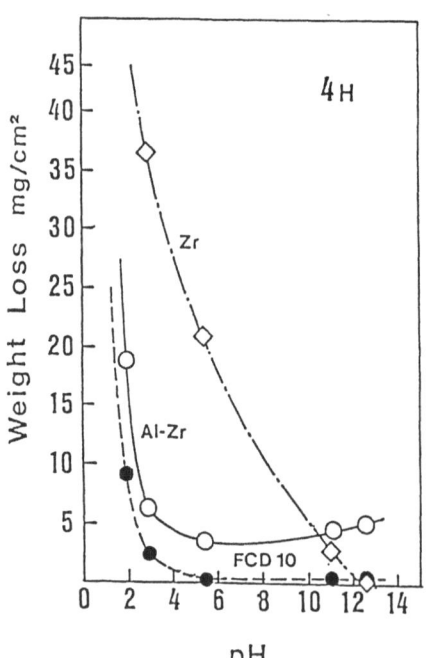

Fig.8B Chemical durability
against pH value.

Table 2 Properties of fluorozirco-aluminate glass compared with fluorozirconate and fluorophosphate glasses.

PROPERTIES	Fluoro-zirco-aluminate	Fluoro-zirconate (ZBLAN)	Fluoro-phosphate (FCD-10)
Optical properties			
Refractive index nd	1.42913	1.49897	1.45589
Abbe value νd	96.39	75.95	90.52
Abnormal partial dispersion			
ΔP_{ig}	0.2637	0.1216	0.2130
Non-linear refractive index			
n_2 (10^{-13}esu)	0.47	0.85	0.57
Transmittance limit			
wavelength (5mm t)			
UV λT=50% (μm)	0.21	0.21	0.31
IR λT=50% (μm)	6.76	7.19	4.07
Thermo-optical constant			
(20~40°C , 633nm)			
(dn/dT)rel. (10^{-6}/°C)	-8.02	-13.45	-6.20
(dn/dT)abs. (")	-9.26	-14.75	-7.47
(ds/dT)abs. (")	-2.57	-6.06	-1.09
Rayleigh's ratio (λ=633nm)			
$R_{90°}$ (10^{-6}/cm)	2.3	5.0	2.1
Physical properties			
Specific gravity S.G.	3.77	4.33	3.63
Chemical properties			
Water durability			
Dw (wt loss%)	0.16	29.2	0.04
Acid durability			
Da (wt loss%)	0.68	32.0	0.29
Thermal properties			
Transformation point Tg (°C)	398	257	440
Sag point Ts (°C)	423	277	465
Thermal expansion coeff.			
(-30~70°C) α (10^{-7}/°C)	154	172	138
Mechanical properties			
Young's modulus E (kg/mm^2)	6300	5380	7260
Shear modulus G (")	2400	2050	2790
Bulk modulus K (")	5600	4770	6080
Poisson's ratio σ	0.31	0.31	0.30
Knoop hardness Hk (kg/mm^2)	320	225	360

ZBLAN composition : ZrF_4 53,BaF_2 20,LaF_3 4,AlF_3 3,NaF 20 (mol%)

FCD 10 : Commercially available optical glass at HOYA.

neutral solution than fluorozirconate glass.

7) Mechanical strength

The physical, thermal and mechanical properties are shown in Table 2. The fluorozirco-aluminate glass has higher mechanical strength than the fluorozirconate glass.

3. FLUOROALUMINATE GLASSES CONTAINING BERYLLIUM FLUORIDE

In order to much more improve the devitrification tendency, BeF_2 was introduced into the fluoroaluminate glass.

1) DTA

Fig.9 shows the DTA curves for bulk glass. The fluoro-beryllo-aluminate glass does not show any surface devitrification. And Tx-Tg are 135, 107 and 89°C for fluoroberyllo-aluminate fluorozirco-aluminate and fluorozirconate glasses. The fluoro-beryllo-aluminate glass is the most stable glass among them.

2) Light scattering

Fig.10 shows the Rayleigh ratio-wavelength relationship. The Rayleigh scattering of fluoroberyllo-aluminate glass is proportional to the $\lambda^{-4.1}$. It means that there is no scattering due to crystallization products. But there is still difference in Rayleigh ratio between synthetic fused silica and the fluoro-beryllo-aluminate glass. The reason is not clear yet. It might be related to impruities in BeF_2, glass composition, melting conditions such as used crucible or atmosphere, or the number and size of scattering particles in density fluctuation.

Table 3 shows Rayleigh ratio for various glasses. Rayleigh ratio at 1.5 and 2.5µm was extrapolated from the experimental values in the visible range. Light scattering is expected to be improved almost by two in fluoroberyllo-aluminate glass compared to the fluorozirconate glass.

Table 3   Rayleigh ratio ($R_{90°}$) for various glasses.

Wavelength (µm)	SiO_2	Fluoro-zirconate	Fluoro-zirco-aluminate	Fluoro-beryllo-aluminate
0.633	6.0×10^{-7}	4.8×10^{-6}	2.2×10^{-6}	1.5×10^{-6}
1.5	1.5×10^{-8}	-	-	-
2.5	-	1.5×10^{-7}	7.3×10^{-8}	5.4×10^{-9}

3) Chemical durability and mechanical properties

The chemical durability and Knoop hardness of the fluoro-beryllo-aluminate glass is almost the same as the fluorozirco-aluminate glass as shown in Table 4.

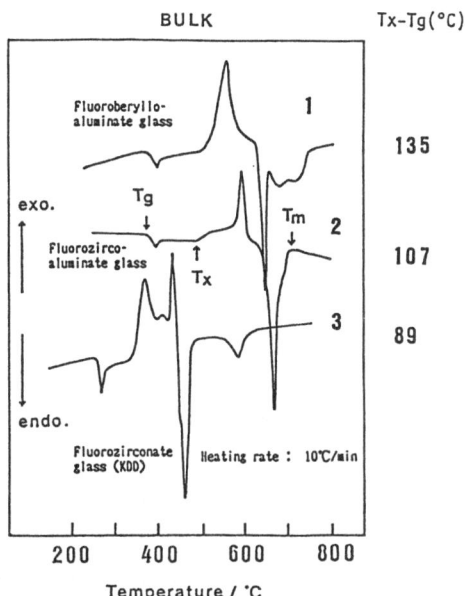

Fig.9 DTA curves of fluoroberyllo-aluminate, fluorozirco-
aluminate and fluorozirconate bulk glasses.

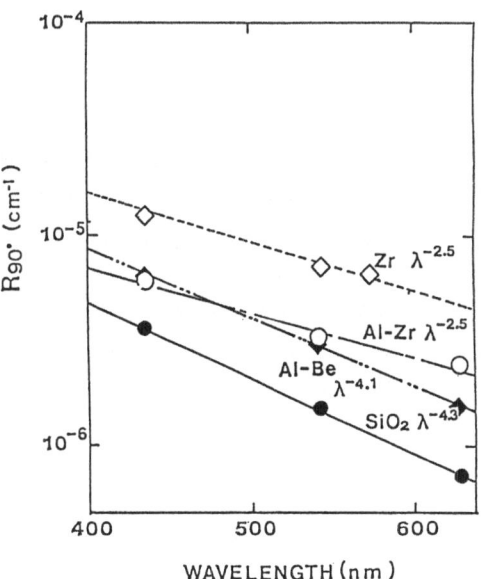

Fig.10 Rayleigh ratio ($R_{90°}$) against wavelength.

Table 4   Chemical durability and Knoop hardness.

PROPERTIES	Fluoro-zirco-aluminate	Fluoro-zirconate (ZBLAN)	Fluoro-phoshate (FCD-10)	Fluoro-beryllo-aluminate
Water durability Dw (wt loss%)	0.16	29.2	0.04	0.18
Acid durability Da (wt loss%)	0.68	32.0	0.29	0.80
Knoop hardness Hk (kg/mm^2)	320	225	360	335

4. CONCLUSIONS

Two kinds of new fluoroaluminate glasses have been developed by introducing zirconium or beryllium fluorides in place of phosphoric oxide in a fluorophosphate glass.

The new glasses are quite stable against devitrification and show lower light scattering, better chemical durability and higher mechanical strength than fluorozirconate glasses.

Especially, fluoroberyllo-aluminate glasses are very promising because of no light scattering due to crystallized particles. But there is still difference in light scattering between synthetic fused silica and fluoroberyllo-aluminate glass. Future investigations will be made on the impurities of raw materials, melting conditions, glass compositions, density fluctuations, etc..

REFERENCES

1. T. Izumitani, M. Tsuru: Japanese Pat. No. 1199772(1984).
2. H. Tokiwa, Y. Mimura, O. Shinbori and T. Nakai: Mater. Sci. Forum 5, 411 (1985).
3. T. Izumitani, T. Yamashita, M. Tokida and K. Miura: Japanese Pat. Application, 61-43065(1986).

DISCUSSION

Q: J.D. Mackenzie. Why do you need CaF_2, SrF_2 and BaF_2? Why all these fluorides?

A: T. Izumitani. The combination and relative proportions of alkaline earth fluorides are very important in order to stabilize the fluoroaluminate glass. And so is the $AlF_3:RF_2$ ratio (R=Ca, Sr, Ba).

Q: J. Lucas. Do you have an idea about the limit of solubility of ZrF_4 in the AlF_3-based melt, without phase separation?

A: T. Izumitani. The addition of too much ZrF_4 to a AlF_3-based melt will cause devitrification, given the crystallization tendency of ZrF_4-based melts. However, there is no phase separation for fluorozirco-aluminate glass, because of the large amounts of modifiers that it contains.

Q: R. Reisfeld. Would it be possible to add BF_3 to ZrF_4 or BeF_2-based glasses in order to reduce the non-linear refractive index?

A: T. Izumitani. The non-linear refractive index increases in the order $BeF_2 < BF_3 < ZrF_4$. Therefore, BF_3 is not always useful to reduce n_2 and it also has a tendency to evaporate.

Q: P.W. France. Have you any idea of the intrinsic minimum losses in these materials and at what wavelength they might occur?

A: T. Izumitani. We do not yet have an idea about these.

SURFACE CHEMISTRY AND SLOW-CRACK GROWTH BEHAVIOR OF FLUOROZIRCONATE GLASSES

CARLO G. PANTANO
Department of Materials Science and Engineering
Pennsylvania State University
University Park, PA, USA, 16802

ABSTRACT
The surface chemical reactions which might contribute to stress corrosion in fluorozirconate glasses have been investigated and the results are related to the slow crack growth behavior of these glasses. The surface studies indicate that aqueous environments are especially aggressive. It is found that the penetration and bulk diffusion of molecular water precedes any hydrolysis or dissolution of the glass structure. The chemisorption of water to produce oxide or hydroxide species is not prevalent except in the case of the basic aqueous environments. The most notable feature of the stress intensity-crack velocity diagrams is a distinct and reproducible stress-intensity threshold at $K_I = .15-.20$ MPa-m$^{1/2}$ in the environments which contain water. This threshold is observed for both gaseous and liquid environments even though the surface chemistry and corrosion in water vapor and liquid water are very different. The wet nitrogen atmospheres also exhibit a velocity plateau whose magnitude increases systematically with the water activity; it approaches the plateau observed in pure liquid water at ~10^{-3} m/s. Altogether, it appears that the crack growth mechanism may be independent of any chemisorption reactions at the crack tip. Perhaps, the formation of a coordinate bond between molecular water and the barium or zirconium cations weakens the coulombic forces of attraction with the fluorine anions, and thereby, influences the energy necessary for crack propagation. This interaction also appears to be the first step in the corrosion reaction. It is further suggested that the stress-intensity thresholds and velocity plateaus may be due to stress-enhanced water diffusion ahead of the crack tip where it modifies both the material and stress distribution.

1. INTRODUCTION

It has already been shown that both fluorozirconate and barium-thorium fluoride glasses exhibit a susceptibility to dynamic fatigue and slow-crack growth (1). These data indicate that water is an especially active environment, but the stress intensity-crack velocity (KV) diagrams have unique characteristics relative to those observed in silicate glasses. Otherwise, very little has been reported on the fracture and stress corrosion behavior of these glasses.

In a more general sense, the stress corrosion mechanisms in ionic materials, whether glassy or crystalline, have not been clarified. In a study of slow crack growth in single-crystal-line alkaline earth fluorides (CaF_2, SrF_2 and BaF_2), both H_2O and HF were found to enhance the crack-growth rate (2). The delayed failure of polycrystalline MgF_2 (3) and ZnSe (4) were also reported, and here too, water-enhanced slow crack growth was observed. Although these studies described the importance of dislocation motion and microstructure in the fracture process, they did not lead to the establishment of a stress corrosion mechanism.

In the most definitive studies of stress corrosion mechanisms in silicate glasses (5-7), the availability of an extensive data base concerning their surface chemistry (adsorption, hydration/dehydration, chemical reactivity, etc.) and corrosion behaviors was of great benefit. A number of papers have appeared recently concerning the corrosion of fluoride glasses (8-15), but the reaction mechanisms are still not well understood. The adsorption and other surface chemical reactions have yet to be reported for these glasses, although some adsorption studies of water vapor on crystalline alkaline--earth fluorides have been reported (16-21).

There is no question that these fluoride glasses are susceptible to stress corrosion, and this will certainly limit many of their potential applications for infrared fiber optics. In this paper we report further on the surface chemistry and slow crack growth behavior of fluorozirconate glasses. The study of chemical effects at fracture surfaces is emphasized where methods including low-energy ion scattering (ISS), Auger electron spectroscopy (AES) and secondary ion mass spectroscopy (SIMS) are used for characterization. The constant moment double cantilever beam technique was employed to determine the slow crack growth behavior in a wide variety of liquid and gaseous environments including water, water vapor, methanol, methanol vapor, ammonia, formamide and water-methanol mixtures. These data are used to discuss the mechanism of stress corrosion in fluoride glasses.

2. EXPERIMENTAL PROCEDURE

All of the double cantilever beam specimens were cut from a single billet of glass. The glass - whose composition is .57ZrF_4. .36BaF_2. .03LaF_3. 04AlF_3 (ZBLA) - was fabricated by Le Verre Fluore. The beams - approximately 6mm x 30mm x 3mm - were cut and oriented so that the crack followed a path of constant thermal history through the glass. A 2mm wide groove - approximately 2mm deep - was machined down the center of the beam to guide the crack. The groove and faces of the glass plates were polished to eliminate residual stresses due to the machining operations. The cracks were initiated by thermal shock with a hot nichrome wire. The crack was then extended under load into the middle two-thirds of the specimen where the crack velocities were recorded. The constant moment double cantilever beam technique (22) was used with the crack elongation being monitored through a 40X traveling microscope. The specimens were enclosed in a sealed chamber which could be purged with a gaseous environments before and/or during the test. The controlled humidity environments were produced by mixing measured amounts of dry nitrogen and water-saturated nitrogen gas. The humidity was measured with a dew point hygrometer.

The clean surfaces used for adsorption studies were created by fracturing pre-notched rods of the glass in an ultra-high vacuum analytical chamber. The chamber could be naked and ion--pumped to a base pressure of less than 1 x 10^{-9} torr before the fracture to eliminate water vapor and other residual gases. The species present in the vacuum environment were monitored with a quadrupole residual gas analyzer. The chamber incorporated a low-energy ion scattering spectrometer (ISS) and an Auger electron spectrometer (AES) for analysis of the 'clean' surfaces before and after controlled-amounts of water vapor were introduced through a leak valve. This system was also used for the chemical analysis of fracture surfaces created externally in air and liquid environments. It is important to point out that a careful evaluation of the electron beam effects which occur during Auger analysis of these glasses was carried out, this evaluation showed that the electron beam current density must be maintained at or below 6 $\mu A/cm^2$ to prevent beam damage to the specimen.

The glasses were hydrated at elevated temperatures in a tube furnace under flowing Ar or O_2 gases which were first saturated with water vapor at 30C or 75C. These glasses, as well as those hydrated at room temperature in water-methanol liquid mixtures, were depth profiled with secondary ion mass spectroscopy (SIMS).

3. CRACK-GROWTH STUDIES

The K-V diagrams measured for ZBLA in a wide variety of environments are summarized in Figures 1 and 2. In Figure 1, the presence of a well-defined threshold - followed by a velocity plateau - appears to characterize the behavior in humid gas and in liquid water. The velocity plateau shifts to higher values with increasing water activity and reaches its maximum value in liquid water. The K-V diagrams in Figure 2 show that ammonia vapor behaves very much like a humid gas, whereas the methanol liquid, methanol vapor and formamide extend the region of slow crack growth to lower stress-intensity. The absense of a threshold or velocity plateau for the nonaqueous environments indicates that they are each active in promoting crack growth.

It is interesting to note that decreasing the activity of water vapor in the inert nitrogen gas phase systematically lowers the velocity plateau, whereas decreasing the activity of liquid water by dilution with methanol yields a straight line whose slope depends upon the water activity (1). The slope of this line can be used to define the stress corrosion susceptibility (n) which is ~77 in pure methanol and is ~40 in methanol containing 1-5% water. The absence of thresholds and plateaus in the water-methanol mixtures suggests that these features may be related to the formation of a surface reaction product between adsorbed or liquid water and the glass. The reaction products can be neutralized or carried away by the methanol 'solvent' in the liquid mixtures but not in the humid gas. In pure water, these reaction products probably accumu-

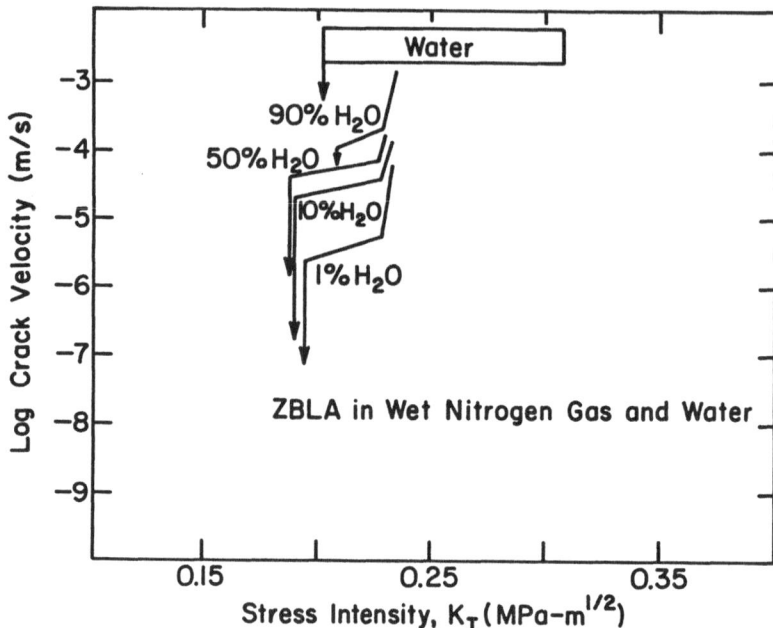

FIGURE 1. The stress intensity - crack velocity diagrams for a fluorozirconate glass in aqueous environments; the crack velocity in water varies over the range represented by the box.

late in the crack, whereas in the pure methanol vapor and formamide they may not form. Later, it is suggested that HF may be a surface reaction product which is unique to the environments which exhibit thresholds and velocity plateaus.

It is quite clear that water is an aggressive crack growth agent. This is consistent with the aqueous corrosion behavior of fluorozirconate glasses where rapid and extensive surface hydration, dissolution and crystalline reaction product formation are evident (8-15). This corrosion reaction is greatly accelerated in acidic solutions, but is practically non-existent in basic solutions. For this reason, additional crack growth measurements were made in a pH 11 ammonium hydroxide solution. The K-V diagram in this solution was comparable to the curve for liquid water. Nonetheless, these fracture surfaces were free of any visible degradation whereas in the pure water environments a highly corroded surface is typically observed.

Additional crack growth measurements were made in mixtures of methanol and either acid or base. The pH of the mixtures was systematically varied between 1 and 12. These data are shown in Figure 3 and they, too, indicate an insensitivity of the crack velocity to pH. The data points are scattered about a line whose slope gives an n value of about 40. In view of the obvious pH dependence of the corrosion and dissolution reactions (15), these observations suggest that the stress

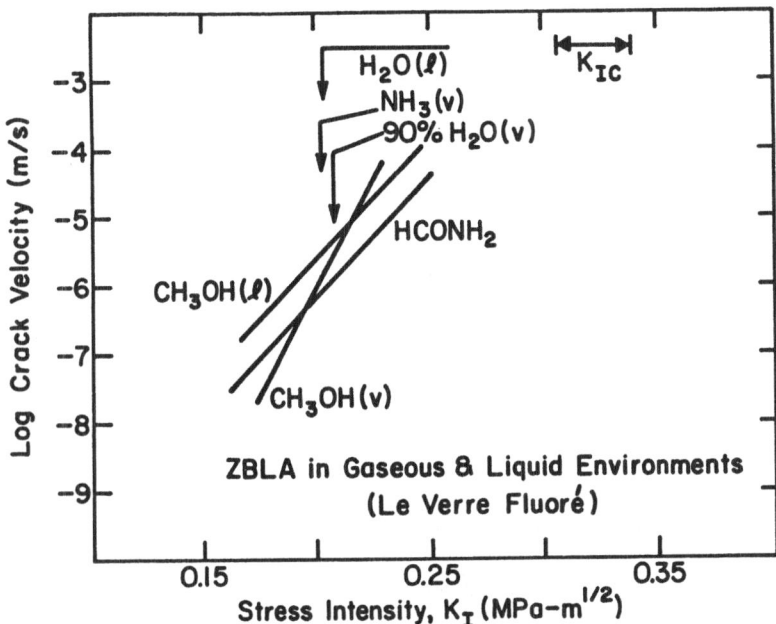

FIGURE 2. The stress intensity - crack velocity diagrams for a fluorozirconate glass in aqueous and non-aqueous environments.

corrosion mechanism in fluorozirconate glasses may be distinct relative to the corrosion mechanism.

4. SURFACE STUDIES

The reaction of clean fracture surfaces exposed to water vapor, as well as the composition of fracture surfaces created in various environments, have been investigated with ion-scattering spectrometry (ISS) Auger electron spectroscopy (AES) and secondary ion mass spectroscopy (SIMS).

The ISS technique is sensitive to the elemental composition of the outermost atomic layer of a solid. It has been used here to determine whether oxidation and/or the chemisorption of water are prevalent reactions in the surface monolayer of clean fracture surfaces. The upper spectra in Figure 4 represents a clean surface created by fracturing a rod of ZBLA glass under ultra-high vacuum conditions. The peaks due to F and Al are clearly evident, whereas the peaks due to Zr, Ba and La exhibit some overlap. The fact the fluorine signal is strong - in spite of its lower scattering cross-section relation to zirconium and barium - indicates that the fluorine ions shield the cations at the surface, and thereby, dominate the sites in the surface monolayer. Clean fracture surfaces of this type have been exposed in-situ to air and water vapor,

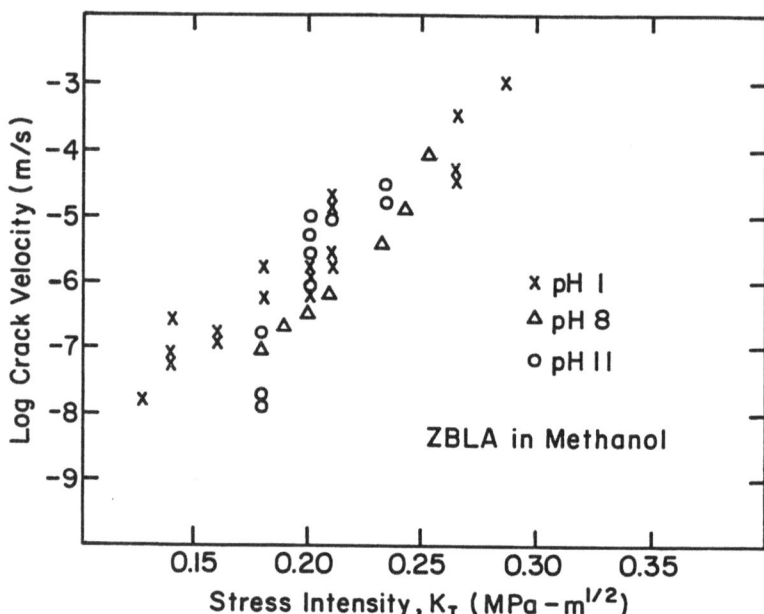

FIGURE 3. The sub-critical crack velocities measured in mixtures of methanol and either water, acid or base; in all cases, the water content is less than 1%.

and in both cases, no oxygen signal due to oxide, hydroxide or water adsorption was detected. In contrast, creation of the fracture surface in air leads to the adsorption of some oxygen (or hydroxyl) as shown in the second spectra of Figure 4. This suggests that active surface adsorption sites may be generated by the stress which is present during the fracture event. In general, though, the extent of oxygen chemisorption in the outermost monolayer of the fracture surfaces is minimal.

A similar ISS analysis of water adsorption was reported by Strecker (21) for clean surfaces of single-crystal CaF_2. He found that water adsorption was practically non-existent until the temperature was lowered to the point where an epitaxial condensation of water was observed. His data also provided an indication of the monolayer sensitivity of this method; i.e., condensation of a single monolayer of water was sufficient to completely extinguish the Ca and F peaks in the ISS spectra, and correspondingly, to generate a strong oxygen signal. Thus, the oxygen or water coverage represented in the middle spectra of Figure 4 is small indeed.

The lower spectra in Figure 4 represents a surface created by fracture in water. Due to the aggressive reaction between fluorozirconate glass and water , though, one must recognize that the surfaces have been corroded in the water after the fracture. Nonetheless, the absence of an oxygen signal indicates that oxide and/or hydroxide reaction products are not abundant at the glass/solution-interface, nor within any associated surface layer. Rather, the spectra reveals the presence of only Zr and F. This is probably due to the local precipitation and accumulation of ZrF_4 in the surface. Others have shown the eventual formation of ZrF_4 crystals at the surface of ZBLA after an extended exposure to water (12).

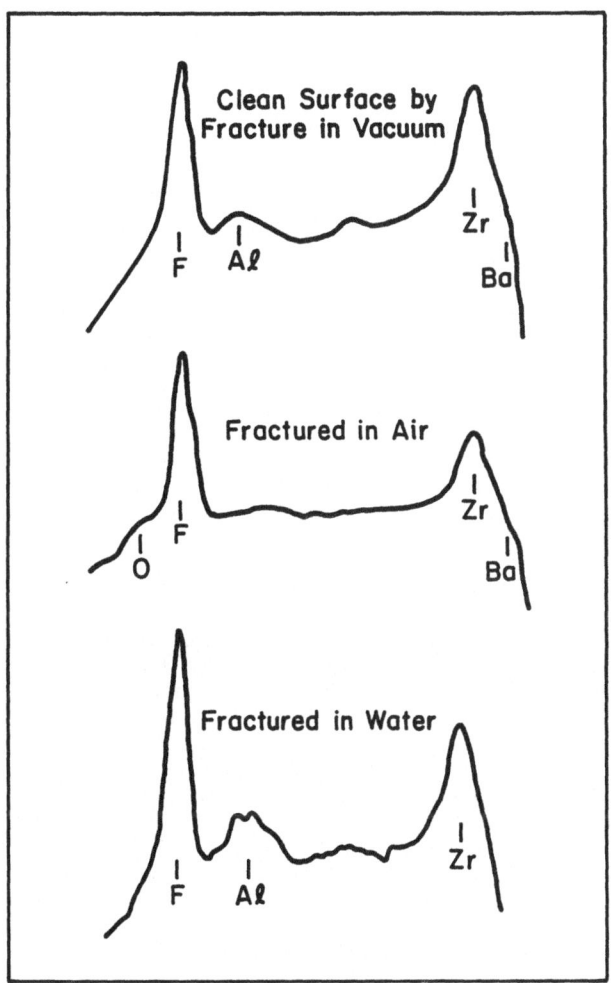

FIGURE 4. Low-energy ion scattering spectra (^{4}He at 1.5 KeV) for fracture surfaces of ZBLA.

The ISS studies have shown that neither the chemisorption of water, nor the formation of oxides and hydroxides, are

prevalent reactions at the surface of ZBLA. However, the heavy
Zr and Ba cations are responsible for the high background in
the low energy region of the ISS spectra, and for this reason,
the sensitivity to oxygen is less than ideal. And since the
low energy ion scattering occurs almost exclusively in the
outermost atomic layer, it is also possible that ion scater-
ring by oxygen, hydroxyl or water species is physically
shielded at the surface by the heavier cations. Thus, a corres
ponding set of surface analytical studies were attempted with
Auger electron spectroscopy (AES). The AES technique averages
the elemental composition in the outermost 3.0-5.0 nm of the
surface; moreover, the AES method is exceedingly sensitive to
oxygen.

The Auger spectra in Figure 5 show that oxygen can be
detected by AES even on the fracture surfaces created in vacuum, but it is not yet known whether this oxygen is intrinsic
to the glass or was adsorbed from the residual vacuum via elec
tron-beam stimulated adsorption (23). In any event, it corresponds to a relatively low concentration (~5-6 atomic percent).
After in-situ exposure to ~1800 Langmuirs of water vapor, the
Auger spectra showed no substantial increase in oxygen, but
the zirconium signal was eliminated and some increase in the
fluorine signal was observed. The attenuation of the zirconium
signal could be due to shielding by some adsorbed species, but
it is unlikely that this species is water because it would
have produced a significant increase in the oxygen signal. The
fact that hydrogen is undetectable by AES suggests that a hydrogen species may be involved. The observed increase in fluorine concentration leads one to consider HF surface reaction
products. The HF could form due to a reaction between ZBLA and
H_2O with the resulting hydroxyl specie diffusing further into
the glass.

The other spectra in Figure 5 was obtained on a fracture
surface created in air. One notes again that little oxygen is
present on the surface, but a carbon signal-presumably an
adsorbed hydrocarbon-is observed. After the analysis of six
facture surfaces created in air, it became apparent that the
oxygen content could vary between 5% and 10%. The C and O
ratio was not constant, and so, hydrocarbons, rather than CO
or CO_2, may be involved. In general, though, the spectra in
Figure 5 further support the contention that these fracture
surface are not especially reactive in humid atmospheres. The
oxygen species - whatever they are - do not cover the surface,
but rather, are present at a relatively low concentration (at
least relative to the anion content of the ZBLA glass).

The Auger spectra in Figure 6 represent the influence of
some liquid environments upon the surface composition. In all
cases, the glasses were fractured under the liquid, and then,
were immediately extracted and dried in a stream of dry nitrogen. The specimen fractured under water shows a significant
attenuation of the barium and zirconium signals, some carbon,
and background levels of oxygen; the obvious decrease in the
[Ba,Zr] - to - [F] ratio, again, points to the possible formation of HF reaction products on the surface. In contrast, the
glasses fractured in methanol and ammonium hydroxide show

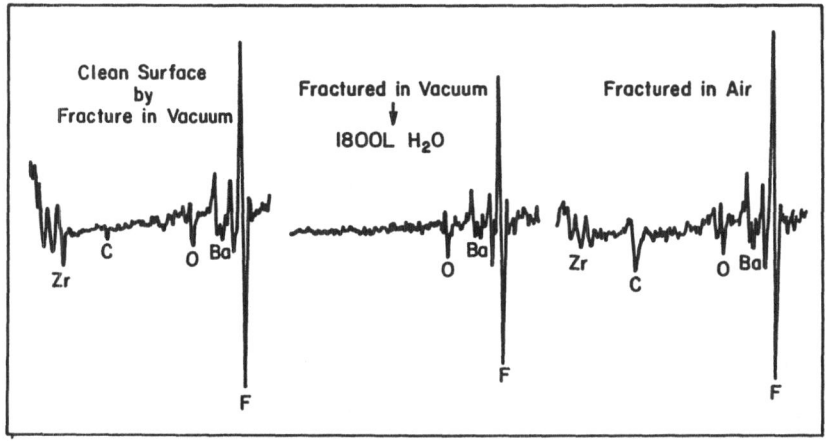

FIGURE 5. Auger spectra for fracture surfaces of ZBLA; the center spectra represents the clean vacuum-fractured surface after exposure to 1800 Langmuirs of water.

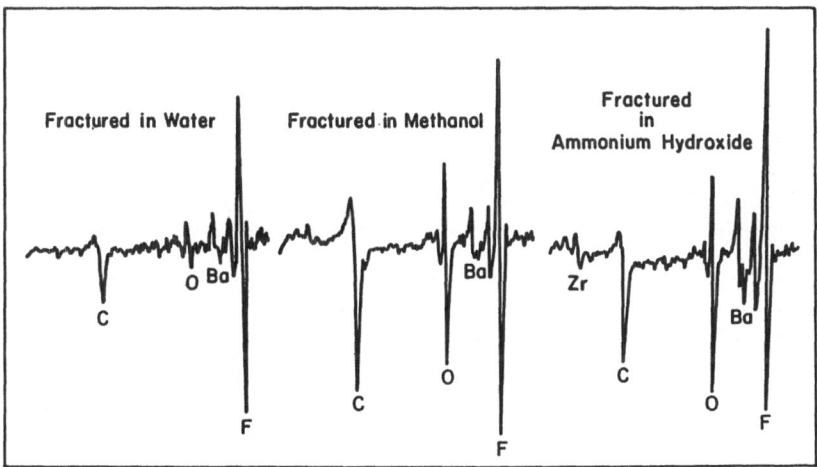

FIGURE 6. Auger spectra for surfaces of ZBLA which were created by fracture in liquid environments; the spectra in Figures 5 and 6 were obtained using a 3KeV electron beam at ~6μA/cm².

substantial increases in oxygen content. The oxygen concentration represented in these two spectra is approximately 20 atomic percent. The fact that nitrogen is not detected at the surface created in the ammonium hydroxide indicates that specific adsorption of oxygen or hydroxyl has occurred. The presence of these oxygen levels at the surface has not attenuated the fluorine signal, and so, the exchange of F and OH seems unlikely even in the highly basic hydroxide solution. A noticeable increase in carbon content was always observed on fracture surfaces created in the ammonium hydroxide which may be due to enhanced adsorption of hydrocarbon, or possibly CO/CO_2 adsorption, at oxidized or hydroxylated surface sites. The surfaces created under methanol consistently showed increases in both the oxygen and carbon, and in general, the C to O ratio was usually larger than that observed for all other surface conditions. This suggests a strong physical adsorption of the methanol species, especially since the exposure to ultra-high vacuum has not triggered their desorption. In the case of alkali-chlorides, bromides and iodides, the physical adsorption of methanol has been observed and attributed to hydrogen bonding with the surface anions (24).

The SIMS technique has been used to depth profile a wide variety of ZBLA glass surfaces after exposure to humid and aqueous environments. In contrast to AES and ISS, this method is relatively insensitive to the outermost atomic layer(s) of the solid, but is ideally suited for profiling .1 to 10.0 μm into the surface. The fact that these glasses do corrode in aqueous environments, but show little reaction at the glass/ /solution interface, suggests that the reactants penetrate the material and breakdown the structure from within. Since it was already known that the hydration of ZBLA is very slow at ambient temperature - even when the relative humidity is as high as 95% - hydration experiments were carried out at 200C and 275C in water-saturated Ar or O_2.

Figure 7 shows that both oxygen and hydrogen are found to penetrate nearly 5,000 nm into the surface within 4 hours at 200C. It is especially noteworthy that the distribution of glass constituents has not been influenced by the water penetration. In multicomponent silicate glasses, hydration in water vapor usually leads to a redistribution of modifier species and the growth of a distinct surface layer. The temperature, time and carrier gas (Ar vs O_2) were varied, and in all cases, only the apparent penetration of H, O and in some instances C, were revealed in the profiles. This is clearly consistent with the ISS and AES surface studies in air and water vapor which showed only minimal surface reactions. A quantitative analysis of the SIMS data has not yet been completed, but the surface concentrations of water appear to be quite low after the exposure to humid gases (or dilute water- -methanol mixtures - see below); thus, the oxygen associated with the molecular water observed in SIMS is probably at, or below, the detection limit in AES and ISS (~100 ppm).

SIMS was also used to examine ZBLA surfaces after exposure to liquid environments. Figure 8 summarizes the hydrogen profiles obtained after ten minute exposures at 30C to acid,

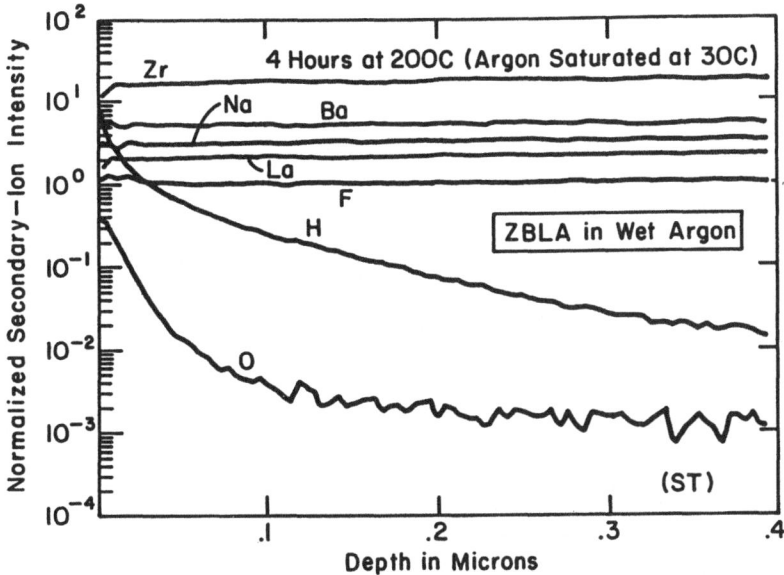

FIGURE 7. The in-depth concentration profiles observed after
an elevated temperature (200C) exposure of a fluorozirconate
glass to argon-which was first saturated with water at 30C;
profiles were obtained using secondary ion mass spectroscopy.

water and base. The absence of any in-depth hydration (or any
other in-depth effect) in ammonium hydroxide is clearly
evident and is consistent with dissolution studies that show
the relative inertness of these glasses to high pH environments. In contrast, the acid solution - within only ten minutes - generated a visible film on the surface, and the hydrogen profile in Figure 8 indicates that this layer is about 4 μm
thick. The uniform hydrogen content, and the corresponding pro
files for other constituents of the glass, reveal that this is
a layer or film of surface reaction products. The scanning
electron microscope (SEM) was also used to examine this surface; a layer containing dense agglomerates and clumps of fluoride crystals were observed. The glass which was exposed to
water also exhibited a visible film on the surface. The corresponding hydrogen-profile in Figure 8 shows that this layer
is about .5-1.0 μm in thickness. Although the profile suggests
that the hydrogen content is greater after the water corrosion,
than after the acid exposure, this is more likely due to a
density difference between the films produced in water versus
acid. The 'acid-generated' surface layer is loosely adherent
and often flakes-off during drying of the sample, whereas the
'water-generated' surface layer is usually more tenacious.
   Since the mechanism of aqueous attack - rather than the for
mation of surface layers - is of primary relevance to stress
corrosion, it again became necessary to lower the water activity by dilution with methanol so that the earliest stages of
the reaction could be investigated. Figure 9 shows two sets of

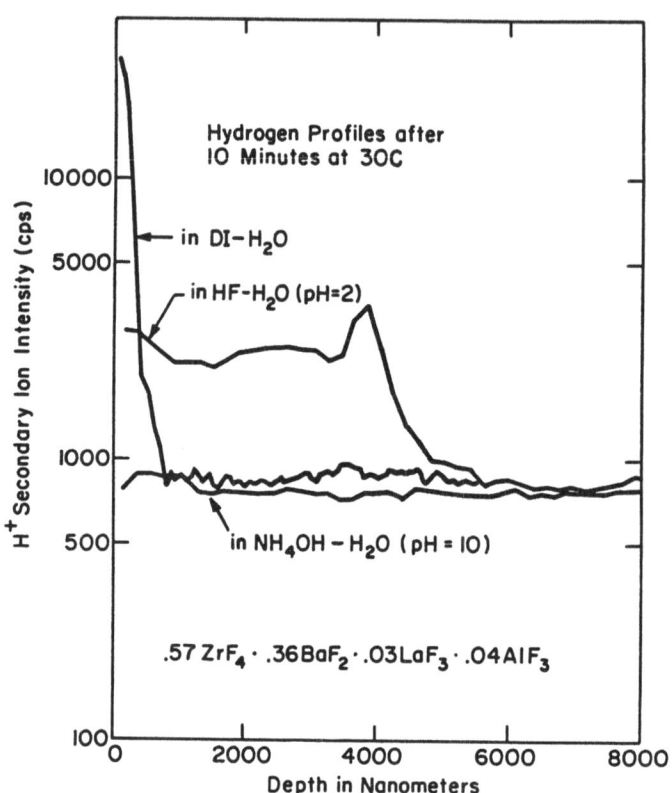

FIGURE 8. The in-depth concentration profiles for hydrogen in the surface layer of fluorozirconate glasses after exposure to water, acid and base; the profiles were obtained with secondary ion mass spectroscopy.

SIMS depth profiles; one set was obtained after a 4 hour hydration of ZBLA in 50% H_2O - 50% CH_3OH and the other after a 4 hour hydration in 70% H_2O - 30% CH_3OH. Although other mixtures and conditions were examined, these two samples bracket the critical range of water activity. In addition to profiling the hydrated surfaces, the solutions were analyzed spectrochemically for dissolved constituents of the glass after the hydration. The profiles in Figure 9 look very much like those due to water vapor hydration at elevated temperatures (Figure 7). Namely, there has been little or no attack of the glass structure - only penetration of hydrogen and oxygen; the corresponding infrared spectra obtained here, as well as those reported elsewhere (13), indicate that these species are associated with molecular wat r. The solution analyses showed negligible quantities of dissolved fluorine in the 50% H_2O -

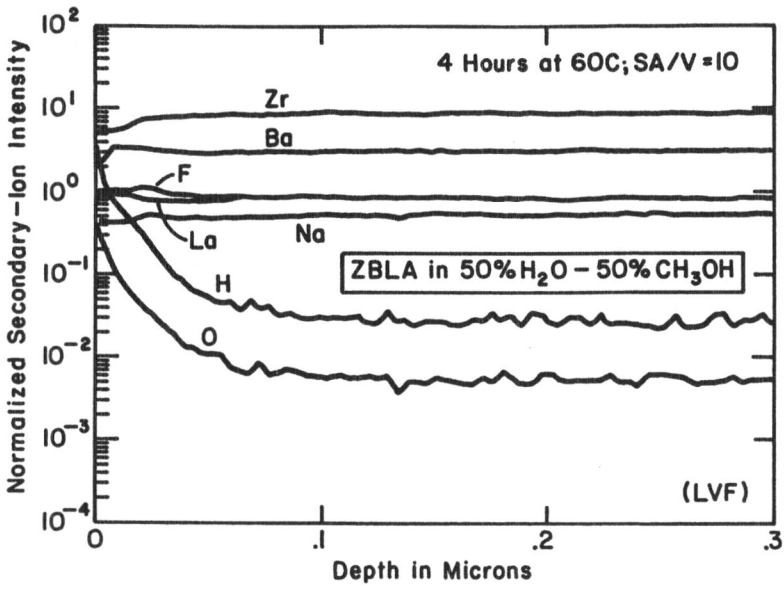

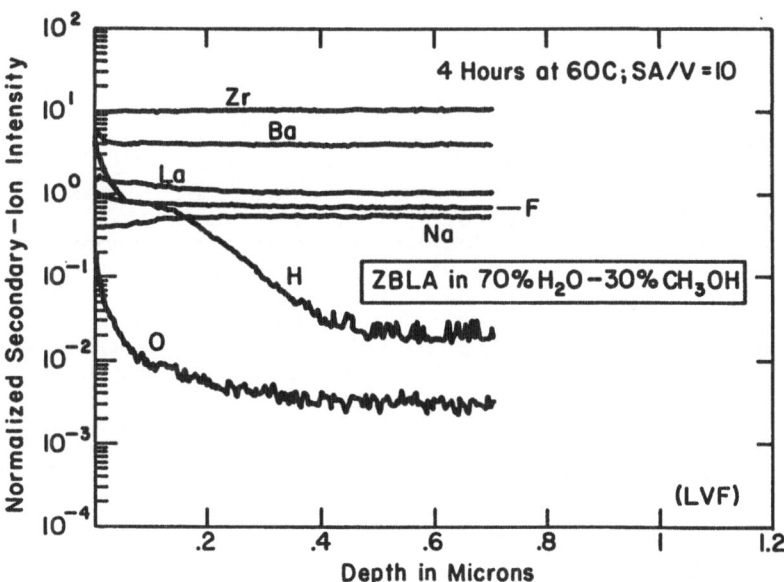

FIGURE 9. The in-depth concentration profiles for a fluorozirconate glass after exposure to water-methanol mixtures; the
profiles were obtained with secondary ion mass spectroscopy.

- 50% CH$_3$OH mixture. This situation changes dramatically once
the water concentration reaches 70% (at least for a 4 hour exposure). In this case, a twenty-fold increase in dissolved
fluorine is observed, and the profiles show the initial stages
of a distinct 'surface layer' formation due to this dissolution of the glass structure. This layer contains substantial
levels of hydrogen and fluorine, but the oxygen profile is
still reminiscent of an underlying diffusion process.

It seems clear that the ISS and AES analyses of water or
acid corroded glasses detected surface reaction products, but
these were fluorides rather than oxyfluorides, oxides or hydro
xides. They are probably precursors to the ZrF$_4$ and BaZrF$_6$
crystals observed at the surface after more extensive corrosion in water and acid (12). The SIMS analyses show that these
fluoride reaction products will form layers at the surface
whose rate of formation increases with decreasing pH. In addition, though, the SIMS mass spectra (not presented here) show
peaks at m/e=20, 40 and 60 whenever the surface has been
exposed to gaseous or liquid water environments. These peaks
can only be attributed to HF$^+$, (HF)$_2^+$ and (HF)$_3^+$ and so it is
likely that these surface layers contain hydrogen fluoride
(hydrogen cannot be detected with AES or ISS). The important
point, though, is that these surface layers are not relics of
the original glass structure, but rather, are the reaction
products of its hydrolytic decomposition, dissolution and
local precipitation. Due to the non-uniform nature of these
surface layers, it is not possible to sputter-profile through
to the underlying 'glass' and clearly define the interaction
that occurs at the interface. But, if the layer formation is
limited by lowering the water activity (e.g., in the water
vapor or water - methanol mixtures), the penetration of molecular water into the glass can be seen.

These data show that in the case of both water vapor (at
200-300C) and liquid water the in-depth penetration of molecular water is more prevalent than hydrolysis or chemisorption
at the glass surface. Tregoat, et al.(25) exposed a barium-
-ytterbium-zinc-thorium fluoride glass to water vapor at 344°C,
and in contrast to the behavior of this ZBLA glass, they
observed HF gas evolution due to this interaction. Their infrared spectroscopic analyses showed the absence of molecular
water vibrations in the hydrated surface, and thus, they proposed an interdiffusion reaction between F and OH. Loehr, et
al. (13), also using infrared spectroscopy, did find molecular
water - and little or no hydroxide species - in the surfaces
of glasses hydrated in liquid water. The origin of these apparent discrepancies is not known, but may be related to tempe
rature and composition differences.

5. DISCUSSION

Altogether, the surface studies indicate that the penetration and bulk diffusion of molecular water precedes any chemisorption, hydrolysis or breakdown of the glass structure.
This appears to be true in humid atmospheres as well as in
liquid environments. Once the water reaches some critical concentration in the glass surface, the glass structure begins to

disintegrate. It is likely that the molecular water - being somewhat nucleophilic - forms coordinate bonds with the Ba and Zr cations. This weakens the coulombic forces of attraction with the fluorine anions, and thereby, 'loosens' the structure. In the case of pure zirconium fluoride, mono-hydrates and tri--hydrates can be generated by exposure of ZrF_4 (ρ=4.65 g/cc) to steam at 100°C-450°C. These hydrates are lower in density than the original ZrF_4 (ρ=3.54 g/cc and 2.81 g/cc, respectively), and perhaps of greater significance, they form with little or no HF evolution (26). Since there is little or no evidence for ion-exchange or preferential leaching of any components in the fluorozirconate glasses, it seems certain that in-depth hydration by molecular water is fundamental to their hydrolytic decomposition and dissolution, i.e. it is the first step in the corrosion reaction.

In H_2O-CH_3OH mixtures and humid atmospheres, the water activity at the surface is low, and thus, the rate at which molecular water concentrates in the surface is limited by its diffusion into the bulk of the glass. Here, the penetration of water occurs with little or no hydrolysis, dissolution or surface layer formation (see Figures 7 and 9). In liquid water, the water activity at the glass/solution boundary is high. Thus, breakdown and dissolution of the glass commences rapidly, and as a result, the reaction products can accumulate and form surface layers. The pH determines the solubility of the glass constituents - once they are released from the glass structure - and so pH is critical to the overall dissolution rate and extent of surface layer formation. At low pH, for example, the dissolution rate is greatly enhanced because the solubility of fluorine increases. The extensive surface layer in this case due to the insolubility of ZrF_4 at fluorine concentration greater than 10-100ppm; this condition is easily achieved in acid near the glass surface, and so the local precipitation of ZrF_4 is extensive. In contrast, the dissolution rate at pH 10 is 100-1000 times less than at pH 2 (15), and under these conditions the penetration of water is not observed. The fact that oxidation and/or hydroxylation of ZBLA was observed in the surface monolayer only for the pH 11 ammonium hydroxide solution suggests that these surface reaction products may 'block' the penetration of water and/or are exceedingly insoluble.

Some additional - albeit qualitative - evidence for a base--catalyzed 'passivation' of the ZBLA surface was provided by examining the corrosion of fracture surfaces which were pre--treated in the base. The fracture surfaces were created and aged in pH 10 ammonium hydroxide solution and were then placed in water. A set of control samples were fractured in air, as well as under the water, for comparison. There was no question that the fracture surfaces created and aged at pH 10 are more resistant to visible surface film formation than the glasses fractured in air or under water; i.e., a reproducible 'incubation time' can be observed before the glasses treated at pH 10 begin to haze in the water.

The surface studies have provided considerable insight to the stress corrosion mechanism in ZBLA. The fact that the crack-growth behavior is relatively insensitive to pH is consistent with the idea that molecular water is primarily res-

ponsible for breakdown of the glass structure in aqueous environments. It is also clear that the crack velocities in acid,
base, water and water vapor, and water-methanol mixtures depend in a consistent way upon the water activity. But since
the extent of surface film formation and dissolution varies
considerably amongst these environments, it can be concluded
that surface film formation and/or dissolution are not important features of the stress corrosion reaction.

In the case of silicate glasses, the mechanism of stress
corrosion has been atributed to a stress-enhanced dissociative
chemisorption at the crack tip (7). This interaction occurs at
the crack-tip between strained bonds and chemical species
in the environment. The theory and experimental data support
the idea that chemical species which possess both electron
donor and proton donor sites are a prerequisite for this reaction. This property of chemical species has been characterized
by Gutman (27) in terms of so-called donor numbers and acceptor
numbers. Thus, chemical environments with high values for both of these num
bers would be expected to enhance the crack-growth rate to the
greatest extent. In the case of these fluoride glasses the
crack-velocity at ~.2 $MPa-m^{1/2}$ (above the threshold for
aqueous environments) shows an excellent correlation with the
sum of those two numbers in all of the environments (28). This
suggests that dissociative chemisorption may also be responsible for stress corrosion in these fluoride glasses. However,
the surface studies of adsorption on clean fracture surfaces
provided little evidence for the chemisorption of water. This
indicates that the presence of strained-bonds may be required
for the chemisorption reaction in this system. Although the
ISS data in Figure 4 support this idea, many of the other sur
faces prepared by fracture in gaseous and liquid environments
did not show chemisorbed oxide or hydroxide species at the surface. It is to be emphasized, though, that these were all
'fast-fractured' in the environment, and so, these surfaces
may not be representative of those prepared under sub-critical
crack growth conditions. Clearly, the in-situ analysis of
fracture surfaces created at low velocities would be a worthwhile endeavor.

On the other hand, the surface studies have indicated that
the penetration of molecular water is the most prevalent interaction between aqueous environments and the fluorozirconate
glasses. If a weakening of the ionic bonds due to the coordina
tion of molecular water with the Zr and Ba cations is responsi
ble for the sub-critical crack growth, then the crack velocity
should scale with the dielectric constant or polarizibility of
the environment. The relationship between crack velocity at
$K_1 = .2$ $MPa-m^{1/2}$, and the dielectric constant of the various
environments, does not support this concept though. Thus,
while hydration by molecular water may be fundamental to the
corrosion and dissolution reactions, it may not be responsible
for the bond-breaking event in stress corrosion.

Of particular significance are the thresholds (or stress-
-corrosion limits) – at about .20 $MPa-m^{1/2}$ – which are exhibited by the H_2O environments and the NH_3. These environments
also exhibit velocity plateaus whose magnitude scales with the

water activity. Since the rapid penetration and diffusion of molecular water is prevalent in these environments, one is led to consider the possible effects of 'stress-enhanced' water diffusion ahead of the crack-tip. This phenomena could influence the stress distribution at the crack-tip and/or modify the glass through which the crack-tip must propagate. Thus, the stress intensity and/or activity of water would determine the extent of water diffusion ahead of the crack-tip, and quite possibly, this water of hydration might dissociate - under stress - as the crack propagates. And in a qualitative sense, an interaction between the glass and environment 'ahead of the crack-tip' would explain the erratic behavior of the cracks in these glasses. In contrast to the behavior of silicate glasses during crack-growth experiments, the cracks in even well-annealed fluoride glasses seldom follow a path which is perfectly orthogonal to the applied moment. It is also found that once the crack has arrested - e.g. in and below the threshold region of stress intensity - the crack will not re--propagate when the load is increased unless the specimen is 'tapped' to get it running.

6. SUMMARY

A detailed interpretation of the stress corrosion behavior in fluorozirconate glasses is only just beginning to emerge. The origin of the stress-intensity thresholds and velocity plateaus is not obvious, but clearly, an understanding of these features could contribute to the development of a model for stress corrosion in these materials. In this regard, it is noteworthy that the environments where velocity thresholds and plateaus are observed also happen to be those where the potential formation of HF reaction products can be proposed. Thus, the dependence of crack growth rates upon the ionic strength of $HF-H_2O$ mixtures would be useful information. In general, though, a more substantial data base concerning the surface chemistry and slow-crack growth characteristic of these glasses will be required, before a definitive model can be proposed and validated. This understanding will be necessary when evaluating the fatigue data yet to be reported for fluorozirconate optical fibers, and will be of great benefit in the design and development of more fatigue-resistant systems. And since these fluorozirconate materials represent a new and unique class of ionically-bonded glasses, the ability to compare their slow crack growth behavior with that of the more covalently-bonded silicate glasses could provide a more fundamental insight to stress-corrosion reactions in brittle solids.

ACKNOWLEDGEMENTS

The author thanks the Air Force Office of Scientific Research for their financial support (AFOSR-82-0013) and acknowledges his co-workers for their contributions: Armando Gonzalez, Cheryl Houser, Theresa McCarthy, Jack Mecholsky and Andrew Phelps.

REFERENCES

1. Pantano, C. G., "Surface Chemistry and Fracture of Fluoride Glasses", Materials Science Forum, 6, 285 (1985).
2. Becher, P.F. and Freiman, S.W., "Crack Propagation in Alkaline Earth Fluorides", J. Appl. Phys. 49. 3779 (1978).
3. McKinney, K.R., Mecholsky, J. J. and Freiman, S. W., "Delayed Failure in Chemically Vapor Deposited ZnSe", J. Am. Ceram. Soc., 62, 336 (1979).
4. Mecholsky, J.J., "Intergranular Slow Crack Growth in MgF_2", J. Am. Ceram. Soc., 64, 563 (1981).
5. Hillig, W.B. and Charles, R. J., "Surfaces, Stress-Dependent Surface Reactions and Strength", in High Strength Materials, (W. F. Zackey, Ed., Wiley, NY, 1965) pp. 682-701.
6. Wiederhorn, S. M., "A Chemical Interpretation of Static Fatigue", J. Am. Ceram. Soc., 55, 81 (1972).
7. Michalske, T. A. and Freiman, S.W., "A Molecular Mechanism for Stress Corrosion in Vitreous Silica", J. Am. Ceram. Soc., 66, 284 (1983).
8. Simmons, C. J., Sutter, H., Simmons, J. H. and Tran, D.C., "Aqueous Corrosion Studies of a Fluorozirconate Glass", Mat. Res. Bull. 17, 1203 (1982).
9. Mitachi, S., "Chemical Durability of Fluoride Glasses in the BaF_2-GdF_3-ZrF_4 System", Phys. Chem. Glasses, 24, 146 (1983).
10. Barkatt, A. and Boehm, L., "The Corrosion Process of Fluoride Glass in Water and the Effects of Remelting and of Glass Composition", Mat. Lett., 3, 43 (1984).
11. Frischat, G. H. and Overbeck, I., "Chemical Durability of Fluorozirconate Glasses", J. Am. Ceram. Soc., 67, C-238 (1984).
12. Doremus, R.H., Murphy, D., Bansal, N.R., Lanford, W. A., and Burman, C., "Reaction of Zirconium Fluoride Glass with Water: Kinetics of Dissolution", J. Mater. Sci., 20, 4445 (1985).
13. Loehr, S. R., Bruce, A. J., Mossadegh, R., Doremus, R. H. and Moynihan, C.T,, "IR Spectroscopy Studies of Attack of Liquid Water on ZrF_4-Based Glasses" Mater. Sci. Forum, 5, 311 (1985).
14. Simmons, C.J. and Simmons, J. H., "Chemical Durability of Fluoride Glasses" to appear.
15. McCarthy, T. A., Houser, C. A. and Pantano, C. G., "Dissolution and Surface Layer   Formation in Fluorozirconate Glasses" to appear.
16. Amphlett, C. B., "The Adsorption of Water Vapour by Calcium Fluoride", Trans. Faraday Soc., 54, 1206 (1958).
17. Hall, P.G. and Tompkins, F. C., "Adsorption of Water Vapour on Insoluble Metal Halides", Trans. Faraday Soc., 58, 1734 (1962).
18. Barraclough, P. B. and Hall, P.G., "Adsorption of Water Vapour by Calcium Fluoride, Barium Fluoride and Lead Fluoride", Trans. Faraday Soc., 71, 2266 (1975).
19. Palik, E. D., Gibson, J. W., Holm, E. T., "Internal Reflection Spectroscopy Study of Adsorption of Water on CaF_2 Surfaces", Surf. Sci., 84, 164 (1979).

20. Wojciechowska, M. and Fiedorow, R., "Surface Chemistry of Porous Magnesium Fluoride", J. Fluorine Chem., 15, 443 (1980).

21. Strecker, C. L., "Determination of the Heat of Adsorption for Water Vapor on Clean Calcium Fluoride Surfaces by Ion Scattering Spectroscopy", Ph.D. Dissertation, University of Dayton (1985).

22. Freiman, S. W., Mulville, D. R. and Mast, D. W., "Crack Propagation Studies in Brittle Materials", J. Mater. Sci., 8, 1527 (1973).

23. Pantano, C. G., and Madey, T. E. "Electron Beam Damage in Auger Electron Spectroscopy", Appl. Surf. Sci., 6, 115 (1981).

24. Chikazawa, M. and Kanazawa, T., "The Adsorption Mechanism of Methanol on Alkali Halides", Bull. Chem. Soc. Japan, 50, 2837 (1977).

25. Tregoat, D., Fonteneau, G., Lucas, J. and Moynihan, C. T., "Corrosion of HMFG by H_2O Vapor", Mater. Sci. Forum, 5, 323 (1895).

26. Waters, T. N., "Some Investigations in the Zirconium Tetrafluoride-Water Systems", J. Inorg. Nucl. Chem., 15, 320 (1960).

27. Gutman, V. The Donor-Acceptor Approach to Molecular Interactions, Plenum Press, NY, 1978.

28. Gonzalez, A., Mecholsky, J. J. and Pantano, C. G., "Stress Corrosion Mechanisms in Fluorozirconate Glasses", to appear.

DISCUSSION

Q: P. W. France. A few years ago, very high values of stress corrosion susceptibility were reported, in the range of n=15-100. Have you actually observed these values and how do you account for them?

A: C. Pantano. These values were obtained by indentation techniques where the observed crack growth was driven by the residual stresses around the indentation. The theory associated with this method assumes the usual K-v relationship, where the slope in region I gives the stress corrosion susceptibility. Our measurements of the K-v diagram for ZBLA show a threshold at $K_I = 0.2$ MPa m$^{1/2}$, which is not typical of silicate glasses. Thus the present indentation method underestimates the susceptibility to sub-critical crack growth because, once the residual stress decays to $K_I < 0.2$ MPa m$^{1/2}$, the crack growth stops.

C: J. Lucas. Our observations of corrosion of fluoride glasses in water are consistent with your results. We also observed H_2O molecule formation during the corrosion process by water vapor. The 6.1 µm absorption is often present even well below the surface of the glass.

Q: C. Moynihan. Do your results indicate that hydrogen penetrates further into the glass surface than does oxygen? IR measurements of hydrated surfaces extending > 10 µm below the glass surface indicate that both oxygen and hydrogen (as OH$^-$ or H_2O) penetrate simultaneously.

A: C. Pantano. No. The hydrogen accumulates in the surface to a concentration higher than in OH or H_2O. Our results indicate that there is a certain amount of the oxygen that can be incorporated in the form of OH$^-$ or O_2^{2-}. We believe that water molecules penetrate the surface to form HF and metal hydroxides. Once the limiting hydroxide level is reached (perhaps ∿ 5-10% of the fluoride concentration), the hydrogen continues to concentrate in the surface (probably through protonation of F_{nb} species) and the released OH$^-$ diffuses further into the glass. However, the scale of depth profiles (∿ 0.1-0.5 µm) obscures the shallow diffusion profile associated with oxygen, which IR measurements have shown to extend to 10-40 µm.

OPTICAL PROPERTIES OF FLUORIDE GLASSES

MARTIN G. DREXHAGE

Solid State Sciences Division
Rome Air Development Center (ESMO)
Hanscom AFB, Massachuscetts 01731 USA

ABSTRACT

The basic optical phenomena responsible for the transparency of heavy metal fluoride glasses are examined in terms of theory and experiment. The preparation and properties of the most widely used zirconium tetrafluoride-based compositions are emphasized, and their optical behavior contrasted with that observed in silicate glasses and other vitreous fluorides. The topics addressed include ultraviolet and infrared absorption, light scattering, refractive index, and the effects of ionizing radiation. The role of impurities such as metallic oxides, hydroxyl species, and carbon dioxide, and the merits and disadvantages of various reactive atmosphere treatments are discussed.

1. INTRODUCTION

The intent of this paper is to briefly review and assess those aspects of the optical behavior of heavy metal fluoride glasses ("HMFG") which are relevant to their application as fiber lightguides and bulk optical components. The broad scope of the term "optical properties" and the contributions of other authors to this volume necessitate that discussion be limited to selected topics. These include a brief introduction to the most widely used HMFG compositions and their preparation, an examination of the various mechanisms which produce high transparency (and optical absorption) in the materials, and a summary of work in the areas of refractive index/dispersion and ionizing radiation effects. For a more detailed understanding of optical phenomena in the vitreous fluorides and their relationship to glass chemistry, the reader is encouraged to examine review papers such as References 1-3.

2. STABLE HMFG COMPOSITIONS AND THEIR PREPARATION

The recent technical literature suggests that many laboratories have focused their research efforts on several specific fluorozirconate compositions. Among them are the formulations shown in Fig. 1, which also reports the density, glass transition (T_g) and crystallization (T_x) temperatures, refractive indices, and expansion coefficients for the materials. A typical fluorozirconate composition, such as the "ZBLA" glass, is seen to consist of zirconium, barium, lanthanum and aluminum fluorides. In the related "ZBLAN" composition, about 20 mol % sodium fluoride is included at the expense of barium (4, 5). To achieve the requisite core/clad

MOL %	HBLA	ZBLA	ZBLAN	BZnLuT
ZrF_4	—	57	55.8	—
HfF_4	57	—	—	—
BaF_2	36	36	14.4	19
LaF_3	3	3	5.8	—
AlF_3	4	4	3.8	—
NaF	—	—	20.2	—
ZnF_2	—	—	—	27
LuF_3	—	—	—	27
ThF_4	—	—	—	27
DENSITY (gm/cm³)	5.88	4.61	~4.52	6.45
T_g (°C)	325	310	263	353
T_x (°C)	426	403	~386	429
REFRACTIVE INDEX n_D	1.504	1.521	1.480	1.534
EXPANSION COEFFICIENT (°C⁻¹)	173×10^{-7}	187×10^{-7}	175×10^{-7}	151×10^{-7}

FIGURE 1. Fluoride glass compositions and selected properties.

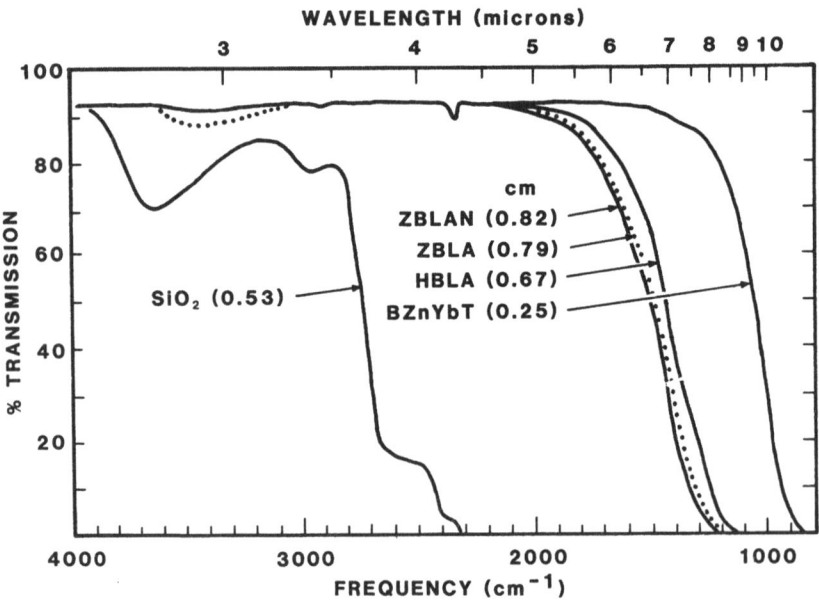

FIGURE 2. Transmission spectra from 2.5 to 10 microns for the glasses of Figure 1 and fused silica.

index difference in optical fibers, the aluminum or sodium
content may be altered slightly or hafnium tetrafluoride may be
partially or completely substituted for zirconium (e.g., "HBLA"
in Fig. 1) Other substitutions/additives commonly employed in
fluorozirconate compositions include the replacement of
lanthanum by gadolinium fluoride, the use of LiF and or lead
fluoride in place of NaF, and the addition of small amounts
(0.1-5 mol %) of indium trifluoride to reduce light scattering
effects (6). Moderate loss optical fibers (50-100 dB/km in the
mid-IR) and bulk glass articles (plates, domes, rods) made from
the fluorozirconate compositions discussed above are now
commercially available (7).

Worthy of more detailed examination are glasses of the
"BZnXT" type (Fig. 1), prepared from the fluorides of barium,
zinc, and thorium, with X representing either lutetium (Lu),
ytterbium (Yb) or yttrium (Y) (8,9). These materials offer
extended transparency in the mid-IR, better chemical durability
and radiation tolerance than the fluorozirconates, and serve as
an excellent matrix for rare-earth ions in potential "active"
optical applications such as laser hosts and frequency
upconversion devices. While generally more difficult to
prepare, their glassforming ability may be enhanced by addition
of sodium or indium fluoride; large fractions (near 30 mol %)
of aluminum trifluoride may also be incorporated with a
concomitant decrease in mid-IR transmission (Sec. 6).

Preparative techniques for the HMFG are now well
established, with most workers favoring anhydrous fluorides as
starting materials (see, e.g., Refs. 10, 11). By contrast,
early explorations employed the in situ fluorination of
metallic oxide starting materials with ammonium bifluoride.
Rigid atmospheric control to exclude moisture during all phases
of the glass fabrication process is a prerequisite; several
laboratories have established dedicated glove box facilities to
minimize this source of contamination (11).

3. TRANSPARENCY RANGE AND LOSS MECHANISMS

Properly prepared (i.e., low hydroxyl, transition metal
and rare earth content) fluorozirconate glass plates in
thickness near 1 cm exhibit approximately 92 % transmission at
wavelengths between 0.3 and 5 microns. Reflection losses at
the front and rear surfaces, proportional to the index of the
materials, account for the majority of the remainder. The
behavior of various compositions in the 2.5 to 10 micron region
is illustrated in Fig. 2, which also shows transmission data
for fused silica of low OH content. Vibrational modes
associated with Si and O limit the transmission of fused silica
to wavelengths of about 4 microns or less. By contrast,
glasses of the ZBLA and ZBLAN type are transparent to roughly 5
microns; the substitution of hafnium tetrafluoride (which has
twice the atomic weight of Zr) extends this range slightly.
Wavelengths of about 7 microns are accessible with the
BZnYbT-type materials; a strong absorption line in these
glasses near 1 micron may be eliminated by substituting
lutetium fluoride for the ytterbium. All the HMFG in Fig. 2
exhibit a small (1-2%) decrease in transmission near 2.8
microns due to bulk and/or surface absorbed hydroxyl species.

The sharp peak near 4.25 microns is the result of carbon dioxide within the materials (12, 13), and is further discussed in Sec. 7.

Optical absorption in transparent materials is the result of both extrinsic and intrinsic factors. Certain extrinsic sources of absorption, such as oxide, hydroxyl, transition metal and rare earth impurities are examined in Sections 4-7. Predictions concerning the potentially high transparency of HMFG are modeled on a consideration of three largely independent intrinsic loss mechanisms:

a) Rayleigh/Brillouin light scattering from microscopic density and composition fluctuations "frozen in" during glass formation. The magnitude of this effect should decrease in proportion to the reciprocal fourth power of the wavelength.

b) Absorption due to electronic transitions in the ultraviolet at wavelengths below about 0.2 microns. In theory, the intensity of such absorption should decrease exponentially with increasing wavelength, making it a neglible factor at long wavelengths in the mid-IR.

c) Infrared absorption due to fundamental vibrations of the glass "lattice" at wavelengths of 25-50 microns. The overtones and convolutions of these vibrations (e.g., Zr with F) give rise to the so-called infrared or multiphonon vibrational "edge" whose absorption intensity (in theory) decreases exponentially with decreasing wavelength.

Summation of the three mechanisms results in a "vee"-shaped intrinsic loss curve when absorption coefficient is plotted against wavelength. In the case of HMFG, this theoretical model (and the extrapolation of certain experimental data) suggests that a loss minimum will occur in the vicinity of 2-3 microns. In this region, losses of 0.01-0.001 dB/km are predicted. In the sections which follow experimental data concerning each mechanism are examined in view of this model, with due consideration to the oftentimes strong influence of extrinsic impurities.

4. LIGHT SCATTERING IN BULK HMFG

Since the details of light scattering in fluoride optical fibers are covered elsewhere in this volume (14), this section emphasizes only those phenomena and results which pertain to an understanding of the overall intrinsic transparency/wavelength curve for HMFG. Due in part to their low glass transition temperatures, various theoretical models suggest that HMFG should exhibit intrinsic Rayleigh scattering behavior with a magnitude lower than that attainable in silica-based glasses (see, e.g., Refs. 1-3). Large ZBLA ingots (3.5 cm diameter) in which the scattered light intensity in the visible region is 0.5-0.6 that of fused silica have been prepared (15-18). These and similar results obtained on long lengths of fiber (14), when extrapolated to the mid-IR region, imply that Rayleigh scattering-limited losses of close to 0.001 dB/km can in principle be achieved. In practice however, extrinsic sources of scattering such as crystallites, bubbles, inclusions, or waveguide imperfections often lead to large deviations from the desired Rayleigh behavior. The reproducible elimination of crystallites in particular is likely the most difficult obstacle confronting HMFG technology; some progress has been made via use of improved melting techniques, raw materials, and additives such as indium trifluoride or sulfates (6). It has recently been suggested that trace oxide impurities introduced from starting materials are also a primary source of light scattering (19).

In addition to providing information about the magnitude and wavelength dependence of the scattered light intensity, scattering measurements of the type described in Refs. 15-18 can also yield data on various optical and elastooptic constants in HMFG. Figure 3 illustrates this point for several fluorozirconate compositions and fused silica, and reports values for the Brillouin linewidths (in the VV polarization), Pockels' coefficient, Poisson's ratio, and elastic constants. Longitudinal and transverse sound velocities may also be obtained (15). Of particular relevance in Fig. 3 is a preliminary assessment of the magnitude of Stimulated Brillouin Scattering (SBS) effects in HMFG (17). This nonlinear phenomena has recently been observed as an additional loss parameter in single mode silicate fibers (20). SBS can occur if the optical power launched into a single mode fiber exceeds some critical threshold level; it causes a significant fraction of that power to be converted into a second light wave, shifted in frequency, traveling backwards towards the transmitter. This critical threshold is proportional to the Brillouin linewidth and inversely proportional to the square of the Pockels' coefficient. It has been suggested that HMFG fibers would have a lower threshold for SBS than silicate waveguides, thus limiting their usefulness in long-length communications systems (20). The fourth column in Fig. 3 shows calculated values for the SBS threshold in fluoride glasses (1.4-4.5) normalized to that of fused silica, and are in contrast with the earlier predictions. Recent observations that the Pockels' coefficients in HMFG appear dependent on thermal history (e.g., annealing after preparation) could allow "tailoring" to a minimum value, thereby attaining a maximum in the SBS threshold (17).

5. ULTRAVIOLET ABSORPTION

Although the HMFG offer considerable transparency in the near-UV, they are not as transmissive as high purity fused silica or many fluoroberyllate glasses. Typical cut-off wavelengths (defined, e.g., as the wavelength at which a few millimeter thick sample exhibits 0% transmission) for most compositions are in the 0.21-0.25 micron (5.9-5.0 eV) region; values near 0.16 microns (8.0 eV) are observed in fused silica and multicomponent beryllium fluoride-based glasses (21). The position of intrinsic electronic transitions in HMFG which manifest themselves as an ultraviolet absorption edge at longer wavelengths have been investigated via reflection spectroscopy by Izumitani and collegues (22,23). Figure 4, adapted from this work, shows representative data for a zirconium-barium-gadolinium-aluminum fluoride glass and fused silica. The peaks in such reflectance spectra correspond closely to the location of absorption bands. In fused silica, UV absorption is caused by transitions to an exciton state

LIGHT SCATTERING: DERIVED PARAMETERS

Glass	$\Delta v_B(vv)$ (MHz)	P_{12}	E_t^h/E_t^s	σ	C_{11}(GPa)	C_{44} (GPa)
ZBL	213.6	.255	1.6	0.30	——	——
ZBLA	153.9	.128	4.5	0.25	74.7	25.1
ZBLAN	159.5	.231	1.4	0.31	——	——
SiO$_2$	153.8	.270	1.0	0.17	77.9	31.0

-BRILLOUIN LINEWIDTHS, Δv_B

-POCKELS COEFFICIENTS P_{11}, P_{12}, P_{44}

-POISSONS RATIO, σ

-THRESHOLD FOR SBS: $E_t \approx \Delta v_B / P_{12}^2$

-ELASTIC CONSTANTS C_{11}, C_{44}

-SOUND VELOCITIES V_T, V_L

FIGURE 3. Parameters derived from light scattering measurements (after Refs. 15-18).

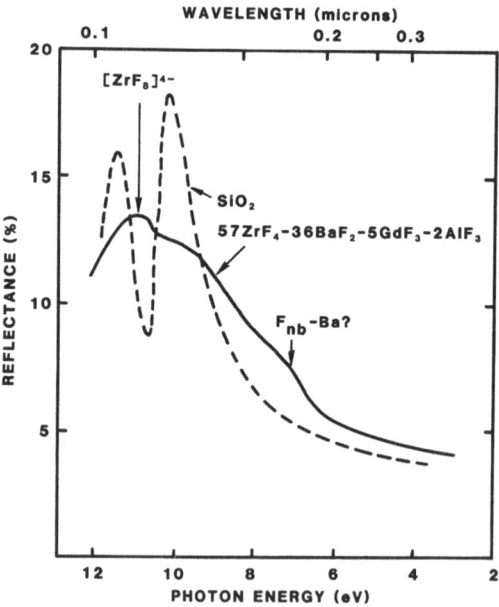

FIGURE 4. UV reflectance for a ZBGA glass and fused silica (adapted from Ref. 23).

(see, e.g., Ref. 24). In the fluorozirconate glasses, tentative assignments for certain bands have been made, while the origin of others is unclear. The 11 eV band in Fig. 4 is thought to arise from transitions within $[ZrF_8]^{4-}$ ions, while the shoulder at 7.5 eV is likely due to non-bridging fluorines combined with Ba^{2+} ions (23). Although it is difficult to define the "onset" of the intrinsic ultraviolet absorption edge from data of the type presented in Fig. 4, the differences in the slopes of the curves in the 7-9 eV region suggest that at longer wavelengths in the UV, the HMFG will have higher absorption coefficients than fused silica. Figure 5 provides confirmation of this behavior, showing absorption data derived from transmission measurements on silica (25) and several HMFG compositions prepared in the author's laboratory. The curve for the ZBLA and ZBLAN glasses represents results obtained from 15 high quality ingots prepared via techniques outlined in Ref. 10. Between 4.5 and 6.0 eV, this data is very reproducible, but becomes more scattered (shown by the error bars) as an absorption band near 3.5 eV due to Fe^{3+} contamination is encountered. The shaded area of Fig. 5 shows the range of absorption coefficients observed in about 10 specimens of BZn(Y or Yb)T glass of the composition noted in Fig. 1. As is the case with many silicate glasses, Fig. 5 suggests that HMFG containing large amounts of relatively "light" cations (e.g., Zr) will exhibit better UV transmission than those containing "heavy" species like thorium or ytterbium. More importantly, the position of the UV edge in HMFG and its closely exponential dependence on wavelength indicates that intrinsic electronic transitions should contribute negligibly to the optical losses observed in the 2-3 micron region of the mid-IR.

6. INFRARED EDGE ABSORPTION

Infrared active vibrational modes of the glass "lattice" are the third mechanism governing the transparency behavior of HMFG. Reflection spectra in the 25-50 micron region reveal the presence of these fundamental absorptions, as shown in Fig. 6 for a variety of compositions (2). In fluorozirconate glasses not containing aluminum, the strong peak near 20 microns is presumed to result from stretching vibrations of Zr complexes with fluorine, while the smaller peak near 40 microns may be due to Ba-F stretching vibrations and/or bend modes of Zr with F. Hafnate glasses (not shown in Fig. 6) exhibit a spectra almost identical to that of the fluorozirconates, with peaks of slightly lower intensity shifted to longer wavelengths. The origins of the bands in the remaining glasses in Fig. 6 are undefined, but certain qualitative compositional trends may be observed. Materials containing "heavy" cations like Ba and Th exhibit longer wavelength (i.e., lower frequency) fundamental vibrations with lower oscillator strengths. When aluminum fluoride is added to these materials, it introduces a new vibrational peak near 625 reciprocal centimeters (16 microns).

The onset of the infrared or multiphonon vibrational edge may be roughly defined as the region in Fig. 6 where the reflectivity goes to zero. At shorter wavelengths, the overtones and harmonics ("multiphonon absorption processes") of the fundamental vibrations produce a limiting absorption

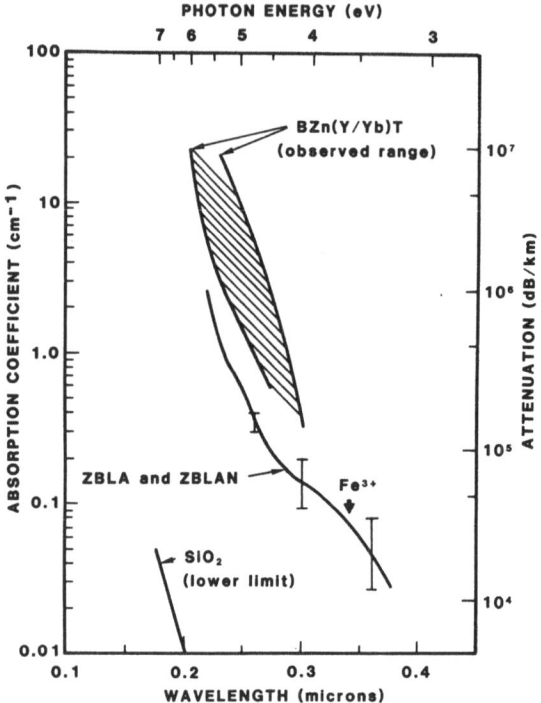

FIGURE 5. The UV absorption coefficient for several fluoride glasses and fused silica.

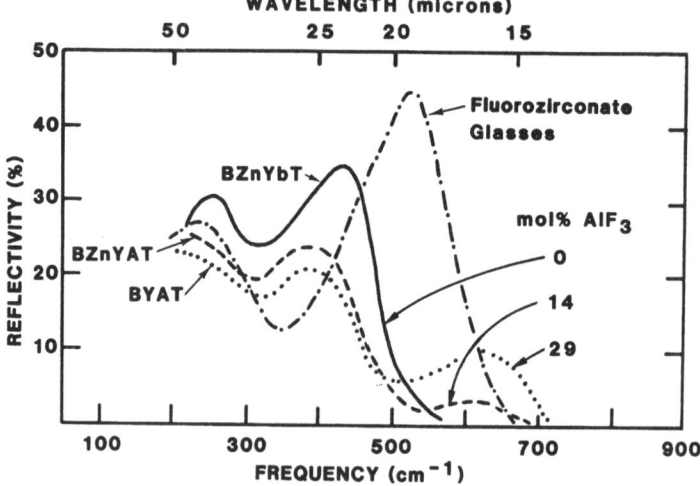

FIGURE 6. Reflection spectra between 15 and 50 microns for several fluoride glasses and fused silica (after Ref. 2).

characteristic of the material's composition whose intensity
decays in a roughly exponential fashion with decreasing
wavelength. In the absence of extrinsic impurities this IR
edge will, at some wavelength, intersect with the intrinsic
scattering curve to produce the "vee"-shaped transparency
alluded to in Sec. 3. Simultaneous plotting of attenuation
data derived from optical fibers at short wavelengths and bulk
glasses at longer wavelengths show that a rudimentary "vee"
curve does result, although it is distorted by the presence of
impurities and (oftentimes) non-Rayleigh scattering behavior
(1-3).

The data of Fig. 6 also serves to confirm the general
trends shown in the transmission curves of Fig. 2. Plots of
absorption coefficient versus wavelength in the 6-10 micron
region manifest themselves as a series of roughly parallel
lines ordered in accordance with the IR edge onset defined in
Fig. 6 (2,9). Thus at a given mid-IR wavelength, BZnYb(or
Lu)T exhibit the lowest absorption coefficients, while those of
the fluorozirconates and fluorohafnates are about one order of
magnitude higher. The absorption coefficient in glasses
containing substantial amounts of aluminum (e.g., BZnYAT and
BYAT, Fig. 6) is higher yet, due to the shift in IR edge
position onset caused by the 16 micron Al-F vibration.

7. SELECTED MID-IR IMPURITIES AND THE ROLE OF REACTIVE
ATMOSPHERES

The detrimental (and sometimes beneficial!) effects of
transition metal and rare earth impurities on optical
properties of fluoride glasses are addressed by other
contributers to this volume (14, 26). Here we will briefly
consider several equally "difficult to remove" sources of
extrinsic absorption which effect the transparency of HMFG.
The wavelength position of these impurities is schematically
shown in the transmission curve of Fig. 7, which represents a
"worst case" scenario assuming all were present in a given
sample of ZBLA glass. The band near 2.8 microns is due to the
stretching vibrations of hydroxyl species, either on the
surface or in the bulk of the glass. The small sharp peaks in
the vicinity of 3.4 microns are the result of C-H vibrations
from trace amounts of polishing oils remaining on the glass
surface. The assymetric stretching vibration of carbon dioxide
dissolved in the glass produces the band at 4.255 microns; not
visible in Fig. 7 are several weak combination bands from the
same species near 2.74 microns. The latter are of particular
concern as they fall near the loss minimum in HMFG fibers (13).
Carbon dioxide contamination may be the result of the combined
effects of reactive atmospheres (e.g., carbon tetrachloride),
crucible materials (e.g., vitreous carbon), and raw materials
purity. Severe hydration of fluoride glasses by liquid water
can cause an absorption band at 6.1 microns. The overtones of
heavy metal-oxygen vibrations (whose fundamentals occur in
further in the IR) give rise to "oxide impurity" shoulders in
the IR edge at wavelengths of about 7.4 microns in
fluorozirconate glasses, and 8.7 microns in BZnYbT-type
materials.

The elimination of hydroxyl contamination from HMFG has

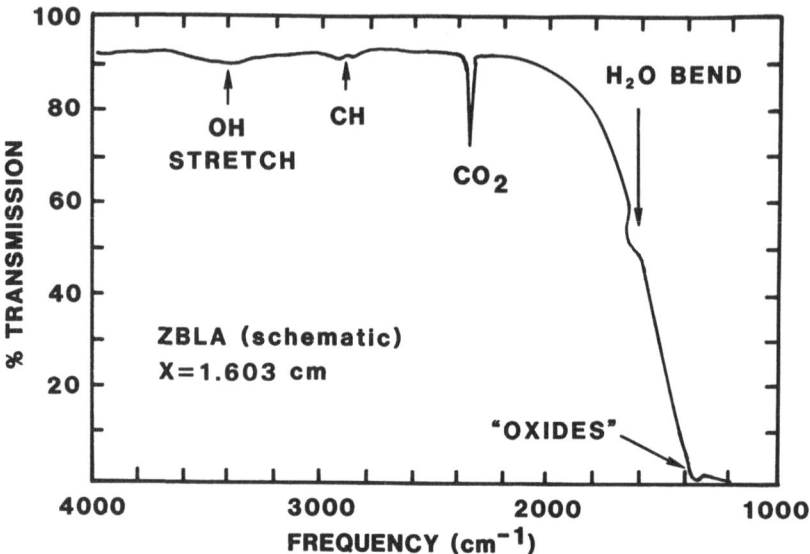

FIGURE 7. Schematic illustration showing location of IR impurity bands which could be observed in a fluorozirconate glass.

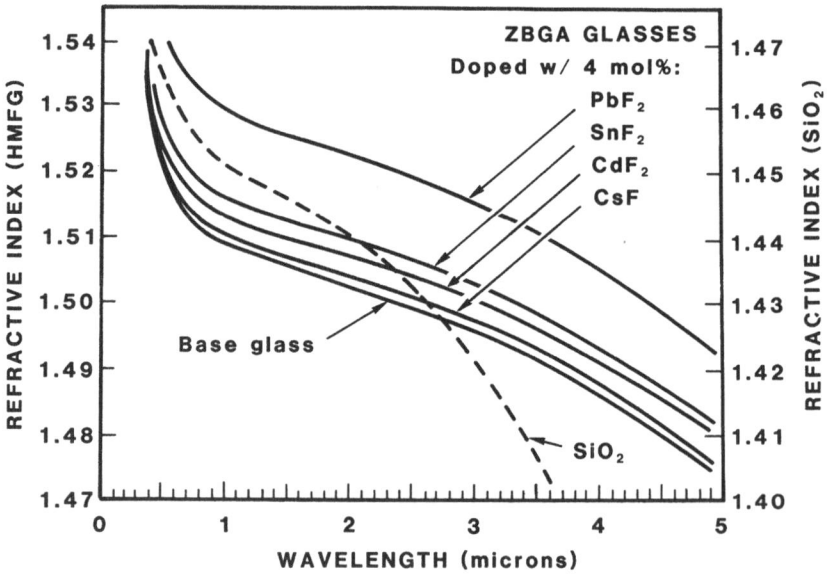

FIGURE 8. Refractive index for several doped ZBGA glasses and fused silica (courtesy Seiko Mitachi).

been the subject of considerable research activity since their discovery. To some extent, these investigations have been hampered by a lack of certainty concerning the extinction coefficient for hydroxyl species in HMFG, i.e., what concentration of OH produces a given 2.8 micron peak intensity. Three broad approaches to the reduction of OH content have been explored. Melting of the glass under extremely dry inert atmospheres has been shown to produce fibers with reduced hydroxyl peaks relative to materials produced under more humid atmospheres (27). The dehydrating effects of relatively high melting temperatures (above 900 Centigrade) and long melting times (several hours) have also been documented (28, 29). The latter method has yielded glasses with OH peak intensities of 20 dB/km (compared to several thousand dB/km in earlier work); multiple-Gaussian fits to this peak suggest contributions of less than 0.01 dB/km from hydroxyl species near 2.55 microns, where an impurity-free transmission "window" appears to exist (28). A possible drawback to the high temperature melting approach is the very high volatility of zirconium tetrafluoride above 850 Centigrade; the initial glass compositions must be adjusted accordingly.

A third approach to hydroxyl removal/reduction involves the use of reactive gas atmospheres during melting, a technique first described by Robinson and co-workers who employed carbon tetrachloride entrained in argon (30). Since then, a wide variety of other gaseous species have been examined for their dehydrating ability, including sulfur hexafluoride (31), nitrogen trifluoride (19), and carbon disulfide (8). Reactive atmosphere techniques have advantages and drawbacks, and quantitative studies of the efficiencies of the various candidates remain to be done. There is good evidence that the intensities of hydroxyl and "oxide" impurity bands are reduced. Improvements in light scattering behavior (19) and extension of IR transmission (8) have also been claimed. On the other hand, too generous a use of reactive atmospheres can lead to sulfur or carbon deposits in the glass, and decomposition products (e.g., chlorine) can effect the refractive index, radiation tolerance, carbon dioxide content, and possibly the crystallization behavior of HMFG. Alternatives to the use of halide-containing gases include treatment of the melt with dry oxygen (to alter the oxidation state of transition metal impurities) and/or the addition of small amounts of ammonium bifluoride whose decomposition products can convert residual oxide impurities to fluorides.

8. REFRACTIVE INDEX AND DISPERSION

The refractive index and dispersion characteristics of HMFG are particularly relevant to the design of optical waveguides and bulk optical components. The sodium-D line indices of fluoride glasses (Fig. 1) are generally in the range of 1.48-1.54, and thus compare favorably with values observed in many multicomponent silicate glasses. Detailed dispersion data for many compositions are now available (e.g., Refs. 32, 33), though care must be exercised when comparing similar glasses prepared with or without reactive atmospheres containing chlorine, whose inclusion tends to increase the index.

Many fluorozirconates and fluorozirconate glasses have high Abbe numbers (typically 75-90) compared with fused silica (appx. 65). The Abbe number describes the ratio of refractivity to dispersion, and a high value indicates more or less equal refraction at all wavelengths. Izumitani and co-workers (22) have investigated the reasons for the "abnormal" partial dispersion tendencies (i.e., high Abbe number) of HMFG by examining their absorption spectra in the extreme ultraviolet, thus extending the data of Fig. 4 to photon energies of 35 eV (0.035 microns). In addition to those shown in Fig. 4, they observed strong peaks in the reflection spectra near 12.9, 18.9 and 22.8 eV. The refractive index of a material may be modeled by the Lorentz equation, the individual terms of which reflect the contributions to the observed index of various fundamental absorptions (and their oscillator strengths) in the far to near UV and the far IR (see, e.g., Refs. 22, 23). The Abbe number, when expressed in terms of the Lorentz equation, shows that high values are favored in materials with short UV wavelength fundamental absorptions. High oscillator strengths at very short wavelengths also contribute to large Abbe numbers in the visible. The reflectance peaks observed by Izumitani and co-workers in the extreme UV, which in general are stronger and at shorter wavelengths than those found in silica, are thus thought to account for the "abnormal" Abbe values.

In contrast to fused silica, the high Abbe numbers for HMFG manifest themselves as relatively "flat" index versus wavelength curves. This is illustrated in Fig. 8 for several zirconium-barium-gadolinium-aluminum (ZBGA) type glasses (courtesy of Seiko Mitachi, NTT Corp.). At long wavelengths, the term in the Lorentz expression which accounts for the far-IR vibrational absorption bands becomes important, and the index of silica decreases rapidly. The fluoride glasses, whose "lattice" vibrations are farther in the IR exhibit a proportionate "lengthening" of the index versus wavelength curve.

The alterations in index which accompany the inclusion of various dopants in the ZBGA base glass composition are also shown in Fig. 8. All the species shown increase the index of the base glass, in good agreement with the molar refractivities determined for these materials by Sun (34) in his studies of fluoroberyllate glasses. To prepare optical waveguides, it is desirable to have dopants which also decrease the index, particularly in comparatively stable glasses such as the ZBLAN composition. This can be accomplished either by partially replacing Zr with Hf, or by introducing NaF at the expense of barium fluoride (4). The effect of NaF is stronger, producing a change in index of about 1 part in the third decimal place per mol %, while hafnium substitution alters the index by approximately 2.5 parts in the fourth decimal place/mol %.

9. RADIATION EFFECTS IN FLUORIDE GLASSES

In certain applications envisioned for heavy metal fluoride glasses and fibers, the materials may be exposed to varying levels of ionizing radiation. Radiation effects have been extensively studied in silicate optical waveguides, and the observed behavior exhibits a complex dependence on composition, impurity content, radiation dose, dose rate, and temperature, among other factors. With these provisos in mind, it has been hypothesized that HMFG should be more tolerant to the effects of ionizing radiation than other glasses as their mid-IR operating wavelengths are far removed from the (radiation-induced) UV/visible color centers which give rise to optical absorption. This indeed appears to be the case in fibers and bulk glasses at wavelengths in excess of about 3 microns. Figure 9, for example, shows the loss spectrum for a fluorozirconate fiber (provided by Le Verre Fluore and measured at the author's laboratory) before and after a radiation dose of 20 kilorads. In the vicinity of the hydroxyl absorption peak and beyond, the pre- and post-irradiation spectra are indistinguishable within experimental error. At longer wavelengths (e.g., 6-10 microns), the IR edges observed in bulk HMFG also appear to exhibit a similar degree of radiation "hardness" (35). It has been suggested that the short wavelength radiation tolerance of fluorozirconate compositions can be improved via addition of 3d ions or lead fluoride to the glasses (36).

Some of the defects responsible for radiation-induced absorption in the near-infrared have been identified and characterized by workers at Oklahoma State University (37, 38) and the Naval Research Laboratory (39-41). Some general conclusions from these studies are illustrated by the data of Fig. 10, which shows the optical absorption induced in a ternary fluorozirconate glass (ZBL) and a BZnYbT material at 80K and 300K by a dose of approximately 25 Mrads (37). Both glasses were prepared using a carbon tetrachloride ("Cl") reactive atmosphere. The defects which give rise to UV coloration in the materials are produced by a radiolysis damage mechanism, and the effects are similar to those observed in alkali halide crystals. In the fluorozirconate (or hafnate) glasses, radiation results in the production of Zr^{3+} (or Hf^{3+}) defects which cause absorption near 0.5 microns. The long tail associated with this absorption extends to 2 microns in the IR, where its magnitude is near 1000 dB/km for the dose of Fig. 10 (38). It is this band which accounts in part for the short wavelength losses observed in Fig. 9. At room temperature, the damage in BZnYb(or Lu)T glasses is quite weak compared to that in fluorozirconate materials, though significant absorption can arise at low temperatures. Annealing of the radiation damage occurs in both types of glasses, with the BZnYbT-type recovering at much lower temperatures.

Figure 10 also shows the deleterious effects of carbon tetrachloride reactive atmospheres on radiation effects; its use can lead to the incorporation of about 2-5 wt % chlorine into the glasses. The defects produced by this species have been extensively studied using optical spectroscopy and resonance techniques (37-41); their long absorption tails can also significantly effect mid-IR transparency. It is clear from the above discussion that the radiation tolerance of HMFG is improved significantly by utilizing compositions free of zirconium or hafnium tetrafluoride, and that it is best to

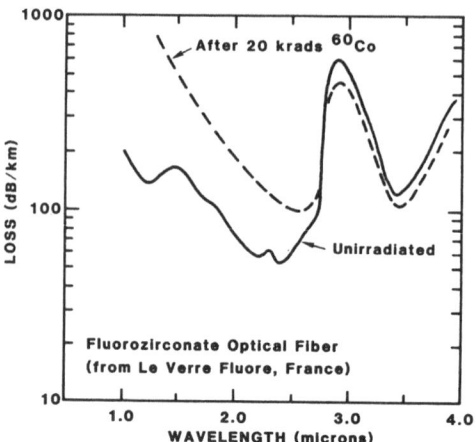

FIGURE 9. Pre- and post irradiation spectra of a fluorozirconate fiber; the radiation dose was 20 kilorads.

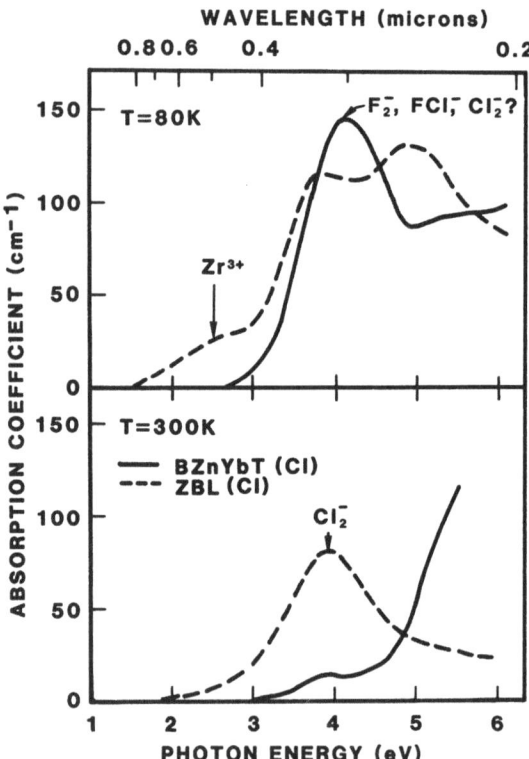

FIGURE 10. Radiation induced absorption in bulk fluoride glasses (after Ref. 37).

avoid the use of chlorine-containing atmospheres for hydroxyl removal. A definitive understanding of the radiation effects "question", however, awaits the preparation of long lengths of extremely low loss fiber.

10. AREAS FOR COOPERATIVE RESEARCH

The promise of high transparency in HMFG offers a host of applications, and also suggests productive areas for cooperative research among university, industry and government laboratories in the NATO countries. It would be of interest, for example, to clearly define the position of the intrinsic ultraviolet edges in these materials, understand in more detail the role played by impurities, and examine the dependence of UV absorption on glass structure and composition. In the mid-IR, quantitative determination of the extinction coefficient for hydroxyl species would allow a better definition of the effects of various glass processing procedures. A related effort could involve a systematic evaluation of the merits and efficiencies of various reactive atmospheres, including their relationship to the refractive index. The preliminary results of Sec. 4 concerning Stimulated Brillouin Scattering and thermal effects on various light scattering parameters are also worthy of more extensive examination.

REFERENCES
1. Miyashita, T. and Manabe, T., IEEE J. Quantum Electron. QE-18, 1432 (1982).
2. Drexhage, M.G., pp. 151-243 in "Treatise on Materials Science Science and Technology Vol. 26: Glass IV" (Academic Press, NY, 1985).
3. Tran, D.C., Sigel, G.H., and Bendow, B., J. Lightwave Tech. LT-2, 566 (1984).
4. Tokiwa, H., Mimura, Y., Shinbori, O. and Nakai, T., J. Lightwave Tech. LT-3, 569 (1985).
5. Tokiwa, H., Mimura, Y., Shinbori, O. and Nakai, T., J. Lightwave Tech. LT-3, 574 (1985).
6. Ohsawa, K. and Shibata, T., J. Lightwave Tech. LT-2, 602 (1984).
7. SpecTran Corp., Sturbridge, MA, USA; Le Verre Fluore, St. Erblon, France; Furukawa Electric Co. Ltd., Tokyo, Japan.
8. Lucas, J., Tregoat, D., El-Houari, A. and Fonteneau, G., Proc. SPIE 484, 36 (1984).
9. Drexhage, M.G., El-Bayoumi, O., Moynihan, C.T., Bruce, A.J., Chung, K.H., Gavin, D.L. and Loretz, T., J. Am. Ceram. Soc. 65, C168 (1982).
10. Drexhage, M.G., Mater. Sci. Forum 5, 1 (1985).
11. Suscavage, M.J., Hutta, J.J., Drexhage, M.G., Perrazo, N., Mossadegh, R. and Moynihan, C.T., Mater. Sci. Forum 5, 35 (1985).
12. Drexhage, M.G., Hutta, J.J., Suscavage, M.J., Mossadegh, R. and Moynihan, C.T., Mater. Sci. Forum 6, 509 (1985).
13. Moore, M.W., Carter, S.F., France, P.W. and Williams, J.R., Mater. Sci. Forum 6, 525 (1985).
14. see Tran, D.C., "Light Scattering in Halide Glass Fibers", this volume.
15. Schroeder, J., Tsoukala, V., Staller, C., Bruce, A.J.,

234

Moynihan, C.T., Hutta, J.J., Suscavage, M.J. and Drexhage, M.G., Electron. Lett. 20, 860 (1984).

16. Schroeder, J., Fox-Bilmont, M., Pazol, B., Tsoukala, V., Drexhage, M.G. and El-Bayoumi, O., Proc. SPIE 484, 61 (1984).

17. Schroeder, J., Floudas, J.A., Stiller, M.A. and Drexhage, M.G., Mater. Sci. Forum 6, 577 (1985).

18. Schroeder, J., Tsoukala, V., Bruce, A.J., Staller, C.O., Hutta, J.J., Suscavage, M.J. and Drexhage, M.G., Mater. Sci. Forum 6 561 (1985).

19. Nakai, T., Mimura, Y., Tokiwa, H. and Shinbori, O., J. Lightwave Tech. LT-4, 87 (1986).

20. Cotter, D., J. Opt. Commun. 4, 10 (1983).

21. Dumbaugh, W.H. and Morgan, D.W., J. Non-Cryst. Solids 38-39, 211 (1980).

22. Izumitani, T. and Hirota, S., Mater. Sci. Forum 6, 645 (1985).

23. Hirota, S., Izumitani, T. and Onaka, R., J. Non-Cryst. Solids 72, 39 (1985).

24. Sigel, G.H., pp. 5-89 in "Glass I: Interaction With Electromagnetic Radiation" (Academic Press, NY, 1977).

25. Pinnow, D.A., Rich, T.C., Ostermayer, F.W., and DiDomenico, M., Appl. Phys. Lett. 22, 527 (1973).

26. see Poignant, H., "The Role of Impurities in Halide Glass" and Reisfeld, R., "Optical Properties of Rare Earth and Transition Metal Ions in Fluoride Glasses", this volume.

27. Ohishi, Y., Mitachi, S. and Takahashi, S., Mater. Res. Bull. 19, 673 (1984).

28. France, P.W., Carter, S.F., Williams, J. and Beales, K.J., Electron. Lett. 20, 607 (1984).

29. Maze, G., Cardin, V. and Poulain, M., J. Lightwave Tech. LT-2, 596 (1984).

30. Robinson, M., Pastor, R.C., Turk, R.R., Devor, D.P., Braunstein, M. and Braunstein, R., Mater. Res. Bull. 15, 735 (1980).

31. Tran, D.C. and Fisher, C., US Patent 4,539,032.

32. Brown, R.N. and Hutta, J.J., Appl. Opt. 24, 4500 (1986).

33. Mitachi, S. and Miyashita, T., Appl. Opt. 22, 2419 (1983).

34. Sun, K.H., Glass Tech. 20, 36 (1979).

35. Rosiewicz, A. and Gannon, J.R., Electron. Lett. 17, 184 (1981).

36. Cases, R., Alcala, R. and Tran, D.C., Mater. Sci. Forum 6, 755 (1985).

37. Yeh, D.C., Sibley, W.A., Suscavage, M.J. and Drexhage, M.G., "Radiation Effects and Optical Transitions in Ytterbium-Doped Fluoride Glass", submitted to J. Appl. Phys.

38. Tanimura, K., Sibley, W.A., Suscavage, M.J. and Drexhage, M.G., J. Appl. Phys. 58, 4544 (1985).

39. Cases, R., Griscom, D.L. and Tran, D.C., J. Non-Cryst. Solids 72, 51 (1985).

40. Friebele, E.J. and Tran, D.C., J. Non-Cryst. Solids 72, 221 (1985).

41. Griscom, D.L. and Tran, D.C., J. Non-Cryst. Solids 72, 159 (1985).

DISCUSSION

C: R. Almeida. You said that the intermediate frequency band of the IR reflectivity spectra may be due to Ba-F stretches. However, I have shown that these same glasses, with Sr or Pb substituting for Ba, and in which the atomic masses and force constants are widely different, still give you the same vibrational frequencies. Therefore, this is a fundamental Zr-F stretching mode, not due to Ba-F vibrations.

Q: A. Bruce. Are the fluorohafnate glasses more resistant to radiation damage than the fluorozirconate glasses?

A: M. Drexhage. No. Their behavior is identical to that of ZrF_4-based materials, except that Hf^{3+} deffects may be formed. Details and spectra are given in ref. 38 of the present paper.

Q: J.D. Mackenzie. You have given an excellent review of the optical properties of these glasses. However there was no comparison with single crystals. Are there any optical data available for these?

A: M. Drexhage. This is a good suggestion, which I will attempt to follow up in the future. Single crystal data are available for some constituents: BaF_2, ThF_4 and LaF_3, for example. Data for the primary ingredient - ZrF_4 - is still unavailable due to lack of means to prepare large single crystals. However, in M. Robinson's paper, he has described a simple way to achieve this and optical data are expected to be available shortly.

OPTICAL PROPERTIES OF RARE-EARTH AND TRANSITION METAL IONS IN FLUORIDE GLASSES.

RENATA REISFELD

DEPARTMENT OF INORGANIC AND ANALYTICAL CHEMISTRY
THE HEBREW UNIVERSITY OF JERUSALEM,JERUSALEM 91904,ISRAEL.

1. ABSTRACT

The absorption spectra of chromium(III) and nickel(II) at octahedral sites in a zirconium barium fluoride glass are analyzed and compared with vitreous and crystalline mixed oxides,suggesting slightly longer Cr-F and Ni-F distances than in crystalline fluorides. Laser properties of rare earth ions in fluoride glass(ZBLA) having composition of $57ZrF_4,34BaF_2,$ $5LaF_3,4AlF_3$, or (PBLA) having composition of $36PbF_2,24ZnF_2,35GaF_3,3YF_3,$ $2AlF_3$ for $^3Nd(III)$ were calculated. Pumping efficiencies, ^{2}lifetimes of excited and terminal levels and stimulated peak cross-sections are presented. It is shown that multiphonon relaxations (in the case of energy difference less than 3000 cm^{-1}) are always orders of magnitude lower in the fluoride glass than in the tellurite glass and the inhomogeneous width is consistently smaller in fluoride glass than in tellurite glass,which was chosen as a model for oxide glass. No significant difference was found for concentration quenching between the two kinds of glasses. Energy transfer between Mn(II) and Nd(III) can increase pumping efficiencies of Nd(III) lasers.

2. INTRODUCTION

Fluoride glasses containing about 50 mole% of ZrF_4 which can be replaced by HfF_4 [1-3], colloquially called ZBLA glass ,have been considered as materials for fiber optics in the range of 0.3-5 µm [4]. Another important category of fluoride glasses containing zinc(II) (or manganese), gallium (III) and lead(II) fluorides was invented [5,6] at the University of Maine, Le Mans (PBLA glass). The absorption spectra and luminescence of $4f^{11}$ erbium(III) and $3d^5$ manganese (II), and the mutual energy transfer between excited states of these two species were studied in such a glass [7]. In ZBLA glass, the luminescence of $4f^2$ praseodymium(III) [8,9], $4f^6$ europium(III) [10,42], $4f^{10}$ holmium(III) [11,12] and erbium(III) [13-15] occurs from more excited J-levels than usual,the lower limit still allowing perceptible luminescence for the energy gap between the emitting J-level and the closest lower-lying J-level at 2000 cm^{-1} (0.25 eV), some 2 to 4 times smaller than in nearly all other glasses and crystals. It may be noted that this rich emission spectrum to several J-levels besides the groundstate is observed also at room temperature.

The superior optical characteristics of fluoride glasses for IR fiber optic applications also provide an ideal medium or host enabling the glass to be integrated into a system acting as a laser light source as well as the actual waveguide material. The spectroscopic and fluorescent properties of Nd(III) containing fluorozirconate glasses have been reported by Weber in work with the Lucas-Poulain group at Rennes [16] and

238

optical absorption of 3d transition metal ions such as Fe, Co, Ni and Cu
by Ohishi et al.[17]. Luminescence and nonradiative relaxation of rare
earths in amorphous fluoride materials has recently been reviewed [18].

3. CHROMIUM(III) AND NICKEL(II) IN ZBLA GLASSES

The absorption spectra of Cr(III) and Ni(II) at room temperature are
given in Fig 1. The observed band peaks, and ligand field and Racah
parameters are presented in Table I.

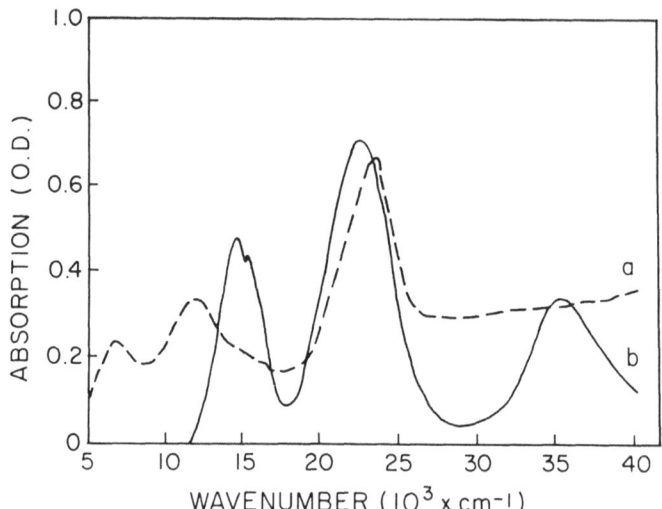

FIGURE 1. Absorption spectra of a) Cr(III) and b) Ni(II) in ZBLA
($56.75ZrF_4$, $34.25BaF_2$,$4.5LaF_3$,$4AlF_3$,$0.5(NiF_2,CrF_3)$) at room temperature.

From the absorption spectra of Cr(III) and Ni(II) in ZBLA glass, both
in the work of Reisfeld et al.[19] and in that of Ohishi et al. [17], as
well as from the absorption spectra of these ions in PBLA glass [5], it
can be demonstrated that the site symmetry for both Cr(III) and Ni(II) in
these glasses is close to cubic octahedral symmetry. Sites of lower
symmetry would have relatively much stronger absorption bands. Such a
situation could be predicted for PBLA glass in view of the feasible
substitution of Zn(II) and Ga(III) by Cr(III) of comparable radius.

Also it is shown that almost all Cr(III) and a large majority of all
paramagnetic Ni(II) complexes in solution as well as solid compounds show
the coordination number N = 6 with octahedral symmetry. However it is not
perfectly trivial that Cr(III) and Ni(II) in ZBLA glass (known from Raman
spectra [20] to have more complicated coordination behaviour) turns out to
be octahedral to a high approximation. The sub-shell energy difference dq
(also designated as Δ) corresponds to the maximum (or strictly to the
center of gravity) of the first spin-allowed transition.The Racah
parameter [21,22] of interelectronic repulsion B is derived from the
diagonal sum rule. Also it is known that:

$$B=(\sigma_2+\sigma_3-3\sigma_1)/15 \tag{1}$$

Such a derivation is rarely possible in Cr(III) because the third spin-allowed transition is usually hidden by electron transfer bands or other intense absorption. In such Cr(III) cases, B can be derived [23] from σ_1 and σ_2 alone, providing B = 850 cm^{-1} for Cr(III) in ZBLA and PBLA. The nephelauxetic ratio β is the ratio between β from Eq. (1) and B_0 for the gaseous ion, 918 cm^{-1} for Cr^{+3} and 1041 cm^{-1} for Ni^{+2}. It is interesting to compare the parameters of Table I with related materials (vitreous and crystalline oxides) which were compiled for 36 Cr(III) cases [24].

The value for dq of Cr(III) in ZBLA glass is distinctly lower than 16100 cm^{-1} reported [25] for the cubic elpasolites $K_2NaGa_{0.95}CrO_{0.05}F_6$ and K_2NaCrF_6 suggesting 1.5 percent (0.03 A) longer average Cr-F distances in the glass than in the crystal, as discussed below for analogous Ni(II) cases. The parameters in Table I are closer to CrF_6^{-3} in solution [21] having dq=15200 cm^{-1} and B=820 cm^{-1} according to Claus Shäffer. They fall inside the intervals dq=14500 to 16400 cm^{-1} and B=620 to 850 cm^{-1} given [26] for Cr(III) in 14 highly different mixed oxide glasses, and may also be compared with dq=17450 and B=725 cm^{-1} for $Cr(OH_2)_6^{+3}$. The first absorption band of the fluorides and many oxide cases, shows a complicated structure because the first two doublet levels ^2E and $^2T_1$ almost coincide with $^4T_2$ providing additional complications of spin-orbit coupling. The most prominent narrow peak occurs at 654 nm (15300 cm^{-1}) in our ZBLA glass, to be compared with 15430 cm^{-1} in a zirconium-barium-thorium fluoride glass [26], which should represent the position of ^2E to a good approximation.

Since $^4T_2$ stretches distinctly well below ^2E, one expects any luminescence to be a broad-band transition between the two lowest quartet levels. Only very weak fluorescence of Cr(III) was seen in fluorophosphate glass [26] and in ZBLA [19]. Lifetimes as well as the peak emissions of Cr(III) in ZBLA glass are presented in Table II. The short lifetimes form a striking contrast, not only to the cubic elpasolites [25] with temperature- dependent lifetimes in the range 0.2 to 0.6 msec, but also to Cr(III) in a lithium-lanthanum phosphate glass [27] with lifetimes around 0.02 msec (0.025 msec at the same low Cr(III) concentration as in the ZBLA glass) and a quantum yield up to 0.23. Much higher quantum yields are observed in glass-ceramics containing crystallites (much smaller than 400 nm) of spinel-type $MgAl_{2-x}Cr_xO_4$ and the isotypic gahnite $ZnAl_{2-x}Cr_xO_4$ [28] and other types [24,29,30]. Such glass-ceramics may be useful as laser materials, conceivably replacing the crystalline alexandrite $Al_{2-x}Cr_xBeO_4$.

The dq value for nickel(II) in ZBLA glass is unusually small when compared with 8800 NiO; 8650 $Ni_xMg_{1-x}O$; 8500 $Ni(OH_2)_6^{+2}$; 7400 NiTiO$_3$; and 7300 $Ni_xMg_{1-x}TiO_3$ (all values [21] in cm^{-1}). It is particularly interesting to compare with crystalline fluorides [21,31] such as 7800 spinel-type Li_2NiF_4, 7700 rutile-type NiF$_2$; 7500 perovskite-type KNiF$_3$ (B=950 and 960 cm^{-1} in the two latter compounds to be compared with 940cm^{-1} in $Ni(OH_2)_6^{+2}$ and 840cm^{-1} in $Ni_xMg_{1-x}TiO_3$). Rudorff, Kändler and Babel [31] pointed out that such variations can be ascribed to slightly varying internuclear distances R. In this perspective ZBLA seems to have Ni-F distances on the average 1.6% (0.03A) longer than crystalline KNiF$_3$. In mixed oxides more dramatic effects can occur, dq of Ni(II) being decreased to 6000 cm^{-1} in ilmenite-type $Ni_xCd_{1-x}TiO_3$ (isotypic with NiTiO$_3$ and MgTiO$_3$) and, as shown by Reinen, to only 4800 cm^{-1} in the perovskite (elpasolite superstructure?) $Ba_2Ca_{1-x}TeNi_xO_6$ [21]. However in such substituted crystals (like in the classical case of ruby $Cr_xAl_{2-x}O_3$) a weak doubt always remains whether the distance M-X between M carrying a partly filled shell and the closest neighbour atoms X fully adapts to the internuclear distances in the closed-shell host lattice. More convincing evidence comes from the hydrostatic high pressure (especially for atoms [32] on special positions) but substituted crystals remain the only technique of significantly increasing R. A direct determination of R (with a precision of about 0.02A) is accessible to EXAFS using the X-ray absorption edge of the substituting M even in low concentration [33].

TABLE I:Optical transitions of Cr(III) and Ni(III) in fluoride glasses

ION	TRANSITION ASSIGNMENT	BAND POSITION OBSERVED[cm^{-1}]		PARAMETERS [cm^{-1}]	
		[17]	[19]	[17]	[19]
Cr(III)	$^4A_{2g}(^4F)$--$^4T_{2g}(^4F)$	14749	14800		
" "	$^4A_{2g}(^4F)$--$^2E_{1g}(^2G)$	15385	-	Dq=1475	Dq=1480
" "	$^4A_{2g}(^4F)$--$^4F_{1g}(^4F)$	22472	22500	B=847	B=850
" "	$^4A_{2g}(^4F)$--$^4T_{1g}(^4P)$	34483	34700	C=3136	
Ni(II)	$^3A_{2g}(^3F)$--$^3T_{2g}(^3F)$	6536	6900		
" "	$^3A_{2g}(^3F)$--$^3T_{1g}(^3F)$	11364	11500	Dq=663	Dq=690
" "	$^3A_{2g}(^3F)$--$^3E_{1g}(^1D)$	14925	-	B=956	B=970
" "	$^3A_{2g}(^3F)$--$^3T_{1g}(^3P)$	22936	23300	C=4006	

Nephelauxetic parameter β in ZBLA glass [19] is 0.926 for Cr(III) and 0.932 for Ni(II)

TABLE II:Lifetimes of Nd(III)(876 nm emission) and Cr(III)(~800nm emission) in ZBLA glass.

DOPANT 1	DOPANT 2	EXCITATION λ ,[nm]	EMISSION λ, [nm]	DECAY,[μsec]		
				τ_1	τ_2	τ_3
—	0.5 Cr	465	797	1.4	3.0	6.0
0.5 Mn	0.5 Cr	414	790	0.4	1.0	3.1
0.5 Nd	0.5 Cr	337	804	0.6	1.2	3.1
0.5 Mn	0.5 Cr	337	804	0.5	1.9	3.9
0.5 Nd	0.5 Cr	450	876	381	410	410
0.5 Nd	0.5 Cr	579	876	380	405	—

The only possibility of luminescence of Ni(II) in ZBLA glass would be at the foot (some 6000 cm^{-1}) of the first absorption band, but we did not detect any. The spin-forbidden absorption band due to the first singlet level ^1E corresponds to the rather broad shoulder at 15000 cm^{-1} (see Fig.1) comparable to the peak [31] of crystalline Ni(II) fluorides between 15000 and 15400 cm^{-1}.

4. NEODYMIUM(III) IN FLUORIDE GLASSES:

4.1. Intensity parameters and radiative transitions of Nd(III)

The absorption spectrum of Neodymium(III) serves as a basis for a complete set of predictions of transition rates within the $4f^3$ configuration of Nd(III). The procedure is based on the theory of Judd - Ofelt and is described in detail elsewhere [14]. Here we describe briefly the main steps of its evaluation.

According to the theory the otherwise forbidden transitions within the f-f configuration of rare-earths become slightly allowed by admixing of wavefunctions of the f-f configuration with odd components of crystal field potential. The intraconfigurational transitions then become subject to a new set of selection rules and oscillator strengths of the transitions depend parametrically on the three phenomenological parameters Ω_2, Ω_4, Ω_6. Reduced matrix elements for the transitions are almost invariant in respect to the crystal field strength [34] and were tabulated for all rare earths ions [35].

The optical transitions of rare earths in solids are predominantly of electric dipole character and their spectral intensities can be described using the treatment of Judd and Ofelt. In this approach the line strength S of a transition between two J states is given by the sum of products of empirical intensity parameters Ω_t and matrix elements of tensor operators $U^{(t)}$ of the form

$$S(J,J') = e^2 \sum_t \Omega_t |\langle aJ||U^{(t)}||bJ'\rangle|^2 \qquad (2)$$

where t=2,4,6

The values of Ω_t are obtained from a least-squares fit of measured and calculated absorption line strengths and typically have an experimental uncertainty of about 10%. The integrated intensities of the absorption bands yield Ω's which are an effective average over the different rare earth environments in the glass.

The most significant factor determining the values is the strength of the odd-order terms in the expansion of the local field at the rare earth sites. These in turn are affected by the nearest-neighbour anion(s) and cations. For a given glass former systematic changes of Ω_t have been observed with the changes in the size and charge of network modifier ions [36].

The three omegas are determined by solving an overdetermined set of linear equations built by equating the measured oscillator strengths with the sum of products of the unknown omegas with the appropriate reduced matrix elements. The three omegas found from the solution are put into a computer program which calculates all the radiative transition rates possible in the system analyzed. The omega parameters for ZBLA and PBLA glasses are given in Table III. The oscillator strengths of Nd(III) are given in Table IV. Generally, Ω_2 is lower in fluoride than in oxide glasses, which is of no consequence for the $^4F_{3/2}$-$^4I_{11/2}$ laser transition for which $U^{(2)}$=0.

Table III: Omega parameters in ZBLA and PBLA glasses.

GLASS	REFERENCE	Ω_2, [pm^2]	Ω_4, [pm^2]	Ω_6, [pm^2]
PBLA	[19]	1.01±0.28	3.73±0.35	6.19±0.43
ZBLA	[19]	1.10±0.25	3.80±0.30	5.53±0.20
ZBLA	[19]	1.95±0.26	3.65±0.38	4.17±0.17

4.2 Fluorescent lifetimes and nonradiative transitions

The calculation of nonradiative transfer rates due to multiphonon decay is accomplished by subtracting the calculated radiative rates from the reciprocals of the integrated lifetimes of as many as possible energy levels of the rare earth ions and plotting the logarithms of the numbers against energy gaps between the levels and their nearest lower neighbour. Then the two parameters of the exponential multiphonon decay rate law are calculated: the exponential parameter α, from the slope of the plot and the electronic factor B, from its intercept [37].

Since only three energy levels of Nd(III) in fluoride glasses have lifetimes long enough to be measurable, the α and B were determined from 3 points only. Fortunately the result, which is α = 0.0053±0.0005 and B = $1.63\pm0.1\times10^{10}$, agrees well with other sets of data which were measured on Ho(III) in ZBLA glass in our laboratory [8]. The parameters are substituted into formula (3)

$$W_{nr} = B\exp[-\alpha\Delta E] \tag{3}$$

where ΔE is the energy gap from the electronic level to its next lower neighbour. The entire set of transition rates is calculated, now with the nonradiative transition rates included [37].

The result of such a procedure is shown in Table IV for Nd(III) in PBLA glass. The calculated oscillator strengths agree well with the measured values. The omegas calculated from our ZBLA glass compare well with the values obtained by Lucas et al.[16] in their study of Nd(III) in ZBLA glass. The last two columns in Table IV compare calculated lifetimes with the measured lifetimes. The outstanding property of fluoride glasses is the relatively long-lived luminescence from levels which are separated only by a small energy gap to the next lower level [12]. Here we are able to record and measure the lifetimes of emissions from two levels: thermalized $^4D_{3/2}$ (361nm) and $^2P_{3/2}$ (387nm). The first three emission lines are identified as belonging to the following transitions: $^4D_{3/2}$-$^4I_{9/2}$ (361nm), $^4D_{3/2}$-$^4I_{11/2}$ (381nm) and $^4D_{3/2}$-$^4I_{13/2}$ (412nm). The next, much weaker group of lines belong mainly to the transitions from $^2P_{3/2}$. The spectroscopic assignments of these lines are given in Table IV for Nd(III) in PBLA glass alone and in Table V for Nd(III) with Mn(II). The integrated lifetimes of these emissions are not influenced by the presence of Mn(II) ions. In both Tables, the first group of transitions having integrated lifetimes varying from 1.4 to 2.1 μsec belong to the transitions from the ^4D manifold. Its predicted lifetime is 0.97 μsec. The second group of lifetimes (15-20μsec) belongs to the transitions from $^2P_{3/2}$. Its predicted lifetime is 13.6 μsec. The predictions which are based on the theory of Judd-Ofelt combined with the exponential multiphonon law are not expected to give better agreement [8].

Table IV: Oscillator strengths and lifetimes of Nd(III) in PBLA glass:
($36PbF_2, 24ZnF_2, 35GaF_3, 2AlF_3, 3YF_3; 2LaF_3, 2NdF_3$)

TRANSITION	WAVELENGTH [nm]	OSCILLATOR STRENGTHS		LIFETIMES, [μsec]	
		CALCULATED	OBSERVED	CALCULATED	OBSERVED
$^4D_{1/2} \to {}^4I_{9/2}$	354}	9.04	4.27	–	0.00007
$^4D_{5/2} \to {}^4I_{9/2}$	357}	9.04	1.27	–	0.00009
$^4D_{3/2} \to {}^4I_{9/2}$	361	–	3.7	1.5	1.0
$^4D_{3/2} \to {}^4I_{11/2}$	381	–	11.5	1.4	1.0
$^4D_{3/2} \to {}^4I_{13/2}$	412	–	1.15	1.2	1.0
$^4D_{3/2} \to {}^4F_{5/2}$	637	–	5.10	1.70	1.0
$^2P_{3/2} \to {}^4I_{9/2}$	387	0.09	0.05	–	13.6
$^2P_{3/2} \to {}^4I_{11/2}$	419	–	1.00	19.0	13.6
$^2P_{3/2} \to {}^4I_{13/2}$	454	0.04	0.04	15.0	13.6
$^2P_{3/2} \to {}^4I_{15/2}$	502	–	0.01	18.0	13.6
$^2P_{3/2} \to {}^4H_{9/2}$	746	–	2.40	18.0	13.6
$^2P_{3/2} \to {}^4F_{9/2}$	886	–	2.40	18.5	13.6
$^2D_{5/2} \to {}^4I_{9/2}$	426-440	0.17	0.04	–	0.0008
$^2P_{1/2} \to {}^4I_{9/2}$		0.17	0.09	–	0.3000
$^4G_{11/2} \to {}^4I_{9/2}$	475}	1.38	0.21	–	0.00003
$^2D_{3/2} \to {}^4I_{9/2}$	478	1.38	0.35	–	0.00005
$^2G_{9/2} \to {}^4I_{9/2}$	483}	1.38	0.42	–	0.1000
$^4G_{9/2} \to {}^4I_{9/2}$	510}	1.38	1.50	–	0.0002
$^4G_{7/2} \to {}^4I_{9/2}$	523	6.98	2.70	–	0.2500
$^2G_{7/2} \to {}^4I_{9/2}$	577}	9.78	2.74	–	0.00015
$^4G_{5/2} \to {}^4I_{9/2}$	577}	9.78	7.30	–	0.0040
$^2H_{11/2} \to {}^4I_{9/2}$	620	0.18	0.18	–	0.0200
$^4F_{9/2} \to {}^4I_{9/2}$	678	0.37	0.66	–	0.0100
$^4F_{7/2} \to {}^4I_{9/2}$	732}	6.41	5.70	–	0.00004
$^4S_{3/2} \to {}^4I_{9/2}$	743}	6.41	0.03	–	0.0009
$^2H_{9/2} \to {}^4I_{9/2}$	797}	7.13	1.40	–	0.00006
$^4F_{5/2} \to {}^4I_{9/2}$	802}	7.13	6.20	–	0.0090
$^4F_{3/2} \to {}^4I_{9/2}$	860	1.92	2.20	190.0	337.00

The integrated lifetime of $^4F_{3/2}$ level of Nd(III) is 190 μsec in $36PbF_2,24ZnF_2,35GaF_3,2AlF_3,3YF_3,2LaF_3,2NdF_3$, 330 μsec in $36PbF_2,24MnF_2,35GaF_3,2AlF_3,3.8LaF_3,0.2NdF_3$ and about 400 μsec in $56.50ZrF_4,34.00BaF_2,4.5LaF_3,4AlF_3,0.5CrF_3,0.5NdF_3$ (Table II). Here the prediction, which is 450 μsec for ZBLA and 337 μsec for PBLA, is near the experimental result, while the short 190 μsec is due to the cross-relaxation mechanism seen below. A summary of lifetimes of Nd(III) ions in various samples is shown in Tables II,IV and V.

TABLE V.Neodymium in PBLA glass;$35PbF_2,24MnF_2,35GaF_3,AlF_3,(4-x)LaF_3,xNdF_3$

TRANSITION	X	WAVELENGTH ,[nm]		LIFETIMES,[μsec]				RISETIME	
		EXCITATION	EMISSION	τ_1	τ_2	τ_3	τ_{in}	SHORT	LONG
$^4D_{3/2}-^4I_{9/2}$	0.2	337	360	1.4	1.5	1.8	1.5	<0.5	-
$^4D_{3/2}-^4I_{11/2}$	0.2	337	381	1.8	2.0	2.3	2.0	<0.5	-
?	0.2	337	810	471	-	-	471	0.60	44
?	0.2	407	810	614	-	-	614	0.60	49
$^4F_{3/2}-^4I_{9/2}$	0.2	337	865	567	507	499	500	1.5	70
$^4F_{3/2}-^4I_{9/2}$	0.2	407	865	632	642	-	637	1.5	85
$^4F_{3/2}-^4I_{9/2}$	0.2	579	865	265	327	345	330	-	7.0
$^4D_{3/2}-^4I_{9/2}$	0.2	337	361	1.3	1.5	1.7	1.5	<0.5	-
$^4D_{3/2}-^4I_{11/2}$	0.2	337	381	1.2	1.4	1.8	1.5	<0.5	-
$^2P_{3/2}-^4I_{11/2}$	2.0	337	419	18.0	20.0	21.0	20.0	<0.5	1.5
$^4D_{3/2}-^4I_{15/2}$	2.0	337	499	19.0	18.0	18.0	18.0	<0.5	1.5
$^4D_{3/2}-^2F_{5/2}$	2.0	337	637	1.8	-	-	1.8	<0.5	-
$^2P_{3/2}-^2H_{9/2}$	2.0	337	743	18.0	18.0	20.0	18.0	<0.5	1.5
$^4F_{3/2}-^4I_{9/2}$	2.0	337	867	176	215	238	220	-	3.0
$^4F_{3/2}-^4I_{9/2}$	2.0	407	867	180	210	240	220	-	3.0
$^4F_{3/2}-^4I_{9/2}$	2.0	579	867	172	192	210	196	-	3.0
$^2P_{3/2}-^2F_{9/2}$	2.0	337	886	18.0	24.0	28.0	25.0	<0.5	2.5

4.3. Cross-relaxation

Special cases of energy transfer are cross-relaxations when the original system loses the energy (E_3-E_2) by obtaining the lower state E_2 (which may also be the groundstate E_1) and another system aquires the energy by going to a higher state E_2 [37]. Cross-relaxation may take place between the same lanthanide (being a major mechanism for quenching at higher concentration in a given material) or between two differing elements, which happen to have two pairs of energy levels separated by the same amount. The cross- relaxation between a pair of rare earth ions is graphically presented in Fig.3 of [37].

The measured lifetime of luminescence is related to the total relaxation rate by

$$1/\tau = \Sigma W_{nr} + \Sigma A + P_{cr} = 1/\tau_o + P_{cr} \qquad (4)$$

where ΣA is the total radiative rate, ΣW_{nr} is the nonradiative rate, P_{cr} is the rate of cross-relaxation between adjacent ions and τ_o is intrinsic lifetime.

The critical radius R_o for cross-relaxation is defined by

$$P_{cr}(R_o) \cdot (1/\tau_o) = 1 \qquad (5)$$

R_o being the critical distance at which the probability for cross-relaxation P_{cr} equals the sum of radiative and multiphonon relaxations.

The cross-relaxation channel for Nd(III) is $(^4F_{3/2}) + (^4I_{9/2}) \rightarrow 2(^4I_{15/2})$. The critical radii in various glasses are presented in Table VI.

TABLE VI. Critical radii for cross-relaxation of Nd(III) in tellurite and fluoride glasses

COMPOUND	CONCENTRATION $[10^{20}/cm^3]$	CRITICAL RADIUS[A]	$\tau_{intrinsic}$ [μsec]	$\tau_{measured}$ [μsec]	EXCITATION λ,[nm]
0.5/ZnTe	1.10	4.74±.11	187	178	579
1.6/ZnTe	3.50	5.07±.36	187	130	579
2.7/ZnTe	5.80	5.81±.25	187	102	579
0.5/ZBLA	.85	3.72±.80	455	443	576
2.0/PBLA	4.02	5.07±.61	345	264	576

ZnTe— $35ZnO, 65TeO_2$

4.4 Laser action :

The formula for peak cross-section is [38]

$$\sigma = \frac{\lambda^4 A}{8\pi c n^2 \Delta\lambda} \quad [cm^2] \qquad (6)$$

where λ - emission wavelength [cm]
$\Delta\lambda$ - full width at half height of emission band [cm]
n - refractive index
A - radiative transfer probability [sec^{-1}]

Threshold power for transverse pumping is

$$P_{th} = \frac{hc(L_o + L_{res}) \cdot 10^{-7}}{2\lambda_p \tau_f lF\sigma\alpha_p} \qquad (7)$$

where L_{res} - resonant power loss due to self-absorption at the laser wavelength which is defined as

$$L_{res} = 2l\sigma N\beta_y/Z \qquad\qquad (8)$$

where N - number density of lasing lines

β_y - Boltzmann factor for the terminal laser level

Z - partition function

l - length of laser

In Nd(III) the terminal level, $^4I_{11/2}$ for the 1060nm luminescence is positioned at ≈ 2000 cm^{-1}, then $E/kT \approx 10$ at room temperature and the Boltzmann factor is $\approx 4.5\times10^{-5}$. For a 1 cm long minilaser at representative values of N and σ:

Lres $\approx$ 0.2-0.1 %.

L_o - Nonresonant loss which is mainly due to the absorption of the medium and loss at mirrors. It is usually taken to be 0 - 1.5%

λ_p - pumping wavelength. In our case it is 806nm of LED, having 25nm bandwidth.

τ_f - lifetime of lasing level.

F - Boltzmann population function of the lasing level. The $^4F_{3/2}$ of Nd(III) is split into 2 main bands in glasses. The fraction of the population at the lasing level (R_1) is 0.64 - 0.72R

α_p - absorption coefficient of the pumped level which in our case is centered at 800nm ($^4F_{5/2} + {}^2H_{9/2}$) and is obtained by dividing the optical density of the sample by its thickness.

Table VII presents the comparison of peak cross-section and threshold power for laser action of Nd(III) in fluoride, oxide and chalcogenide glasses for transverse pumping. From the Table it can be seen that the laser characteristics for Nd(III) in fluoride glasses are quite similar and even better than in ED-2 glass.

TABLE VII:Spectroscopic and Laser Properties of Nd(III) in Fluoride Glasses as Compared to Chalcogenide and Oxide Glasses.

Host	Assignment	Laser wavelen. [nm]	Conc. [cm^3] x10^{20}	Abs.coef. at 806nm [cm^{-1}]	$\Delta\lambda$ [nm]	σ [cm^2] x10^{-20}	P_{th} [W/cm^2] L_o=1% L_r=0.2%	P_{th} [W/cm^2] L_o=1% L_r=0%	τ_f [μsec]	Ref.
ZBLA	$^4F_{3/2}$-$^4I_{11/2}$	1049	2.72	3.14	26.7	2.9	57	-	400	16
PBLA	$^4F_{3/2}$-$^4I_{11/2}$	1039	4.02	3.57	33.0	2.75	112	-	190	19
PBLA	$^4F_{3/2}$-$^4I_{13/2}$	1306	4.02	3.57	65.0	0.85	-	256	190	19
ED-2	$^4F_{3/2}$-$^4I_{11/2}$	1060	1.83	1.27	27.8	2.9	173	-	300	38
ED-2	$^4F_{3/2}$-$^4I_{13/2}$	1340	1.83	3.14	64.4	0.72	-	590	300	38
ZnTe	$^4F_{3/2}$-$^4I_{11/2}$	1060	3.46	4.73	29.0	3.6	93	-	130	19
ZnTe	$^4F_{3/2}$-$^4I_{13/2}$	1340	3.46	4.73	73.0	0.76	-	367	130	19
GLS	$^4F_{3/2}$-$^4I_{11/2}$	1077	2.63	14.50	-	7.95	11.3	-	100	41
GLS	$^4F_{3/2}$-$^4I_{13/2}$	1370	2.63	14.50	-	3.60	-	27.7	100	41

ZBLA - 57.0ZrF$_4$,34.0BaF$_2$,3.0LaF$_3$,4.0AlF$_3$,2.0NdF$_3$ mole%

PBLA - 36PbF$_2$,24ZnF$_2$,35GaF$_3$,2AlF$_3$,3YF$_3$,2LaF$_3$,2NdF$_3$ mole%

ED-2 - 60SiO$_2$,27.5Li$_2$O,10CaO,2.5Al$_2$O$_3$,0.16CeO$_2$ mole%,2.012Nd$_2$O$_3$ wt%

Chalcogenide,GLS - 3Ga$_2$S$_3$,0.85LaS$_3$,0.15Nd$_2$S$_3$

Zinc Tellurite,ZnTe - 35ZnO,65TeO$_2$ mole%,2 Nd$_2$O$_3$ wt%

5. ENERGY TRANSFER FROM MANGANESE(II) TO ERBIUM(III) AND NEODYMIUM(III)

5.1. Energy transfer from Mn(II) to Er(III)

Energy transfer between Mn(II) and Er(III) in PBLA glass having composition $36PbF_2, 24(Mn,Zn)F_2, 35GaF_3, 5Al(PO_3,F)_3$ doped with Er(III) has been studied recently [7]. The emission of Mn(II) in absence of Er(III) consists of a broad band centered around 630nm and an integrated lifetime of 1.4msec. In the presence of Er(III), the intensity and lifetimes are decreased as a result of energy transfer to the $^4F_{9/2}$ level of Er(III). The fluorescence of Er(III) arising from $^4S_{3/2}$ at 543nm has an integrated lifetime of 60μsec in the absence of Mn(II) and is decreased to 10μsec in the presence of Mn(II) as a result of energy transfer to Mn(II). The 666nm luminescence of Er(III) is enhanced in the presence of Mn(II) and has a non-exponential time dependence. The longer component corresponds to the transfer of energy from Mn(II) to Er(III) while the short-lived component is probably due to the cascading down Er(III)-Mn(II)-Er(III) through states above the Stokes threshold of Mn(II).

5.2. Energy transfer from Mn(II) to Nd(III)

The evidence of energy transfer between Mn(II) and Nd(III) in PBLA and ZBLA glass is seen in Figs. 3 and 4 where Nd(III) can be excited via Mn(II) at 337 nm or at 400 nm [43]. The risetime of about 0.1 msec in Figs. 3 and 4 is characteristic of the time at which the energy transfer takes place. Fig.4 shows the approximately exponential decay with lifetime 0.34 msec by excitation in the Nd(III) band at 579nm, the emission being measured at 876nm. When the excitation is done at 404 nm, at the low-energy edge of the narrow $^6S - ^4G$ absorption band of Mn(II), the same Nd(III) emission at 876nm shows a rise-time of about 0.1 msec followed by an exponential decay with the lifetime 1.45 msec Such storage of energy in the lowest quartet of Mn(II) was previously observed [7] for Mn(II) and Er(III) in zinc gallium lead fluoride glass. The energy scheme for transfer between Mn(II) and Nd(III) in such a glass is presented in Fig.2. In the absence of Nd(III), the luminescence of Mn(II) in ZBLA measured at 545nm shows an approximately exponential decay curve with lifetime between 13 and 14 msec(note the concentration of Mn(II) at 1 wt%). This mechanism of energy storage has obvious potential applications in laser materials.

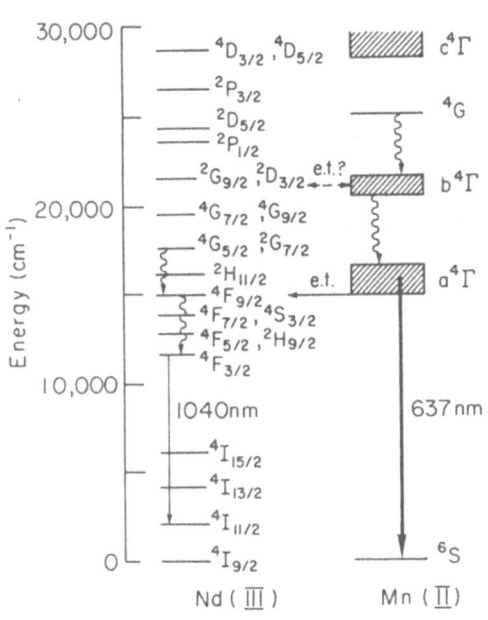

Fig.2 Scheme of energy levels of Nd(III) and Mn(II) in fluoride glass.

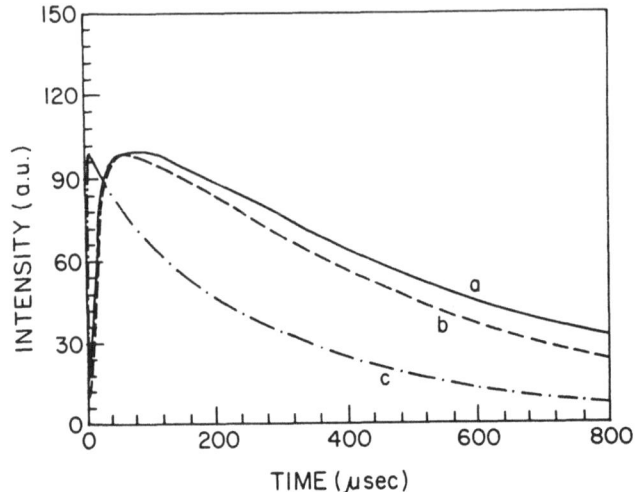

Fig.3 Luminescence decay curves of Nd(III) in PBLA glass(36PbF$_2$,24MnF$_2$, 35GaF$_3$,2AlF$_3$,3.8LaF$_3$,0.2NdF$_3$). a) Excitation 407nm. Lifetime 637 µsec. Risetime 85µsec. b) Excitation 337nm. Lifetime 500µsec. Risetime 70µsec. c) Excitation 579nm. Lifetime 330µsec. Risetime 2.0µsec.

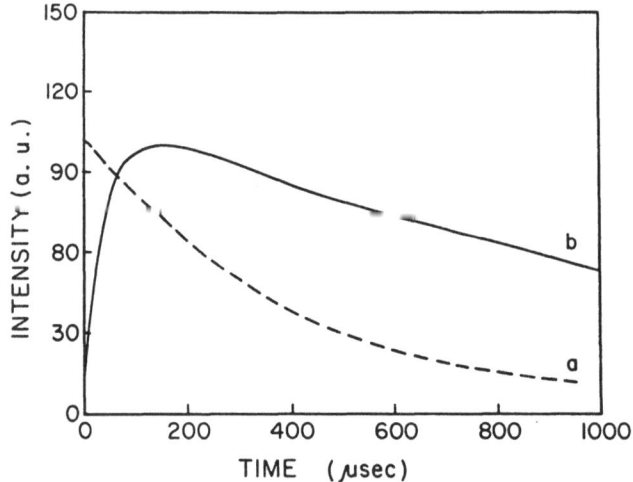

Fig.4 Luminescence decay curves of Nd(III) in ZBLA glass (55.75ZrF$_4$, 33.75BaF$_2$,4.5LaF$_3$,4AlF$_3$,1MnF$_2$,1NdF$_3$). a)Direct excitation of Nd(III) at 579 nm. Lifetime 340µsec. b)Excitation of Mn(II) at 404nm. Lifetime 1450 µsec.

Quantitative calculation of energy transfer between Mn(II) and Nd(III) in PBLA glass was performed by the steady state and dynamic measurements. The results of the calculations using the two methods [43] are presented in Table VIII.

Table VIII. Efficiency of energy transfer from Mn(II) to Nd(III) in PBLA glass ($35PbF_2$, $24MnF_2$, $35GaF_3$, AlF_3, $(4-x)LaF_3$, $xNdF_3$)

| X | EXCITATION WAVELENGTH | EFFICIENCY FROM: | | TRANSFER RATE[sec^{-1}] |
		DECAY TIME	STEADY STATE	
0.2	407	0.52	0.57	1040
0.2	337	0.57	0.56	1550
2.0	407	0.94	0.93	14400
2.0	337	0.92	0.92	12700

A fair agreement between the two series of results using steady state and dynamic measurements indicates that the two methods are mutually consistent in respect to the system studied.

The transfer rates indicate that the transfer proceeds according to a first-order kinetics, the concentration of Mn(II) being constant and concentration of Nd(III) variable. The transfer rates depend strongly on the concentration of Mn(II); the transfer rate for higher concentration of Mn(II) is larger as the Mn-Nd distance decreases. The conclusion being in full accord with intuition nevertheless raises the question whether the kinetics depend on the concentration of ground state Mn(II) ions or rather on the concentration of the excited ions. In our work we observed a systematic trend of faster transfer rates and shorter lifetimes of Mn(II) in presence of Nd(III) when the samples were excited at 337nm. The intensity of the 337nm light from the nitrogen laser is, in our case, a factor of 10 higher than the 407nm light from a dye laser, and the oscillator strength of Mn(II) at 337nm is some 5 times lower than at 407nm. Thus we can expect a twofold increase in excited Mn(II) concentration in the case of excitation at 337nm. If the transfer rates were dependent on the concentration of the excited Mn(II) ions we would then expect an increase in transfer rate, decrease in the lifetime of Mn(II), decrease in the risetime of luminescence of Nd(III) $^4F_{3/2}$ level and decrease in the lifetime of the same level. The experimental results favor the excited state concentration dependence.

6. SUMMARY

Energy transfer between manganese and neodymium ions in the system studied proceeds through at least two distinct channels:

1- Radiative energy transfer from neodymium to manganese ions in the range of 27700 cm^{-1} to 22000 cm^{-1}.

2- Nonradiative energy transfer from manganese to neodymium ions. The energy transfer occurs probably among the lowest manifold of manganese and a number of levels of neodymium in the range of 25000 cm^{-1} to 20000 cm^{-1}.

The transfer rate of the process is almost linear with concentration of neodymium ions, from 0.2 mole% of neodymium to 2.0 mole% at constant 24 mole% of manganese. At low concentration of neodymium the manganese serves as an efficient storage of energy, which results in the lifetime of neodymium with a factor 2-3 longer than its intrinsic lifetime. The phenomena may be utilized in a Q-switched Nd laser for energy storage.

3- Radiative energy transfer from manganese to neodymium ions. The radiation emitted by manganese is absorbed by the $^4G_{5/2}$ band of Nd(III) in the region around 600 nm.

There is evidence pointing towards the role which the intensity of the exciting light plays in respect to the rate of energy transfer between Mn(II) and Nd(III) in the system investigated.

A series of experiments is planned in order to study the dependence of the energy transfer rate on the intensity of exciting light, as well as a number of computer simulations of the apropriate model.

The systems studied are particularly suitable for a theoretical analysis; the PBLA samples represent an almost ideal case of Nd(III) acceptors which do not interact strongly one with another ($36PbF_2$, $24MnF_2$, $35GaF_3$, $2AlF_3$, $3.8LaF_3$ $0.2NdF_3$) and a similar case but with a moderately strong interaction among the acceptors ($36PbF_2$, $24MnF_2$, $35GaF_3$, $2AlF_3$, $2LaF_3$, $2NdF_3$).

On the other hand the $55.75ZrF_4$, $33.75BaF_2$, $4.5LaF_3$, $4AlF_3$, $1MnF_2$, $1NdF_3$ glass represents a system in which there is a moderate interaction between the donors and weak interaction between the acceptors at equal concentrations of both. Finally, by examining Table VII, we come to the conclusion that Nd(III) in fluoride glasses has good laser qualities and could be incorporated into fiber optics systems as an integrated light source.

ACKNOWLEDGEMENT: Research partly supported by US Air Force Weapon Laboratory AFB New Mexico, under Contract No. F29601-81-C-0012.
The author is very grateful to Prof C.K. Jørgensen and Dr. M. Eyal for discussion and to Mrs. Esther Greenberg for her help in preparing this work.

REFERENCES
1. Poulain, M., Poulain, M. and Lucas, J., Mater. Res. Bull.10, 243, (1975).
2. Poulain, M., Chanthanasinh, M. and Lucas, J., Mater. Res. Bull. 12, 151 (1977).
3. Poulain, M. and Lucas, J., Verres Refract. 32, 505 (1978).
4. Tran, D. C., Sigel, Jr. G. H. and Bendow, B., J. Lightwave Technology, LT-2, 566 (1984).
5. Miranday, J. P., Jacoboni, C. and De Pape, R., J. Noncryst. Solids 43, 393 (1981).
6. Miranday, J. P., Jacoboni, C. and De Pape, R., Rev. Chim. Miner. 16, 277 (1979).
7. Reisfeld, R., Greenberg, E., Jacoboni, C., De Pape, R. and Jørgensen, C. K., J. Solid State Chem. 53, 236 (1984).
8. Eyal, M., Greenberg, E., Reisfeld, R. and Spector, N., Chem. Phys. Lett. 117, 108 (1985).
9. Adam, J. L. and Sibley, W. A., J. Noncryst. Solids 76, 267 (1985).
10. Reisfeld, R., Greenberg, E., Brown, R. N., Drexhage, M. G. and Jørgensen, C. K., Chem. Phys. Lett. 95, 91 (1983).
11. Tanimura, K., Shinn, M. D., Sibley, W. A., Drexhage, M. G. and Brown, R. N., Phys. Rev. B 30, 2429 (1984).
12. Reisfeld, R., Eyal, M., Greenberg, E. and Jørgensen, C. K., Chem. Phys. Lett. 118, 25 (1985).
13. Reisfeld, R., Katz, G., Jacoboni, C., De Pape, R., Drexhage, M. G., Brown, R. N. and Jørgensen, C. K., J. Solid State Chem. 48, 323 (1983).
14. Reisfeld, R., Katz, G., Spector, N., Jørgensen, C. K., Jacoboni, C.

and De Pape, R., J. Solid State Chem. 41, 253 (1982).

15. Shinn, M. D., Sibley, W. A., Drexhage, M. G. and Brown, R. N., Phys. Rev. B. 27, 6635 (1983).

16. Lucas, J., Chanthanasinh, M., Poulain, M., Poulain, M., Brun, P. and Weber, M. J., J. Noncryst. Solids 27, 273, (1978).

17. Ohishi, Y.,Mitachi, S., Kanamori, K. and Manabe, T., Phys. Chem. Glasses 24, 135 (1983).

18. Reisfeld, R., Proc. Int'l. Symp. on Rare Earth Spectroscopy, Wroclaw, Poland, 1984. World Scientific Pub. Co. PTE. Ltd. 587 (1985).

19. Reisfeld, R., Eyal, M., Jørgensen, C.K., Guenther, A.H. and Bendow,B., Chimia (in publication).

20. Bendow, B., Banerjee, P. K., Lucas, J., Fonteneau, G. and Drexhage, M. G., J. Am. Ceramic Soc., 68, C92 (1985).

21. Jørgensen, C. K., Oxidation Numbers and Oxidation States, Springer-Verlag, Berlin (1969).

22. Jørgensen, C. K., Modern Aspects of Ligand Field Theory, North-Holland Amsterdam (1971).

23. Jørgensen, C. K., Pappalardo, R. and Rittershaus, E., Z. Naturforsch. 20a, 54 (1965).

24. Reisfeld, R., Rep. Ser. Swedish Acad. Engineering Sciences in Finland (Proc. Advanced Summer School on Electronic Structure of New Materials Loviisa, Finland, 1984) 40, part I, 7 (1985).

25. Ferguson, J., Guggenheim, H. J. and Wood, D. L., J. Chem. Phys. 54, 504 (1971).

26. Andrews, L. J., Lempicki, A. and McCollum, B. C., J. Chem. Phys. 74, 5526 (1981).

27. Reisfeld, R. and Kisilev, A., Chem. Phys. Lett. 115, 457 (1985).

28. Reisfeld, R., Kisilev, A., Greenberg, E., Buch, A. and Ish-Shalom, M., Chem. Phys. Lett. 104, 153 (1984).

29. Kisilev, A., Reisfeld, R., Greenberg, E., Buch, A. and Ish-Shalom, M., Chem. Phys. Lett. 105, 405 (1984).

30. Bouderbala, M., Boulon, G., Reisfeld, R., Buch, A., Ish-Shalom, M. and Lejus, A. M., Chem. Phys. Lett. 121, 535 (1985).

31. Rüdorff, W., Kändler, J. and Babel, D., Z. Anorg. Chem. 317, 261 (1962).

32. Smith, D. W., J. Chem. Phys. 50, 2784 (1969).

33. Cramer, S. P. and Hodgson, K. O., Inorg. Chem. 25, 1 (1979).

34. Reisfeld, R. and Jørgensen, C. K., Lasers and Excited States of Rare Earths, Springer-Verlag, Berlin (1977).

35. Nielson, C. W. and Koster, G. F., Spectroscopic Coefficients for the p^n, d^n and f^n Configurations (MIT Press, Cambridge, 1964).

36. Weber, M. J., Proc. Int'l. Conf. on Lasers '82, New Orleans, Dec. 13-17 (1982).

37. Reisfeld, R. and Eyal, M., J. de Physique, Colloque 7, 349 (1985).

38. Stokowski, S. E. and Weber, M. J., Laser Glass Handbook M-95, Lawrence Livermore National Laboratory, California 1979.

39. Reisfeld, R., Kisilev, A. and Jørgensen, C. K., Chem. Phys. Lett. 111, 19 (1984).

40. Kisilev, A. and Reisfeld, R., Solar Energy 33, 163 (1985).

41. Bornstein, A. and Reisfeld, R., J. Noncryst. Solids 50, 23 (1982).

42. Blanzat, B., Boehm, L., Jørgensen, C. K., Reisfeld, R. and Spector N., J. Solid State Chem. 32, 185, (1980).

43. Reisfeld, R., Eyal, M., Jørgensen, C. K., Jacoboni, C.and De Pape, R., Chem. Phys. Lett. (1986) (submitted).

DISCUSSION

Q: P.W. France. Would you care to comment on the relative opti‐
cal efficiencies of rare-earth doped fluoride glasses and
silicate systems, in terms of their potential use for fiber
lasers?

A: R. Reisfeld. The efficiencies of fluoride glasses should be
comparable to or may be better than silicates.

Q: R. Almeida. Is it possible to prepare Ho-containing fluori‐
de glass lasers for single mode emitters at 2.55 µm?

A: R. Reisfeld. I hope that it will be possible to prepare a
2.5 µm emitter by using Er or Tm and we plan to study these
glasses now.

C: R. Almeida. One should not use expressions such as "the
structure of fluorozirconate glasses", since I do not ex‐
pect the structure of ZBLA glass to be equal to that of
ZBLAN, e.g.. The chain structure applies strictly to the
2:1 ZrF_4-RF_2 composition (R=Sr, Ba, Pb).

Q: M. Drexhage. Have you examined other energy transfer phe‐
nomena, such as frequency doubling, in heavy metal fluoride
glasses? If so, how does the conversion efficiency compare
with the results reported by Angell et al. for Yb^{3+}/Er^{3+}
doped silicate glasses?

A: R. Reisfeld. We have not yet studied the Yb^{3+}/Er^{3+} system,
but we did separately study Er^{3+} and Yb^{3+} in fluoride
glasses. From the calculated matrix elements and Judd-Offelt
parameters   we   can accurately estimate that such con‐
version should be possible. One should be able to obtain
green (and possibly blue, though less intense than the
green) emission from Er^{3+}, as a result of energy transfer
from Yb^{3+} followed by upconversion.

Q: M. Drexhage. The spectroscopic results just presented are
interesting from a fundamental viewpoint, but have you ac‐
tually placed a polished doped glass rod in a pump cavity
and made it lase?

A: R. Reisfeld. T. Izumitani, at Hoya Corporation, is now pre‐
paring such kind of rod which will be tested soon.

LOSS MECHANISMS IN ZrF$_4$ BASED IR FIBRES

P. W. FRANCE, S. F. CARTER, M. W. MOORE and J. R. WILLIAMS

British Telecom Research Laboratories
Martlesham Heath
Ipswich
U.K.

ABSTRACT

Loss mechanisms in ZrF$_4$ - based infrared fibres are reviewed with particular emphasis on intrinsic losses, extrinsic absorption losses and extrinsic scattering. This information is then used to evaluate the status of low-loss fibres (0.9 - 4 dB/km) that are currently being reported from several laboratories, and a breakdown of losses and impurity concentrations is given. Finally an estimate is made of the likely eventual impurity concentrations in the starting materials and the information is used to estimate the potential realistic losses that might be expected in infrared fibres made by melting techiques.

1. INTRODUCTION

Current losses in ZrF$_4$ - based infrared fibres are rapidly approaching 0.1 dB/km in the wavelength range 2.5 to 2.7 μm. Although they cannot yet directly compete with silica fibres made by MCVD, where typical losses are 0.2 dB/km at 1.55 μm, they are beginning to fulfill some of their potential and show signs of achieving losses well below those of silica. Even now, fibre can be manufactured with losses in the region 1 - 5 dB/km [1,2,3] which could be of use in military telecommunication systems requiring the use of radiation hard fibre. Moreover many non - telecommunication applications can already be met, eg. infrared imaging, temperature sensing and gas detection, where infrared transmitting fibre is required in moderate lengths (< 100m) with losses in the range 10 - 100 dB/km. The aim of this work is to review the loss mechanisms that have so far been identified in ZrF$_4$ - based fibres and use this information to assess the losses that exist in typical fibres that are currently being produced.

Fig. 1 compares the intrinsic losses of several candidate materials to be considered IR fibres. The curves have been derived from theoretical studies of Rayleigh scattering by Lines [4] and estimates of the IR edge compiled from several sources [5,6,7]. The IR edge for the ZrF$_4$ glass (ZBLAN) is taken from this work discussed below. BeF$_2$ and ZnCl$_2$ clearly have the lowest potential losses (0.005 and 0.003 respectively) but suffer form the disadvantages of being very toxic or hygroscopic. ZBLAN has the next lowest intrinsic loss with a value close to 0.011 dB/km and at present is probably the best candidate for low loss fibre.

2. LOSS MECHANISMS

2.1. Intrinsic Losses

The curves in Fig. 1 show that the minimum intrinsic loss of a material is given by the combined loss due to Rayleigh scattering and the multiphonon absorption edge. Several authors have attempted to estimate

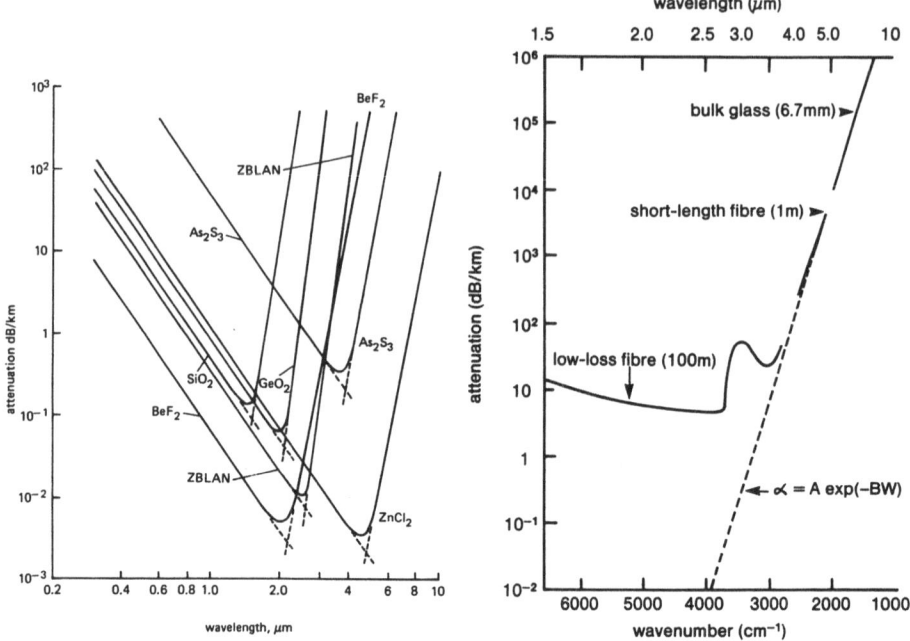

Fig. 1 *Intrinsic losses*

Fig. 2 **Multiphonon edge in ZBLAN**

and measure Rayleigh scatter in ZrF_4 glasses and some of these results are given in Table 1. Here R is used as the scattering coefficient in dB km^{-1} μm^4, so that the scattering loss $\propto$ is given by:

$$\propto_R = R / \lambda^4 \qquad (\lambda = wavelength) \qquad \dots (1)$$

TABLE 1   Rayleigh scattering in ZrF_4 glasses

Reference	Material	R	
[4]	SiO_2	0.66	}theoretical
[8]	SiO_2	0.74	
[4]	ZrF_4	0.336	}theoretical
[9]	ZrF_4	0.69	
[10]	ZrF_4	0.14 - 0.30	
[11]	ZrF_4	0.567	

Measured values of C vary from 0.14 to 0.69, whereas Lines has estimated a theoretical value of 0.336. Since this estimate is mid range it is taken to be a reasonable figure for the expected intrinsic scattering.

Rupprecht [12] first suggested an approximate formula for the IR edge in glasses:

$$\propto_{IR}(w) = A exp(-Bw) \qquad \dots (2)$$

where $\propto_{IR}$ is the loss in dB/km, w is the wavenumber in cm^{-1} and A and

B are material constants. We have been able to estimate values of A and B by determining the infrared edge for a ZBLAN glass. This was done by measuring the infrared absorption in bulk glasses (0.64 cm path length), short length fibres (1m) and low loss fibres (100m), and therefore determining the edge over 5 orders of magnitude as shown in Fig. 2. The results indicate that over this range the IR edge obeys equation (2) and gives vlues of:

$$\log_e A = 23.222, \quad B = 7.164 * 10^{-3}$$

The results also allow us to extrapolate reasonably confidently out to 2.55 μm where the IR edge loss is expected to be 0.007 dB/km. The discrepancy between the fibre data and the projected IR edge is due to extrinsic scatter, OH and Ce absorption in the fibre. Combining equations (1) and (2) gives a minimum loss of 0.011 dB/km close to 2.5 μm.

2.2. Extrinsic Absorption Losses

The requirement to produce ultra-low-loss IR fibres will put strict limitations on the impurity content of the starting materials, especially since the predicted minimum losses are so low. It is likely that the eventual loss will also be determined as much by extrinsic absorptions as by any other source and so in this section an attempt will be made to identify some of the major impurities.

2.2.1. OH Absorption:

Absorption due to OH has always been a problem in silica based fibres operating at 1.3 or 1.55 μm and will continue to be a problem in fluoride glasses made by melting techniques. There is even more cause for concern in longer wavelength systems since the OH fundamental is expected at 2.8 μm and is close to minimum loss wavelength at 2.5 μm. A measured OH spectrum is in a ZrF_4 glass is shown in Fig. 3, taken from [13]. The fundamental lies at 2.87 μm, with combination bands at 2.24 and 2.42 μm. Levels of OH can be reduced to of order 2 ppb (assuming the same absorption coefficient as silica) corresponding to a fundamental peak

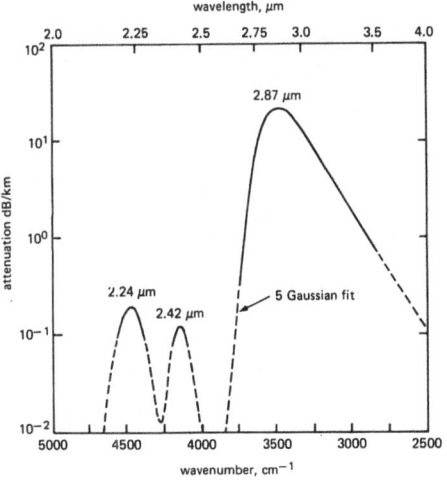

Fig. 3 OH absorption   Fig. 4 Fe absorption

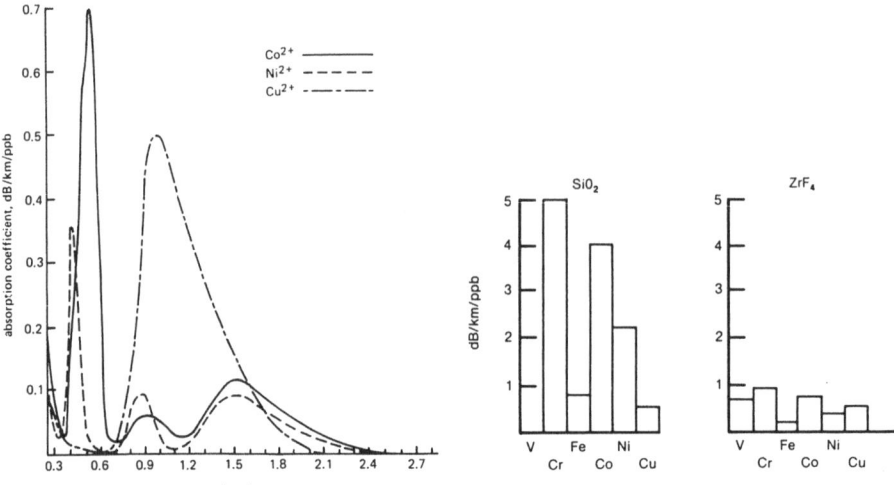

Fig. 5 Absorption due to Co, Ni and Cu

Fig. 6 Peak absorption coefficients in SiO₂ and ZBLAN

height of less than 20 dB/km. This has been achieved using dry processing conditions and refining the glass at high temperatures, since it appears that under these conditions OH can be liberated from the glass according to:

$$- Zr - OH \; + \; F - Zr - \; \longrightarrow \; - Zr - O - Zr - \; + \; HF$$

Fig. 3 has been drawn assuming a peak height of 20 dB/km and Gaussian extrapolations of the tails suggest a low-loss window near 2.55 μm. This fortunately coincides with the minimum intrinsic loss near 2.5 μm.

2.2.2. 3d Transition Metals:

In a glass matrix the free ion energy levels of the 3d transition metals are split by the action of the ligand field surrounding each ion. Electron transitions between these levels generally occur in the visible region and consequently these elements can cause loss. The most common impurities are Fe and Cu, and together with Co and Ni these ions also absorb in the NIR. Fortunately many of transition metals exist in more than one oxidation state with different absorption spectra and for example Fe can exist as Fe^{2+} and Fe^{3+}. Only the divalent form is absorbing at 2.5μm [14]. Consequently oxidation of Fe^{2+} to Fe^{3+} can remove the Fe absorption and is illustrated in Fig. 4 .

Unfortunately the oxidised form of Cu (Cu^{2+}) also absorbs close to 2.5 μm but its absorption coefficient is less than Fe at this wavelength, and so oxidised glasses give overall lower losses [15]. Co and Ni have only one stable oxidation state, the divalent form, and the absorption spectra for Cu^{2+}, Co^{2+} and Ni^{2+} are illustrated if Fig. 5. Gaussian fits have also been made to the absorption tails of the transition metal ions and Table 3 lists the extrapolated absorption coefficients at 2.55μm.

One point of interest is that the 3d transition metals generally sit in octahedral coordination in a ZrF_4 based glass, whereas in silica glass they tend to favour a mixture of octahedral and tetrahedral [14].

This leads to lower extinction coefficients than in silica and this point is illustrated in Fig. 6.

2.2.3. 4f Rare Earth Metals:

The rare earth metals exhibit electronic transitions amongst the 4f shell, which also lie in the visible and NIR regions of the spectrum. Since the lanthanide ions have 5s and 5p electrons lying outside the the 4f shell, they are largely shielded from the effects of the ligand field. Consequently the transitions are between the free ion levels and are only slightly perturbed by the ligand field. This leads to sharp absorption bands that are approximately an order of magnitude less intense than the 3d transition metals.

The most stable state for the rare earth ions is generally the tripositive form and only Eu has been observed by us to exist in more than one oxidation state. Ce might be expected to exist in both the Ce^{3+} and Ce^{4+} forms but again we have only observed the trivalent state as is the case for the other lanthanides. Fig. 7 shows the absorption spectra for Eu in a ZBLAN glass made under both oxidising and reducing conditions. Eu^{2+} and Eu^{3+} were identified according to the melting conditions, Eu^{3+} having a spectrum typical of the trivalent lanthanide ions, with sharp absorption peaks extending from the UV to the NIR. Nd^{3+} is the most absorbing ion at 2.5 μm since it has an absorption centred at 2.51 μm due to the $^4I_{9/2} \longrightarrow {}^4I_{13/2}$ transition. Other ions that give significant loss close to 2.5 μm are Pr^{3+}, Sm^{3+}, Eu^{3+}, Tb^{3+} and Dy^{3+}. Gaussian fits have been made to these absorption tails and Table 3 lists the expected absorption coefficients at 2.55 μm.

2.2.4. Other Impurities:

Several other impurities have been identified as causing loss in ZrF_4 glasses and can be classified into three groups. The first group is that of dissolved gases and in particular CO and CO_2 have been observed to be present in the glass [16] arising from carbon and oxide impurities in the starting materials. Only CO_2 has absorptions which lie close to 2.5 μm, notably at 2.675 and 2.775 μm and these are illustrated in fibre in Fig.8. Fortunately these peaks are very sharp and can readily be reduced by using better quality materials. Dissolved gases are therefore not expected to add significantly to the loss.

The second category of impurities is that of molecular ion absorptions, in addition to OH^- which has already been discussed. In particular NH_4^+ has been identified to absorb at 2.96 and 3.04 μm [17] and generally arises from NH_4HF_2 used as a fluorinating agent. Again the absorptions are small and well away from 2.5 μm so that they should not be a problem. Other identified molecular ions include SO_4^{2-} and PO_3^- [18] but these have negligible absorption below 4.5 μm.

The third category of impurities that have been identified are reduced species of zirconium, including Zr^{3+} and metallic zirconium [19]. These species have been observed in glasses melted under reducing conditions and Zr^{3+} is responsible for the characteristic black phase that is frequently seen in these melts. Under severely reducing conditions metallic zirconium can be found and imparts a yellow colouration to the glass corresponding to an absorption peak at 333 nm. In general these reduced species can lead to extrinsic scattering but can readily be removed by melting under oxidising conditions such as those required to oxidise Fe to Fe^{3+}.

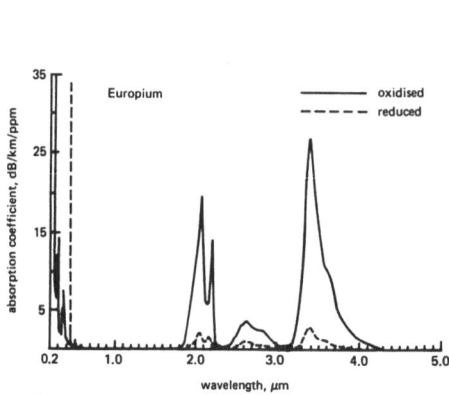

Fig. 7 Eu absorption

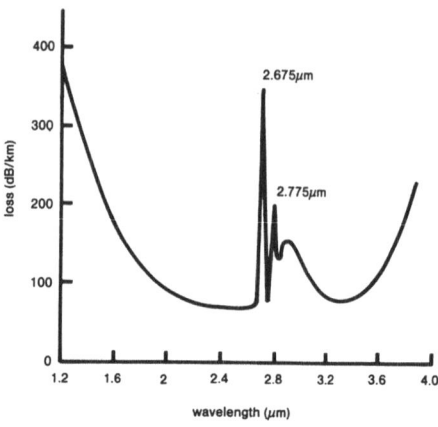

Fig. 8 Fibre loss showing CO_2 absorption

2.3. Extrinsic Scattering

Rayleigh scattering is caused by microscopic changes in glass density and hence refractive index on a scale which is much smaller than the wavelength of the light travelling through the sample. This leads to a wavelength dependence that varies as the fourth power and has been discussed above. Larger scale fluctuations or small imperfections within the fibre core will also cause extrinsic scattering and therefore increase the overall fibre loss. In general extrinsic scattering can be subdivided into two main types: Mie scattering and wavelength independent scattering. Mie scattering is caused by imperfections whose size is of the order of the wavelength of the exciting light and leads to a λ^{-2} dependence, whereas wavelength independent scattering (λ^{0}) is caused by larger particles.

Some examples of possible defects are firstly crystals of LaF_3 that can nucleate within the fibre core during processing causing deformation; secondly bubbles can become trapped in the preform during the casting process; and thirdly we have identified sub micron scattering centres which maybe due to small particles of oxide present in the raw materials. This point has been further amplified by Mitachi et al. [20] who have determined scattering loss as a function of doped oxide content. They conclded that additions of 600 ppm of oxide as Gd_2O_3 could increase wavelength indepentent scattering by 100 dB/km. Large scale diameter fluctuations can also increase wavelength independent losses. Crystallisation can be reduced by the use of more stable glass compositions and in particular ZBLAN has been shown to be fairly stable. Improved processing should also help to reduce bubble formation and better quality raw materials may help to reduce extrinsic scattering.

3. CURRENT STATUS OF LOW LOSS FIBRES

Several laboratories throughout the world have now produced low-loss fluoride fibres in lengths ~100m. In particular NRL have measured 0.95 dB/km [1], NTT have reported 3.0 dB/km [2], BTRL have measured 4.5 dB/km [3] and Le Verre Fluore have made 18 dB/km on 200m [21]. These measurements were made using a full NA launch (BTRL used half NA launch with 60m short length) and are illustrated in Fig. 9.

In order to further classify the total loss into absorption and scattering two approaches have been taken. We have independently measured scatter loss using an integrating sphere, and absorption loss using a calorimetric technique [22]. The results indicate that scatter loss obeys the formula:

$$\alpha = C + R/\lambda^4$$

where C is a wavelength independent offset determined to be 1.0 dB/km, and R was measured to be 1.16. NRL have also analysed their fibre using similar techniques and have determined values of 0.07 for C and about 5.0 for R.

By curve fitting to the absorption data we have also been able to estimate our impurity concentrations in current fibres and the results are tabulated below:

Fe^{2+}	< 0.1 ppb	Nd^{3+}	20 ppb
Co^{2+}	15	other Ln^{3+}	< 5
Ni^{2+}	20		
Cu^{2+}	40		

4. FUTURE POTENTIAL OF LOW LOSS FIBRES
4.1. Estimates of Impurity Concentrations

In order to estimate the total extrinsic losses that might be expected in ZrF_4 fibres it is neccessary to compile information on the impurity concentrations in the raw materials. The impurities of most concern are the 3d and 4f transition metals listed in Tables 2 and 3. 3d transition metals were of course of considerable interest in the fabrication of sodium borosilicate fibres by double crucible techniques [23]. In this case powders of Na_2CO_3, $BaCO_3$, ZrO_2 and Al_2O_3 were used to make the glasses and extensive measurements were made of the impurity contents. By coincidence all of these powders can be used as precursor materials in the formation of NaF, BaF_2, ZrF_4 and AlF_3 used for 95% of the ZBLAN composition and allow us to make provisional estimates of the impurities that might be expected in the fluorides. From a knowledge of the absorption coefficients and spectra of each 3d metal ion and by examining the total absorption losses in double crucible fibre, impurity levels given in Table 2 have been estimated. It will be further assumed here that these estimates include extra loss arising from the processing since the figures quoted are based on double crucible fibre after processing. A further improvement to the Fe contribution can be assumed by using oxidised glasses where the Fe^{2+}/Fe^{3+} ratio is expected to be less than 0.01.

TABLE 2	Impurity	Expected Concentration
	Fe_{2+}	< 10 ppb
	(Fe^{2+})	< 0.1 ppb
	Cu	< 2.0 ppb
	Co	< 0.1 ppb
	Ni	< 0.3 ppb

Rare earth ions have very sharp absorptions over the visible region which would enable them to be detected fairly easily in the absorption spectra of double crucible fibre. On examining the loss plots however all rare earths were below detection limits of approximately 0.1 ppb and it

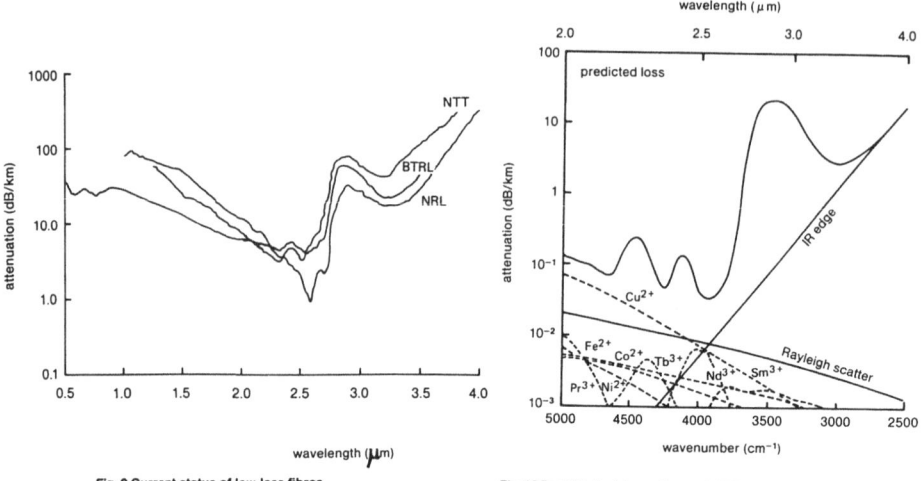

Fig. 9 Current status of low-loss fibres

Fig. 10 Predicted minimum losses in ZrF₄ fibres

can therefore be concluded that for the case of NaF, BaF_2, ZrF_4 and AlF_3:

$$All \quad Ln^{3+} \quad ions \quad < \quad 0.1 \; ppb \quad in \; Na, \; Ba, \; Zr, \; Al.$$

By far the largest source of rare earth impurities will be from the LaF_3. Because the lanthanides are chemically similar they are difficult to separate out. Again because of their characteristic sharp absorptions it has been possible to determine rare earth concentrations in current IR fibre and for example we have determined:

$$Nd^{3+} \quad < \quad 3 \; ppb \quad in \; ZrF_4 \; glass.$$

Since LaF_3 is only a minor component of the glass we can further conclude:

$$Nd^{3+} \quad < \quad 50 \; ppb \quad in \; LaF_3 \; powder.$$

Fortunately it should be possible to reduce these levels using solvent extraction and estimates of the improvement using several complexing agents give a factor of ten. We can therefore further estimate the rare earth content as:

$$Each \quad Ln^{3+} \quad < \quad 5 \quad ppb \; in \; LaF_3$$
$$or \quad < \quad 0.3 \; ppb \; in \; ZBLAN \; glass.$$

The one exception to this may be Ce which is more abundant.

4.2. Estimates of Minimum Loss

The overall minimum loss in ZrF_4 based IR fibres can be estimated by combining the expected impurity concentrations with the absorption coefficients. Table 3 summarises each of these and and lists the expected loss at 2.55 µm. Fig.10 also illustrates these absorptions and shows that when combined with the intrinsic losses the minimum loss wavelength should be at 2.56 µm. Tran et al. have recently reported fibre [1] with scattering losses dominated by Rayleigh type and with only a 0.6 db/km wavelength independent scatter loss. Moreover scattering measurements on bulk glasses have shown that it is possible to reduce scattering down to intrinsic levels and consequently we will assume here that it will be possible to reduce extrinsic scattering in fibre to levels below 10^{-3}

dB/km. Consequently these results suggest that absorption losses of about 0.02 dB/km may be feasible and that overall losses of 0.035 dB/km could be expected at 2.56 µm. This figure compares with a typical loss of 0.20 dB/km at 1.55 µm in silica fibre so that factors of 6 - 7 improvement may be possible with IR fibres.

TABLE 3   Estimated Losses

Impurity	Concentration ppb	Absorp. Coef. 2.55µm dB/km/ppb $* 10^{-3}$	Loss 2.55µm dB/km $* 10^{-3}$
OH^-	2.0	< 1.0	< 1.0
Fe^{2+}	$10\ Fe = 0.1\ Fe^{2+}$	15.0	1.5
Cu^{2+}	2.0	3.0	6.0
Co^{2+}	0.1	17.0	1.7
Ni^{2+}	0.3	2.4	0.7
Ce^{3+}	5.0	–	–
Pr^{3+}	0.3	0.01	0.3
Nd^{3+}	0.3	22.0	6.6
Sm^{3+}	0.3	3.3	1.0
Eu^{3+}	0.3	2.5	0.7
Tb^{3+}	0.3	0.2	–
Dy^{3+}	0.3	1.6	0.5
Rayleigh Scattering	.	.	. .	7.9
IR Edge	.	.	. .	7.6
		TOTAL	.	35.5

5. CONCLUSIONS

The intrinsic losses of ZrF_4 based IR glasses currently used to fabricate IR fibres have values of about 0.015 dB/km near 2.5 µm. By coincidence the OH spectrum suggests the presence of a low-loss window at 2.55 µm where the loss due to OH is expected to be less than 0.001 dB/km. The major absorbing impurities that are expected are Fe, Cu and Nd. An examination of the loss spectrum of double crucible fibre made from materials that could be used as precursors for fluorides suggest that impurities concentrations of 10 and 2 ppb might be expected for Fe and Cu repectively. LaF_3 would be the major source of Nd and by the use of solvent extraction we have estimated that concentrations of Nd might be reduced to 5 ppb. This would correspond to 0.3 ppb in the glass since LaF_3 is only a minor component. These impurities would give a total absorption loss of about 0.02 dB/km at 2.55 µm.

We can therefore conclude that overall losses close to 0.035 dB/km might ultimately be obtainable in IR fibres, a factor of three higher than the intrinsic losses and a factor of seven better than a typical silica fibre. This figure is obviously tentative and assumes that extrinsic scattering losses are negligible, but does at least give a first approximation of losses that might be realised in ZrF_4 based IR fibres.

6. ACKNOWLEDGEMENTS

The authors would like to thank the Director of Research, British Telecom plc, for permission to publish this paper and DTI under JOERS for partial sponsorship of the work.

7. REFERENCES

[1] D C Tran, Proc. OFC; paper TUA1; Atlanta, Georgia (1986).
[2] S Takahashi, SPIE Proceedings 618, Infrared Optical Materials and Fibers IV.
[3] P W France, S F Carter, M W Moore and J R Williams, to be published.
[4] M E Lines; J Appl Phys , 55, 4052 (1984).
[5] T Izawa, N Sibata and A Takeda; Appl Phys Lett, 31, 33 (1977).
[6] L G Van Uitert and S H Wemple; Appl Phys Lett, 33,57 (1978).
[7] J R Gannon; SPIE 266 Infrared Fibres, 62 (1981).
[8] B J Ainslie, K J Beales, D Cooper, S Craig and A J Skeats; 8th ECOC, Stuttgart, 208 (1984).
[9] D C Tran, K H Levin R J Ginther and G H Sigel; Electron Lett, 18, 1046 (1982).
[10] J Schroeder, M Fox-Bilmont, B G Pazol, V Tsoukala, M G Drexhage and O El-Bayoumi; SPIE 484 Infrared Optical Materials and Fibers III, 64 (1984).
[11] D C Tran; Proc. ECOC Venice , 13 (1985).
[12] G Rupprecht; Phys Chem Solids, 34, 2091 (1964).
[13] P W France, S F Carter, J R Williams, K J Beales and J M Parker; Electron Lett, 20, 607 (1984).
[14] P W France, S F Carter and J M Parker; Phys Chem Glasses, in press (1986).
[15] P W France, S F Carter, M W Moore and J R Williams; Electron Lett, 21, 602 (1985).
[16] S F Carter, P W France and J R Williams; Phys Chem Glasses, in press (1986).
[17] P W France, S F Carter and J R Williams; J Am Ceram Soc, 67, C-234 (1984).
[18] M Poulain,and M Saad; J Lightwave Technology, LT-2, 599 (1984).
[19] S F Carter, P W France, M W Moore and E A Harris; Phys. Chem. Glasses in press.
[20] S Mitachi, S Sakaguchi, H Shikano, T Shigematsu and S Takahashi; Jpn. J. Appl. Phys., 24, L827 (1985).
[21] Le Verre Fluore, private communication.
[22] K I White and J E Midwinter; Opto-Electron., 5, 323 (1973).
[23] K J Beales, C R Day, W J Duncan, J E Midwinter and G R Newns; Proc IEEE 6, 591 (1976).

DISCUSSION

C: P. Klein. You dealt with the possibility of seeing HF in
   fluoride glass, rather than being certain that the 2.9 μm
   absorption is solely due to OH. From the solubility of BaF_2
   in liquid HF, leading to the formation of $BaF_2 . (HF)_5$, I
   believe that it is possible to have HF inside the glass. If
   e.g. NaF.HF does not decompose at 500-600°C, why shouldn't
   another hydrofluoride be stable at the melting temperature
   of fluoride glass - at least in mol fractions as low as
   0.01-0.001?

Q: J. Lucas. In order to achieve the lowest theoretical loss,
   what will be the allowed loss for the OH peak?

A: P.W. France. If the extrapolations that we have made for
   the OH peak tails are correct, then we have already
   achieved significantly low OH values that give losses
   $<10^{-2}$ dB/Km. In any case, not much more progress will be
   required in this particular area.

Q: R. Almeida. Do you think that, in order to achieve sufficiently low losses, one has to be able to move to the long
   wavelength side of the OH peak?

A: P.W. France. I don't know. But the OH peak is asymmetric,
   probably due to the simultaneous occurence of free and
   bonded OH species. So it will probably be difficult to go
   to the right hand side of the bonded OH shoulder.

Q: C.T. Moynihan. Why is the OH absorption band so asymmetric?

A: P.W. France. The explanation advanced by Adams was that hydrogen bonding can significantly affect the position of the
   OH fundamental and that H-bonded hydroxyl groups can have
   their absorption peaks shifted from 2.8 μm to ∿ 4 μm and
   beyond. Hydrogen bonding is likely to be very significant
   in fluoride glasses, probably in the form Zr-OH...F-Zr .
   Since the OH is in a glassy matrix, the exact anionic sites
   and the degree of hydrogen bonding will vary, causing a
   distinct asymmetry in the OH peak, with an extended tail on
   the long wavelength side.

Q: C. Moynihan. It was interesting to see that the fiber loss
   data you showed indicated that LiF-containing glasses had
   IR edges shifted to shorter wavelengths compared to NaF-containing glasses. This was predicted by M. Drexhage and my-
   self in around 1980 and was recently verified at RPI on
   bulk glasses. However, the IR shifts are not extremely large. How important will it be to choose NaF over LiF in
   order to get the 2.5 μm fiber losses down?

A: P.W. France. More precise measurements of the IR edge have now been made in ZBLAN glasses and they predict that the multiphonon loss at 2.5 μm will be ∿ 10^{-2} dB/Km. Clearly, any small additions of LiF will increase this loss and significantly affect the intrinsic minimum loss. Therefore, it seems likely that, for ultra-low loss fibers, LiF cannot be tolerated, nor can slight increases in the amount of AlF_3.

C: C. Pantano. As far as the asymmetry of the OH band, I would like to comment that this is probably due to hydrogen bonding interactions which involve molecular water rather than simply metal hydroxide - fluoride interactions.

Q: H. Schneider. Did you, at BTRL, use the double crucible technique for making fluoride glass fibers?

A: P.W. France. We use the rotational casting preform technique. We have only done a few experiments on fluoride glass fiber drawing from crucibles.

FLUORIDE GLASS OPTICAL FIBERS IN FRANCE

H. POIGNANT
CNET LANNION B - ROC/FOG, Route de Trégastel, 22301 LANNION,
France

ABSTRACT
    The state of the art in the fluoride glass optical fiber
technology in France is reviewed here. After presenting the
studied glass compositions and the conventional melting and
casting methods used to achieve both the bulk glasses and the
preforms, the results of characterization studies concerning
the absorption and scattering losses of the glasses, the effect of some dopants such as PbF_2 or NaF on the refractive index, as well as spectral fiber loss, will be given. Future
prospects dealing with fluoride glass single mode fibers and
different approaches for the achievement of glass preforms
will be presented too.

1. INTRODUCTION
    A very strong interest is now appearing in the development
of Zirconium fluoride glass based optical fibers, due to their
potential ultra-low transmission loss in the 2-3 μm wavelength
range. In fact, attenuation values as low as 10^{-2} or 10^{-3} db/km
around 2.5-3.5 μm have been predicted [1], which are the result of both reduced Rayleigh scattering loss [2] and negligible multiphonon absorption [3] at such wavelengths. So, these
new materials really appear as very promising candidates for
achieving long haul optical fiber communication systems. However, the lowest loss reported up to now is about a few db/km
at 2.5-2.6 μm wavelength [4][5][6][7] for relatively short
length fibers (20-40 meters). So, this has to be decreased by
two orders of magnitude to approximate the very low expected
values. This will need a considerable effort to decrease both
the extrinsic scattering and absorption loss, the latter being
asscoiated to the presence of many kinds of impurities in the
glasses (transition metals, rare earth traces, oxides, hydro-
xiles...). The achievement of 10^{-2} db/km fluoride glass fibers
will, in fact, involve new preparation techniques, allowing
impurity levels not exceeding a few ppb or less to be obtained.
Though some work has already started in the research of more
promising routes to reach the ultimate goal [8][9], both glasses and preforms are still prepared using conventional melting
and casting processes in most of the world laboratories working in this new glass field.
    In the present paper, the current status of fluoride glass
optical fiber research and development in France is reviewed.
The glass compositions investigated, their properties and the

methods used for preparing bulk glasses and preforms are reported. Fiber loss characteristics are given too and, finally,
some prospects for the next future studies to be carried out.

2. GLASS COMPOSITIONS

ZrF_4 based fluoride glasses remain quite unstable materials,
compared to silica glass, and the attainment of low loss fibers needs the drawn material to satisfy both the stability
criterion against crystallization and a viscosity/temperature
profile which is appropriate. From these points of view, the
ZBL base glass compositions (made of ZrF_4, BaF_2, LaF_3) are unsuitable and have to be modified by incorporating some additives well known to act as stabilizing agents. For this reason,
metallic fluorides such as AlF_3 [10] or ThF_4, LiF, NaF are involved in the drawable glass compositions. The table 1 lists
the typical base glass compositions investigated at both CNET
and LE VERRE FLUORE S.A., which proved to present the required
properties for the fabrication of homogeneous fibers when adding NaF or ThF_4.

TABLE 1: Typical CNET and LVF base glass compositions.

Glass composition (Mole %)	ZrF_4	BaF_2	AlF_3	LaF_3
CNET	57	35	3.5	4.5
LVF [11]	57	34	4	5

3. BULK GLASS FABRICATION AND CHARACTERIZATION

3.1 Glassmaking

The fluoride glasses are prepared by a conventional melting
technique. The Table 2 shows the typical procedure employed

TABLE 2: Typical melting procedure for a 45g fluoride glass
batch (CNET)

	1/NH_4HF_2 route		2/Fluoride route	
Starting Materials	ZrO_2	Merck Optipur	ZrF_4	R.M.C. or Merck
	BaF_2	Merck Suprapur	BaF_2	Merck Suprapur
	Al_2O_3	Merck Optipur AlF_3	AlF_3	Merck ZLA or RMC
	La_2O_3	5N O.S.I	LaF_3	Merck ZLA or RMC
	NaF	Merck Suprapur	NaF	Merck Suprapur
	NH_4HF_2	Merck Optipur (42g)	NH_4HF_2	Merck Optipur (5g)
Fluorination	3 hours at 300°C		0.5 hour at 300°C	
	1 hour at 400°C (N_2 atm.)		0.25 hour at 400°C (N_2 atm.)	
Melting	0.5 hour at 850°C		0.5 hour at 850°C	
	(N_2/O_2 atm.)		(N_2/O_2 atm.)	

for a 45 g batch fabrication. As can be seen, two routes are possible, which are either the NH_4HF_2 method (starting materials may be both oxides and fluorides, the conversion of oxides being achieved by NH_4HF_2) or the fluoride approach (only anhydrous fluorides are present in the starting glass charge, a small amount of NH_4HF_2 ensuring the fluorination of eventual oxide traces in the starting materials).

Both methods present advantages but also may have negative effects on the optical quality of the synthetised glass, as indicated in Table 3. Of course, a common feature is that each process needs an extremely dry atmosphere around the glass during the whole preparation stage until it is quenched.

TABLE 3: Criteria list for the choice of a fluoride glass melting system.

	NH_4HF_2 ROUTE	FLUORIDE ROUTE
FOR	-High purity oxides available -Lower batch sensitivity to moisture -Water removal easier (careful fluorination)	- Larger size melts possible - Less reaction products - Expected lower scattering loss - Ability of Zr and Al fluorides to be purified [12] [13].
AGAINST	-Large NH_4HF_2 amount implicates large size crucible for a reduced melt -Water and alien species (NH_3, NH_4F..) generated during the fluorination may give excess loss scattering	- High purity fluorides (Zr, Al, La) not readily available. - Fluorides need to be completely anhydrous (hydroxifluorides are very difficult to remove) - Fluorides tend to pick up moisture - Water removal more difficult (RAP)

The figure 1 shows some fluoride glass samples currently obtained at CNET using the NH_4HF_2 route.

In order to find a fairly suitable glass composition for preform making, characterization studies devoted to the thermal and optical properties are necessary. The effect of NaF and PbF_2 on the glass stability and the refractive index have been determined, and calorimetric experiments were performed to evaluate both absorption and scattering loss.

3.2.1 Effect of PbF_2 and NaF on the glass stability [14]

The stable glass character against crystallization may be evaluated from the determination of the reduced parameter $\beta = \frac{T_c - T_g}{T_g}$ (K) [15], where T_c and T_g are respectively the crystallization and vitreous transition temperatures. The figure 2

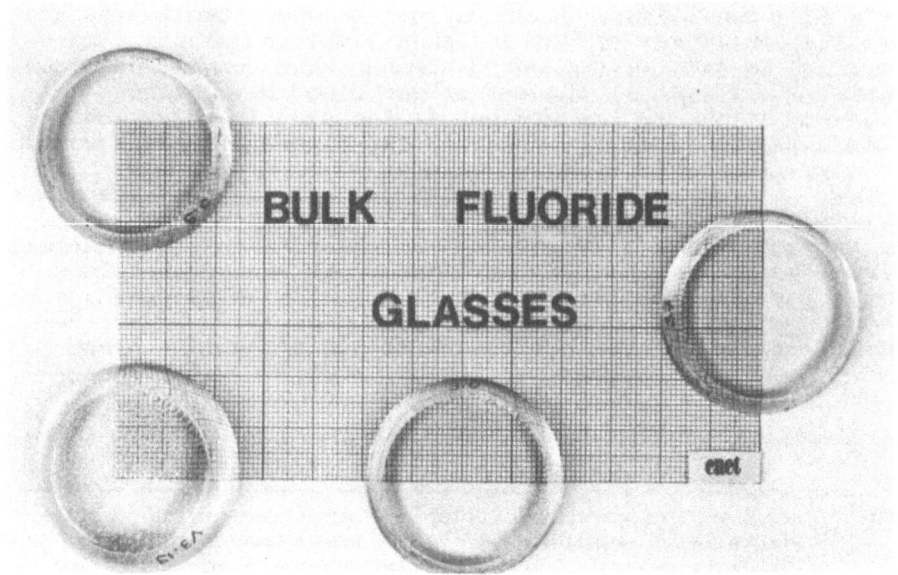

FIGURE 1: Typical fluoride glass samples (CNET)

shows the linear decrease of T_g when NaF or PbF_2 are substituted for BaF_2 in the base glass composition.

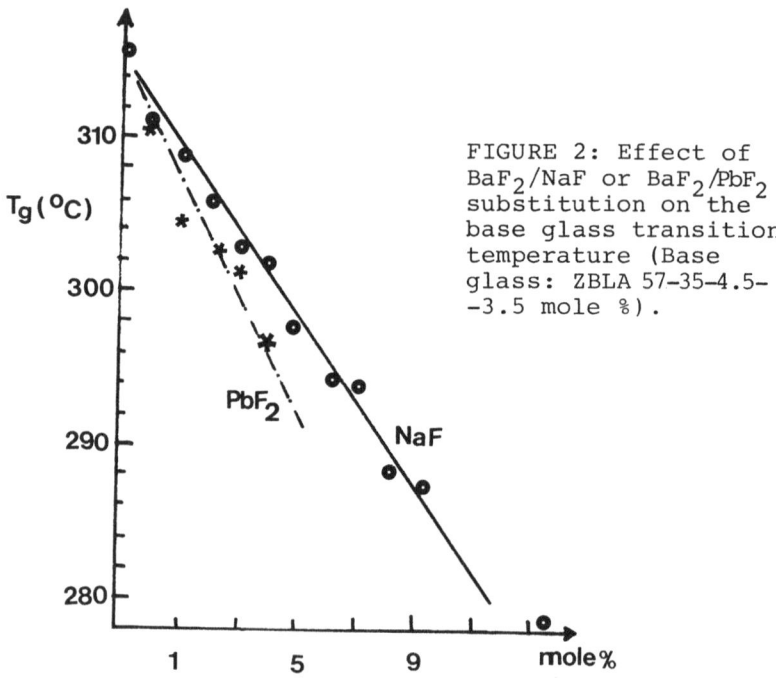

FIGURE 2: Effect of BaF_2/NaF or BaF_2/PbF_2 substitution on the base glass transition temperature (Base glass: ZBLA 57-35-4.5--3.5 mole %).

On figure 3, we have plotted the β values versus the PbF$_2$
and NaF glass content. In the case of NaF/BaF$_2$ substitution,
β increases regularly from 0.10 (base glass) up to 0.18 for a
15 mole % NaF glass. On the other side, it must be pointed out
that the effect of PbF$_2$/BaF$_2$ substitution is quite different.
The curve shows a maximum for a 3 mole % PbF$_2$ content and then
β decreases for higher percentages. It is well known that PbF$_2$
enhances the fluoride glass crystallization, even when enter-
ing the glass composition at quite low level ($\sim$ 4-5 mole %).

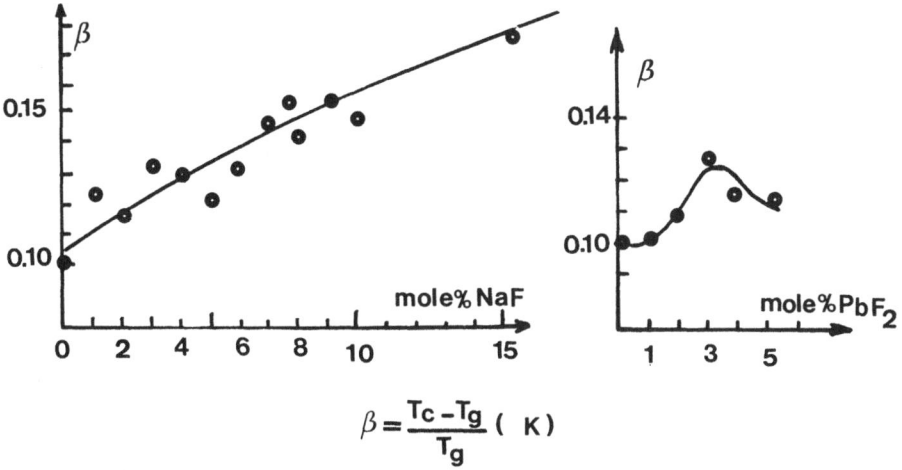

$$\beta = \frac{T_c - T_g}{T_g} \ (\ K \ )$$

FIGURE 3: Value of β parameter versus PbF$_2$ or NaF glass
content (Base glass: ZBLA 57-35-4.5-3.5 mole %)

3.2.2 Effect of PbF$_2$ and NaF on the base glass refractive
       index [14]
    In the design of optical fibers, it is important to control
the refractive index difference between the core and cladding
glasses. The figure 4 shows the base glass refractive index
change when PbF$_2$ or NaF are incorporated.   As can be seen,
Δn values of 10^{-2} are achievable for PbF$_2$ and NaF contents res
pectively equal to 4 and 10 mole %. The PbF$_2$ straight line slo
pe is more important than the NaF one, due to the high polari-
sability of Pb^{2+} cation. However, due to possible induced crys
tallization with PbF$_2$, the refractive index difference which
is needed for preforms is preferably achieved by employing NaF
at high percentages, PbF$_2$ doping being limited usually at the
1-2 mole % level.

3.2.3 Absorption and scattering loss measurements
    The evaluation of both absorption and scattering losses on
the bulk glasses before remelting them when preparing the preform is of great interest, since its allows to qualify the
first melting procedure which has been followed to elaborate
the glass. Really if the absorption losses are dominantly depend

ing on the starting materials purity, the extrinsic scattering losses are strongly associated with the experimental melting con ditions (temperature, time, atmosphere,...). For this reason, calorimetry [16] provides an interesting tool permitting not only to select the glass samples for preform making but also to change some parameters in the melting schedule for improving the glass quality. Table 3 lists both absorption and scatter- ing loss measured at the 0.647 μm wavelength on currently fa- bricated glass smaples (not remelted; size 35 x 5 x 5 mm^3).

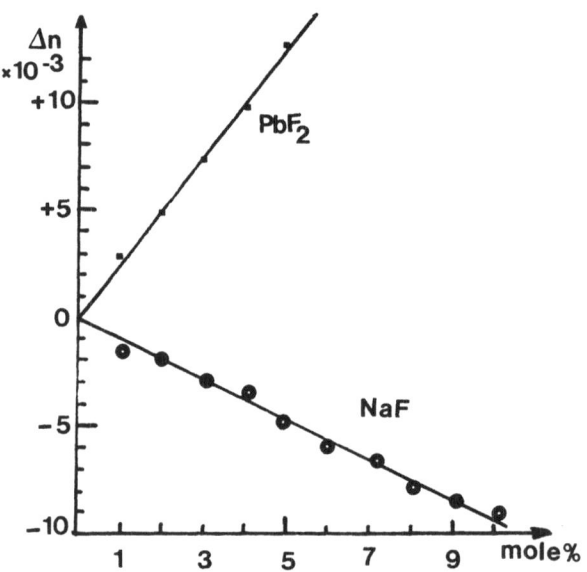

FIGURE 4: Effect of BaF_2/NaF or BaF_2/PbF_2 substitution on the base glass refractive index (measurements performed at 632.8 nm).

The lowest losses are about 20 and 10 db/km for absorption and scattering respectively, which is similar to results repor ted previously [17] [18] for remelted and cast glasses.

3.2.4 Dispersion characteristics

Since it is related to the transmission bandwith of the fi- ber, the material dispersion $M(\lambda) = -\lambda/c \cdot d^2n/d\lambda^2$ is an impor- tant parameter to investigate for fluoride glasses. The figu- res 5.a and 5.b show the dispersion curves versus wavelength measured on bulk ZBLA glass [19] and on a single mode fiber [20] respectively. The results are in good agreement and quite close to those published by [21][22]. The zero material disper‾sion is found at about 1.65 μm (compared to ∿ 1.3 μm for fused‾ silica) and the dispersion curve slope is lower for fluoride glass than for SiO$_2$.

TABLE 4: Absorption and scattering loss calorimetric results on bulk glasses using different starting materials (Zr_1 = ZrO_2 Merck; Zr_2 = ZrO_2 Cezus).

Glass	db/km at 0.647 μm		Starting materials
	$\alpha_{abs.}$	$\alpha_{scatter.}$	
$V_{core\ 1}$	42	20	Oxides (Zr_1) + NH_4HF_2 Merck
$V_{core\ 2}$	38	20	Oxides + ZrF_4 756
$V_{core\ 3}$	32	22	Oxides + ZrF_4 756 sublimated
$V_{clad.1}$	52	26	Oxides (Zr_2) + NH_4HF_2 Merck
$V_{clad.2}$	38	38	" " " "
$V_{clad.3}$	49	25	Oxides (Zr_2) + NH_4HF_2 Prolabo
$V_{clad.4}$	20	10	Oxides (Zr_1) + NH_4HF_2 Prolabo

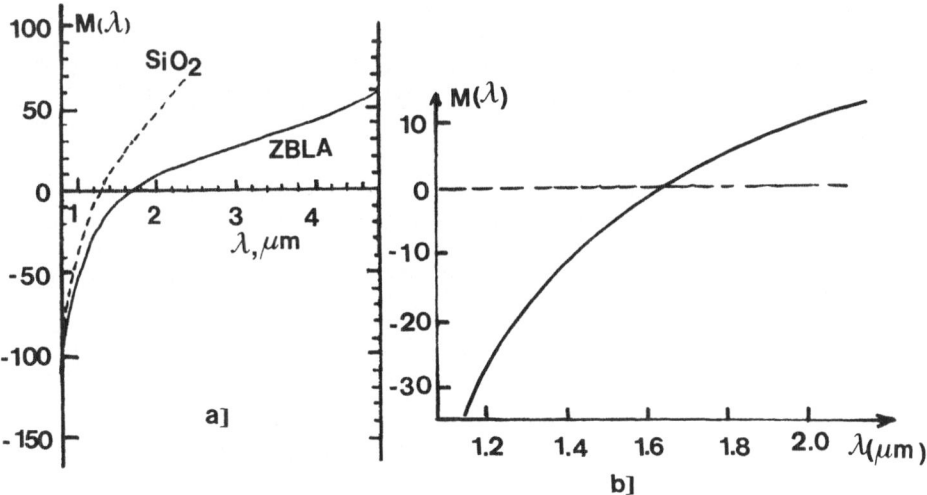

FIGURE 5: Material dispersion spectra for ZBLA glasses
a) on bulk glass   b) on fiber

4. PREFORM FABRICATION

4.1 Preform compositions

Table 5 lists the typically used core and cladding glass compositions for preform fabrication.

TABLE 5: Typical CNET and LVF fluoride glass preform compositions.

GLASS		(Mole %) ZrF_4	BaF_2	AlF_3	LaF_3	NaF	ThF_4	PbF_2	Δn
CNET	Core	56	33	3	3	4	–	1	7×10^{-3}
	Cladding	55	27	4	4	10	–	–	
LVF [5]	Core	56	30	5	5	–	4	–	7×10^{-3}
	Cladding	55	31	5	5	4	–	–	

In the CNET one, the refractive index difference is essentially resulting from a BaF_2/NaF substitution (only 1 mole% PbF_2 enters the core composition) while for LVF fibers, the Δn is achieved by NaF/ThF_4 replacement. In both cases, Δn values of about 7-8 x 10^{-3} are easily obtained and the thermal glass characteristics appear to be rather well matched.

4.2 Preform fabrication

The techniques used for fluoride glass preform manufacturing generally involve melting-casting processes. In fact, preforms are usually obtained by pouring the core glass melt into a cladding tube. This well known built in casting method is illustrated in figure 6. As can be seen, the cladding tube is achieved either by upsetting the metallic mold (previously heated near T_g) in which the cladding glass has just been poured (route 1), or by removing the mold bottom plate (route 2) allowing the glass still liquid to be evacuated. The step index preform is thus obtained by pouring the core glass melt immediately inside this cladding tube.

Such method, which gives quite smooth core-cladding interface appears, however, unsuitable for achieving homogeneous thickness cladding tubes and another process, previously called "rotational technique" by [23], has to be used. As indicated in figure 7, the calculated cladding glass weight is poured into the mold which then rotates at quite high speed (3000-5000 rpm).

Highly concentric tubes are obtained in this way. After a short time (10-20 sec., depending on the glass melt size) the rotation is turned off and the core glass poured into the tube. However, the internal aspect of the core-cladding interface appears different from the corresponding built in casting one and, usually, more bubbles appear at this interface, even when

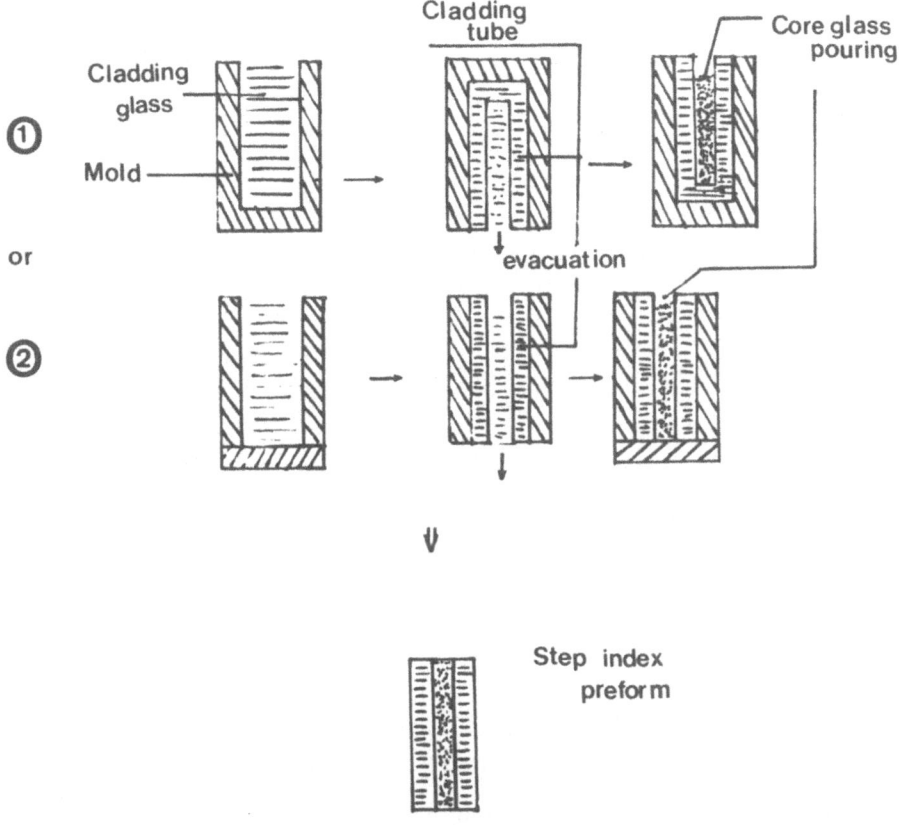

FIGURE 6: Fluoride glass preform built
in casting method.

the mold and pouring temperatures are suitably adjusted. Using
one or the other method, preforms of 12 mm outer diameter and
100 mm length are obtained, with core size within 4-8 mm.

5. FIBER FABRICATION AND CHARACTERIZATION

5.1 Fiber drawing

After polishing, the preforms are drawn into fibers using a
conventional drawing system, involving a resistance heated furnace, especially designed for maintaining a dry atmosphere
around the heated preform, in order to prevent some glass crystallization resulting from ambient moisture attack. CNET and

274

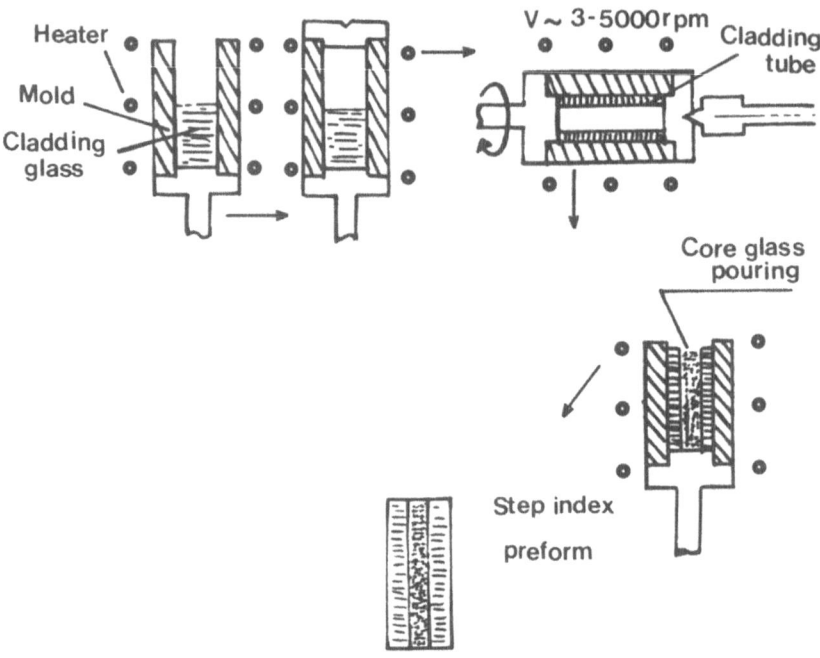

FIGURE 7: Fluoride glass preform
rotational casting.

LVF preforms are typically drawn at 340° and 380° respectively, with drawing speeds between 5 and 15 m/mn. A U.V. curable
coating is usually applied on the fiber to improve its mechanical properties.

5.2 Fiber characterization

The fluoride glass fiber loss measurements are carried out
using a system consisting mainly of a light source, a monochro
mator and a cooled PbS or InSb detector, using the cutback
method. The figure 8 shows the loss spectrum for the best
french fluoride glass fiber up to now.

As can be observed, the minimum loss is 13 db/km at the 2.65
µm wavelength with a 2.9 µm OH obsorption peak of about
110 db/km.

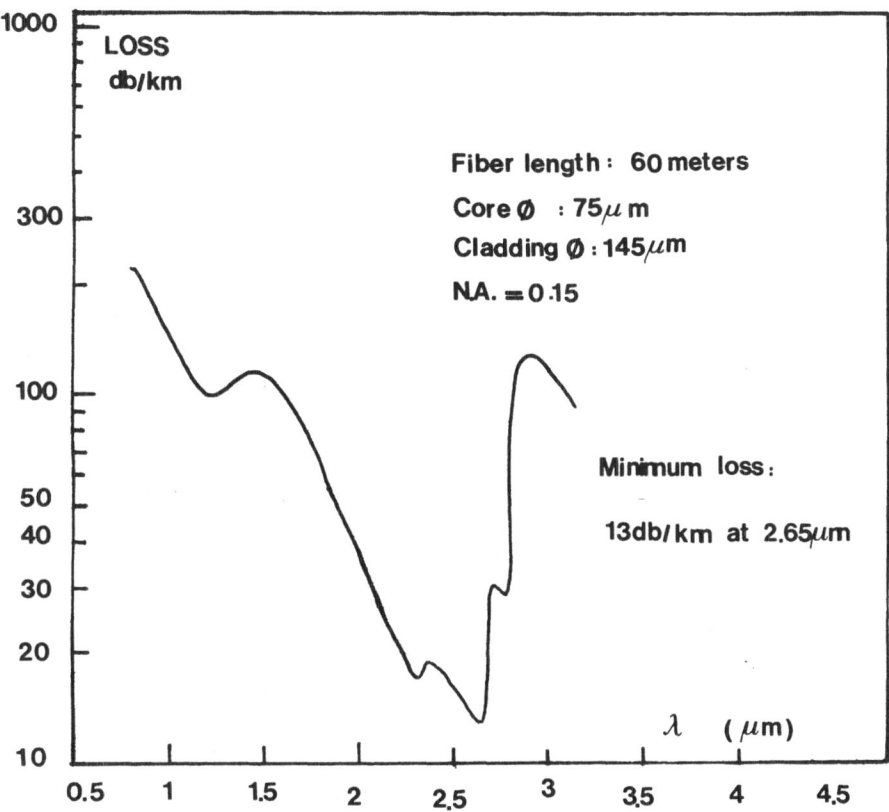

FIGURE 8: Loss spectrum of a 60 meter long
fluoride fiber (LVF).

6. FUTURE PROSPECTS

Though the studies actually performed at CNET and LE VERRE
FLUORE S.A. will still continue, some other research aims appear now, especially devoted to the single mode fiber achievement and the development of fluoride glass making techniques
such as vacuum and vapour processes.

6.1 Single mode fiber fabrication

Some work has already started at both laboratories to attempt making such kind of fibers. The method involved is the
preform in tube one, previously described by [24] and schemati
cally presented in the figure 9.

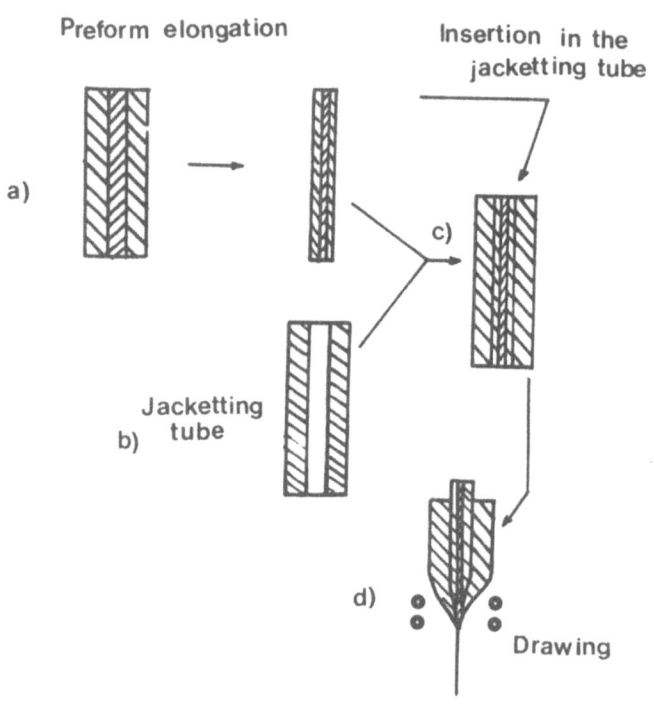

FIGURE 9: Actual single mode fluoride glass
fiber preparation method.

After being elongated, a glass preform is inserted in a
jacketting tube with well defined thickness and refractive index. Figure 10 shows the refractive index profile obtained
with a twice elongated preform. Using such process, core diameters as low as 0.5 mm can be achieved.
    Recently, a fluoride glass single mode fiber with 120 db/km
loss at 2.0 μm was reported [20].

6.2 Fluoride glass vacuum deposition
    Since the first studies dealing with the vacuum fluoride
glass evaporation [9], some improvements have been brought to
the method. Starting with fluoride glasses based on PbF_2-MnF_3-
-GaF_3 or PbF_2-ZnF_2-GaF_3 systems [25], homogeneous fluoride
films were successfully deposited into ZBLA glass substrates
[26]. The figure 11.a shows the weight increase of the substra

te and the film thickness versus the experiment time. The thickness remains practically constant for durations exceeding 0.5 hour (e ∿ 12-13 μm) in the following conditions: evaporation temperature = 765°C, starting glass charge = 1 g of PMG glass).

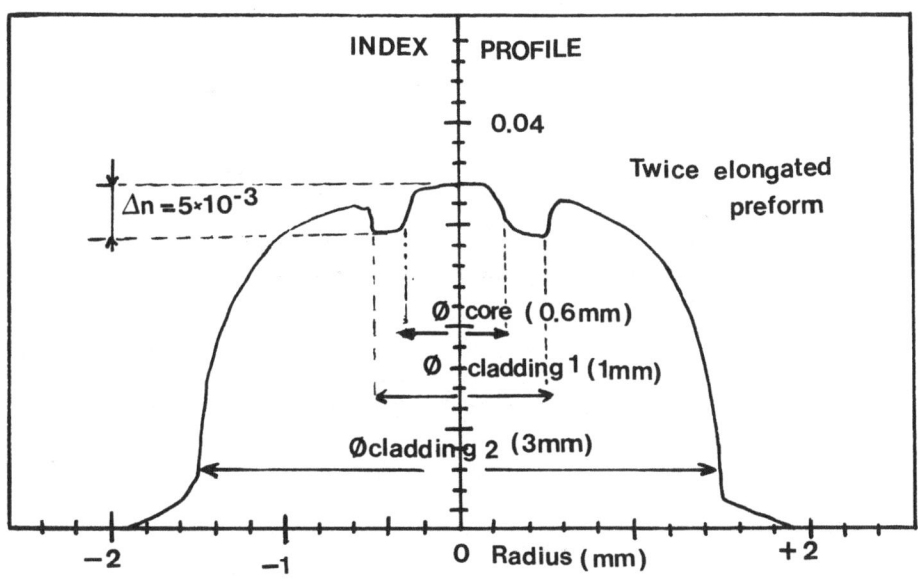

FIGURE 10: Refractive index profile of a
twice elongated preform.

On the figure 11.b, the deposited film thickness is plotted versus the evaporation temperature for different starting glass compositions. The PZG system appeared to be the most evaporable, allowing thicknesses as high a s24 μm to be deposited. For temperatures higher than 810°C, the film tends to crack, owing to too high a deposition rate.

The figure 11.c presents the film thickness versus the starting glass charge. Chemical analysis of the deposited film revealed it consists mainly of PbF_2 and GaF_3, in the 45-55 mole % range. When using PZG glass starting charge, 2 or 3 mole % ZnF_2 were found in the film composition.

6.3 Vapour deposition

Though experiments in this way have not yet really started, studies devoted to the fluoride glass collapsing are currently underway. The first results indicate this operation is quite feasible in special experimental conditions. Fluoride components deposition will be likely undertaken in the very near future.

278

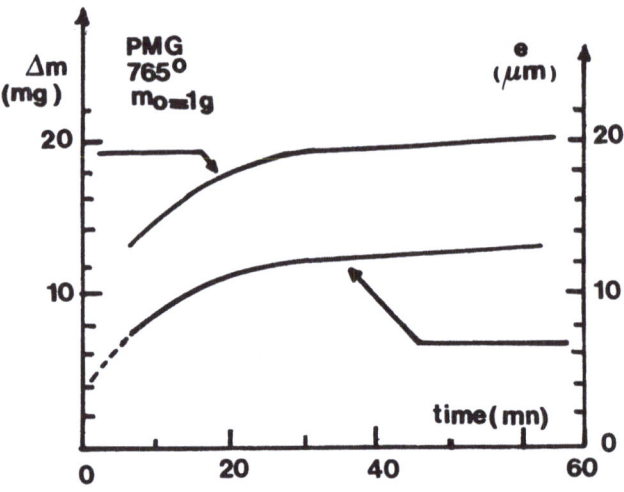

FIGURE 11.a: Plot of film weight and
thickness versus evaporation time.

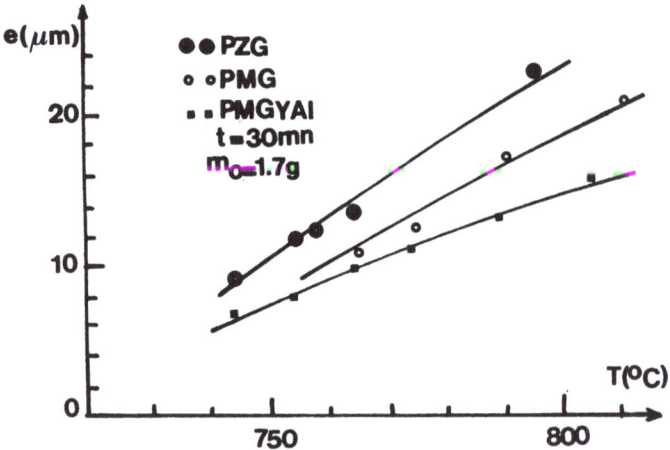

FIGURE 11.b: Plot of film thickness versus
evaporation temperature for
different starting glass
compositions.

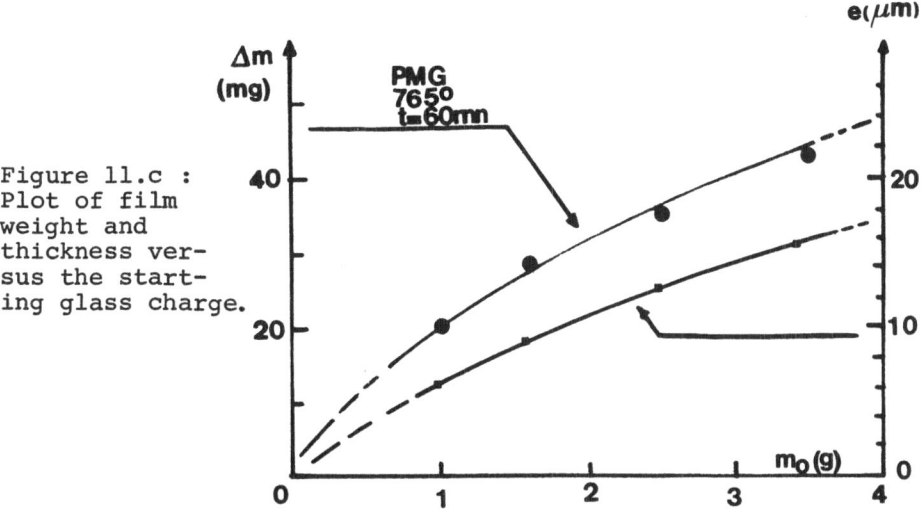

Figure 11.c :
Plot of film
weight and
thickness ver-
sus the start-
ing glass charge.

7. CONCLUSION

In this paper, we have described the present state of the art on fluoride glass fibers in France. Investigations on the properties of ZBLA based glasses have been carried out in order to yield suitable compositions for fiber drawing. Though the lowest loss achieved up to now is somewhat higher than the best published results, some loss decrease is likely to be achieved by careful purification of the starting materials and improvements in the preform casting methods. Loss of about 1 db/km seems feasible. However, this conventional fluoride glass fabrication method will be surely inappropriate for achieving the expected low loss fibers, and other processes in volving vapour deposition need to be studied and investigated.

REFERENCES

1. Shibata S. et al.: "Prediction of loss minima in IR optical fibers", Electron. Lett., 17, (1981), p. 775.
2. Poignant H.: "Dispersive and scattering properties of a ZrF_4 based glass", Electron. Lett., 17, 25/26, (1981), p. 973.
3. Poignant H.: "Multiphonon absorption in IR fluoride glasses", Electron. Let., 18, 5, (1982), p. 199.
4. Tran D.C.: "Fluoride glass optical fibers", Proceedings of the 3rd International Symposium on Halide glasses, Rennes (France/1985).
5. Maze G. et al.: "Reduction of OH absorption in fluoride glasses", J. Lightwave Technol., LT2, 5, (1984), p. 596.
6. Kanomori T. et al.: "Crystallization characteristics of ZrF_4 based glasses in cooling and reheating process", Jap. J. Appl. Phys. 24, 9, (1985), p. L758.

7. Shibata T., et al.: "Fabrication of high strength, low loss fluorozirconate glass optical fibers", Proceedings 3rd Int. Symp. on Halide glasses, Rennes (France/1985).

8. Tran D.C.: "Preparation of heavy metal fluoride glass optical fibers", Proceedings 2nd Int. Symp. glasses, Troy (USA), (1983).

9. Poignant H.: "Fluoride glass evaporation", Proceedings 3rd Int. Symp. on Halide glasses, Rennes (France/1985).

10. Lecoq A. et al.: "Étude phénoménologique du role stabilisa teur de l'aluminium dans les verres au ZrF_4", Verres et Réfractaires, 34, (1980), pp. 333-343.

11. Maze G. et al.: "Fluoride glass fibers for telecommunications" Proc. SPIE, 400, (1983), p. 60.

12. Mitachi S., et al.: "Impurity minimized fluoride glass optical fiber", IOOC 1983, Tokyo (Japon).

13. Poignant H.: "Impurity analysis of fluoride glass starting materials", Proc. 3rd Int. Symp. Halide glasses, Rennes (France/1985).

14. Falcou C.: Internal CNET study, not published.

15. Bansal N.P., et al.: "Effect of composition on the crystal lization behaviour of HMF glasses", Proc. 3rd Int. Symp. Halide glasses, Rennes (France/1985).

16. Zaganiaris A.: "Simultaneous measurement of absorption and scattering loss in bulk glasses and optical fibers by a microcalorimetric method", Appl. Phys. Lett., 25, 6, (1974).

17. Poignant H., et al.: "Viscosity and optical measurements of fluoride glasses", Electron. Lett., 18, 24, (1982), p.104.

18. Poignant H. et al.: "IR fluorozirconate and fluorohafnate glass optical fibers", 8th ECOC, Cannes (France 1982).

19. Jeunhomme L., et. al. "Material dispersion evaluation in a fluoride glass", 7th ECOC, Copenhague (Denmark 1981).

20. Monerie M., et al.: "Fabrication and characterization of fluoride glass single mode fibers", Electron. Lett., 21, 25/26, (1985).

21. Jingui K. et al.: "Material dispersion evaluation in a fluoride glass", Electron. Lett., 18, (1982), p. 164.

22. Bendow B. et al.: Appl. Optics. 20, (1981), p. 3688.

23. Tran D.C. et al.: "Fluoride glasses prepared by a rotational casting process", Electron. Lett., 18, 15, (1982), p. 657.

24. Miyashita T., et al.: "Progress in fluoride glass fiber research and development in Japan", 2nd Int. Symp. Halide Glass, Troy (USA 1983).

25. Jacoboni C. et al.: "Fluoride glasses of 3d transition metals", Glass Technol., 24, 3, (1983), p. 164.

26. Poignant H. et al.: "Vacuum deposition of fluoride films", to be published.

DISCUSSION

Q: C. Pantano. Did you verify or investigate whether the eva-
   porated films were, in fact, amorphous?

A: H. Poignant. The nature of the films depended on the initial glass composition and on the experimental conditions
   (substrate, temperature of the crucible, etc.). When the
   PMGYAl fluoride glass composition was used as the starting
   charge, the deposited films were always amorphous. However,
   further experiments are needed to elucidate the role of the
   starting glass charge upon the vitreous or crystalline nature of the deposited films.

PROGRESS IN FIBER PREPARATION IN JAPAN

S.YOSHIDA

THE FURUKAWA ELECTRIC CO., LTD.
9-15, 2-CHOME, FUTABA, SHINAGAWA-KU, TOKYO 142, JAPAN

ABSTRACT
   The current status of research work on fluoride glass optical fibers in Japan is reviewed. Glass compositions employed, purification of raw materials, technology for preparation of fibers and the transmission loss properties are described. The excess light scattering loss is examined as well. Fluoride glass fibers have attained a level enabling use in short haul applications.

1. INTRODUCTION

Although the main motive of efforts to develop infrared fibers has been to achieve ultra-low-loss properties enabling repeaterless long distance communication systems, there also are demands for short haul applications such as fiber sensing, infrared imaging, remote spectroscopy and laser power delivery. For these potential uses, four general types of infrared transmitting materials have on an experimental basis been formed into fibers; these are heavy metal oxide, halide and chalcogenide glasses, and halide crystals.

Crystalline fibers have the lowest theoretical transmission losses among these materials, but fibers primarily meant for fiber sensing or CO_2 laser light delivery have been prepared using KRS-5, AgCl or CsBr. The lowest loss obtained at 10.6 μm is 90 dB/km, and the graded index fiber has recently been prepared using KRS-5 and KRS-6. Chalcogenide glass fibers typically composed of As-S, Ge-Se or Ge-Se-Te have also been developed. However, as studies have revealed a disadvantage as regards transmission loss, that is a weak absorption tail, their applications are considered to be limited to short haul transmission, such as infrared spectroscopy or thermal imaging, utilizing their broad transparent wavelength region. As for oxide glasses, a germanate glass fiber has been developed using the VAD technique, a production process for high silica fibers. Its transmission loss was reduced before that of other materials in the early years of development, but as its infrared cutoff wavelength is shorter than that of the others, it is expected to find use as a highly efficient non linear fiber optic device, rather than in infrared fibers.

Halide glass, fluoride glass in particular, has been considered to be the most promising material for ultra-low-loss optical fibers in Japan, as well as in other countries. However, before realization of such fibers, there also are demands related to short haul applications.

In Japan, development of fluoride glass fibers has been conducted in the laboratories of three enterprises. Fibers apparently aimed at long distance repeater-less communication have been investigated at NTT and KDD Laboratories. The Furukawa Electric Co. is developing them mainly for short distance applications. In addition, research work on the basic glass structure has been performed in some universities.

2. MATERIALS AND PURIFICATION

Typical glass compositions developed for fiber preparation are shown in Table 1. These glass systems have been chosen on the basis of experimental analysis of stability against crystallization and other changes during preform fabrication and/or fiber drawing.

TABLE 1. Typical fluoride glass compositions developed in Japan.

Lab. Glass	Compositions	
NTT (1) ZBGA	Core	$60.5ZrF_4 - 31.7BaF_2 - 3.8GdF_3 - 4\,AlF_3$
	Cladding	$58.6ZrF_4 - 30.7BaF_2 - 3.7GdF_3 - 7\,AlF_3$
NTT (2) ZBLAYL	Core	$49ZrF_4 - 25BaF_2 - 3.5LaF_3 - 2.5AlF_3 - 2YF_3 - 18LiF$
	Cladding	$48ZrF_4 - 23.5BaF_2 - 2.5LaF_3 - 3.5AlF_3 - 2.5YF_3 - 20LiF$
KDD (3) ZBLAN(H)	Core	$53ZrF_4 - 20BaF_2 - 4LaF_3 - 3AlF_3 - 20NaF$
	Cladding	$39.7ZrF_4 - 18BaF_2 - 4LaF_3 - 3AlF_3 - 22NaF - 13.3HfF_4$
Furukawa(4) ZBLAN(I)	Core	$54.9ZrF_4 - 17.7BaF_2 - 3.9LaF_3 - 3.7AlF_3 - 14.7NaF - 0.2InF_3 - 4.9PbF_3$
	Cladding	$54.9ZrF_4 - 22.6BaF_2 - 3.9LaF_3 - 3.7AlF_3 - 14.7NaF - 0.2InF_3$

ZBGA and ZBLAYL glasses have been developed at NTT's Ibaraki Electrical Communication Laboratory (1-2). Crystallization kinetics of various ZrF_4 based glasses in melt-cooling and reheating processes, which simulated preform casting and fiber-drawing respectively, were investigated by calorimetry. Critical cooling rates for glass formation were estimated, and kinetic parameters of crystallization in reheating were obtained just above deformation temperatures. As the result of these experiments, it was concluded that ZBGA and ZBLAYL glasses are stable against crystallization during fiber-drawing processes. The ZBLAYL glass in particular shows high stability during both cooling and reheating.

The possibilities for ZBLAN(H) glass were suggested by researchers at KDD Laboratory after investigations in which the relationship between reheating conditions and excess scattering was clarified for many core and cladding glass compositions (3). HfF_4 dopant in cladding characterizes this glass system, which is free from crystallization around the fiber-drawing temperature.

Furukawa's Central Research Laboratory has investigated the ZBLAN(I) glass system (4-5). Although a precise understanding has been gained of the effect of NaF in stabilizing fluorozirconate glasses in recent years, the glass was itself discovered relatively early (6). Addition of small amounts of InF_3 have proved effective in reducing the light scattering due to an inhomogeneous microstructure in the fluoride glass.

The purity of materials used in these glass systems greatly affects the optical transmission properties of fibers. Various methods to remove impurities from raw materials have been investigated (7). These include recrystallization, extraction, chemical vapor-phase deposition, distillation and sublimation. Through the evaluation of transmission losses of fibers from materials thus purified, sublimation has been found to be the most effective technique for reducing impurities below 0.1 ppm. Fig. 1 shows loss spectra of purified fibers. ZrF_4 and AlF_3 can be sublimated and separated from impurities; on the other hand, troublesome impurities are themselves sublimated out from BaF_2 and GdF_3.

Techniques for reactive atmosphere processing of the glass melt are also useful for eliminating loss increases due to impurities. Bubbling a glass melt with NF_3 was found effective for reducing mid-infrared losses by oxidizing Fe^{2+} and Cr^{2+} ions and shifting their absorption peaks to shorter wavelengths (8). The estimated loss due to ferric ions (Fe^{3+}) at 2.5 and 3.5 μm is far lower than that due to ferrous ions (Fe^{2+}) as indicated in Table 2.

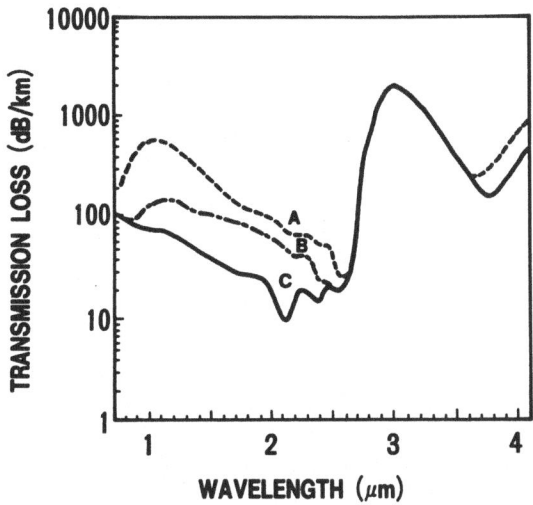

A : Fiber with recrystallized materials.
B : Fiber with sublimated ZrF₄, BaF₂, GdF₃, and AlF₃,
and with untreated NH₄F·HF.
C : Fiber with sublimated ZrF₄, BaF₂, GdF₃, AlF₃, and NH₄F·HF.

FIGURE 1. Transmission loss spectra with sublimated materials (7).

TABLE 2. Estimated absorption loss due to iron ions (8).

Ions State	Loss(dB/km/10^6)	
	2.5μm	3.5μm
Fe^{2+}	33.3	19.5
Fe^{3+}	$<10^{-3}$	$<10^{-3}$

Complex ions such as OH^-, NO_3^-, SO_4^{2-} and CO_3^{2-} contained in the raw materials bring about loss increases in the minimum loss regions of fluoride glass fibers as well. Besides, there is transmission degradation due to NH_4^+, which is added in the form of NH_4HF_2 during the glass preparation. ZBLAN glasses were prepared under Ar or NF_3 atmospheres to investigate the effect of NF_3 processing in eliminating these impurity ions (9). Although CO_3^{2-} and NO_3^- could be thermally decomposed in Ar, oxides were produced in the glasses, causing a scattering increase and crystallization of the glasses. In contrast, the NF_3 processing suppressed such degradation completely. NH_4^+ ions which cause serious absorption peaks in the minimum loss region, have been removed by NF_3 processing. SO_4^{2-} was decomposed under both atmospheres without any scattering increase. Unfortunately, it was found difficult to remove PO_4^{3-} by NF_3 processing.

Reactive atmosphere processing using gases such as Cl_2, CCl_4, HF and SF_6 has been reported to be effective in removing the most troublesome hydroxides in the fluoride glasses. However, there seems to be a problem of scattering loss increases. NF_3 processing has also been found preferable in this respect (10). In order to throw light on the dehydration effect and oxidation resistant properties, glasses doped with NaOH and ZrO_2 were prepared under Ar or NF_3 atmospheres. Measurements were made of their absorption spectra, excess scattering and tendency to crystallize. Under an Ar atmosphere, a part of the hydroxides were decomposed thermally, but oxides were consequently produced in the glasses, which caused excess scattering and crystallization. On the other hand, NF_3 processing was proven superior, in that the hydroxides and the oxides in the glasses were eliminated without any resultant scattering increase.

To estimate the loss due to OH absorption accurately, evaluation of the molar extinction coefficient for OH in fluoride glasses is helpful. Its value has been obtained by a thermo-gravimetric technique in which the change in infrared light absorption and in weights of sample glasses during hydration and dehydration processes were measured (11). The result of 16.4 l/mol/cm for ZBGA glass is very small compared with values for silica glass, 77-181 l/mol/cm.

3. FIBER PREPARATION

Both preform rod drawing and crucible drawing processes have been employed in fiber preparation. So as to obtain highly transparent preforms, melting and casting of fluoride glasses are performed under atmospheric control (1, 12), in order to decompose impurities, oxidize harmful ions and avoid formation of oxide particles and of reduced Zr. To satisfy these many requirements, optimization of atmospheric processing, including selection of gases and flow conditions, is necessary; but a satisfactory process seems not to have been established yet, though various investigations are being performed and NF_3 processing may be part of the answer.

Core-cladding structures of preforms are fabricated by built-in casting (7) or rod-in-tube techniques (4). In the former, a newly designed dry box has been used in which the core glass and cladding glass can be cast in an extremely dry atmosphere (1). In the latter, the core glass has been cast under an atmosphere of dry nitrogen, and the rod has been polished mechanically with grinding materials. Cladding tubes have been prepared by rotational casting (13). The preform is about 9 mm in diameter and 10 cm long. After jacketing a teflon-FEP tube, the preform is drawn into fibers using a platinum ring heater whose heating zone is fairly narrow. Fibers are also drawn under atmospheric control, for instance, in dry helium gas (4). For short haul applications, teflon-FEP cladding fibers are fabricated as well.

One of the causes of excess losses in actual fluoride glass fibers is considered to be drawing-induced scattering. The relation beetween the scattering intensity and reheating conditions in ZBLAN glasses was investigated with precision (14). It has been revealed that the drawing-induced scattering is essentially caused by nucleation of microcrystallites, and begins to increase at a temperature which is fairly lower than that predicted by DTA or DSC. Nevertheless, it has been concluded that fluoride glass fibers can be drawn using the preform method without remarkable drawing induced scattering given optimized glass compositions (ZBLAN) and reheating conditions. An example of reheating-induced scattering characteristics is shown in Fig. 2.

Examples of fluoride glass fibers drawn from preforms are: a) a 230 μm core diameter fiber with teflon-FEP cladding and 160 m in length (5), and b) a fiber with 36 μm and 147 μm core and cladding diameters and 150 m in length (1). A single mode fiber was also prepared with a depressed cladding structure, the core, cladding and jacketing of which were 18, 92 and 170 μm in diameter respectively (15).

As preforms cannot be made large bacause of the necessity of rapid quenching, which is fatal for fluoride glasses, the fibers drawn to date have been limited in length to 100 m or so. Consequently, the crucible drawing technique is expected to enable continuous drawing of long fibers. A simulation of crucible drawing was conducted to determine whether it would be applicable or not (16). After ZBLAN glass melt was held for 15 minutes near the fiber-drawable temperature, it was quenched to room temperature. Its Rayleigh scattering loss did not increase, as shown in Fig. 3. Based on this result, fibers were drawn using a double crucible system, shown in Fig. 4, controlling the glass melt holding time and the temperature of the crucible, and 500 m long single-mode and 700 m long multimode fibers were obtained.

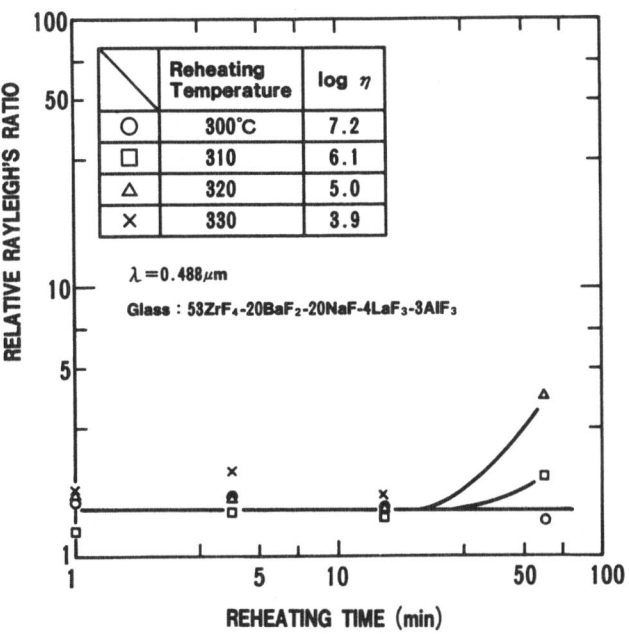

FIGURE 2. Scattering intensity vs. reheating time (14).

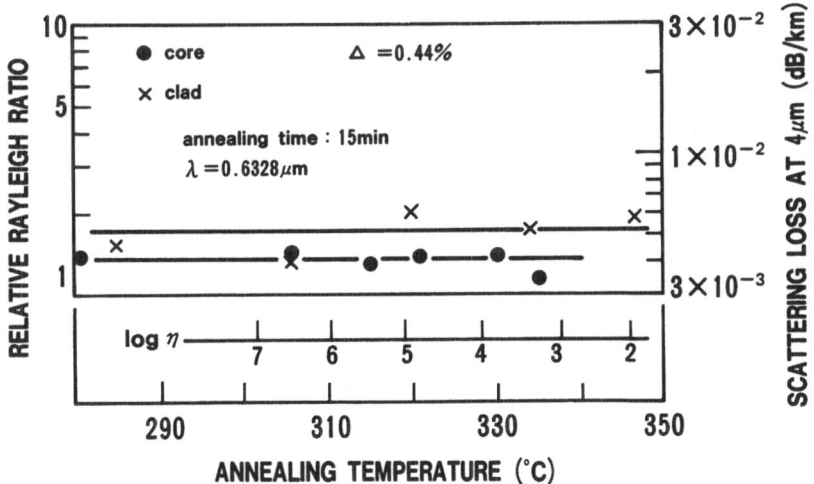

CORE : 53ZrF$_4$-20BaF$_2$-20NaF-4LaF$_3$-3AlF$_3$
CLADDING : 39.7ZrF$_4$-18BaF$_2$-22NaF-4LaF$_3$-3AlF$_3$-13.3HfF$_4$

FIGURE 3. Scattering-annealing characteristics of ZBLAN(H) glass (16).

As the current FEP and ACC coated fluoride glass fibers have low mechanical strengths, reinforced cabling technology has been developed (4). It is also described in another paper of this volume (17).

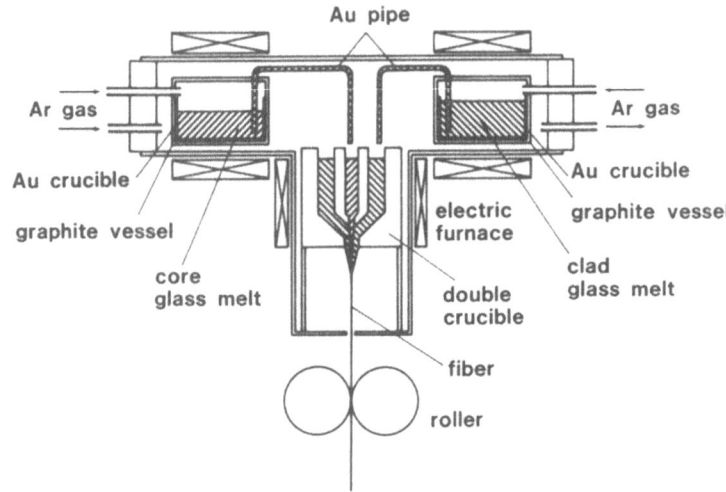

FIGURE 4. Schematic diagram of double crucible fiber-drawing system (16).

4. SCATTERING LOSS ANALYSIS

As the scattering loss comprises a major portion of transmission losses in fluoride glass fibers, analysis and elimination of its causes are the present concerns. To clarify the origin of the excess scattering, the scattering behavior of glass melts upon cooling was investigated (18). Light scattered at 90° to the incident ray from glass melt in a graphite cell was measured while the glass temperature fell from 500°C to 200°C. In stable glasses, there was homogeneous light scattering whose intensity depended strongly on the melting time. As can be seen in Fig. 5, which was obtained at room temperature, when glasses were melted for more than one hour, they no longer had bright scattering centers, and the scattering intensity showed a Rayleigh λ^{-4} dependence. It has been inferred that this type of scattering is due to corpuscles of oxide impurities contained in the raw materials and/or produced during melting by reaction with atmospheric gases. This is because a sufficiently long heating time is necessary to dissolve oxide impurities and to incorporate them into the glass network.

On the other hand, oxygen contents in fibers having various loss values were examined by the activation analysis method using a cyclotron (19). In a relatively high loss fiber, the loss spectrum showed a linear relationship with λ^{-2} and a fairly large λ-independent scattering loss, which was due to structural imperfections. The latter decreases with the oxygen content, as indicated in Fig. 6. This implies that oxides in fluoride glass fibers cause the scattering loss.

The above invenstigations both suggest that the removal of oxides from fluoride glasses is indispensable for obtaining ultra-low-loss fibers.

To examine scattering properties of fluoride glass fibers, a system for measurement of spherical coordinate angular distributions of scattered light from the fiber was developed and used (20). From the data measured and the loss spectrum difference between two kinds of fibers, the scattering losses were analyzed and interpreted as follows. Non-Rayleigh scatterng with λ^{-2}-dependent and λ-independent properties appears as strong forward scattering, is due to defects in fibers, and is dominant in scattering losses measured to date. The Rayleigh scattering term has

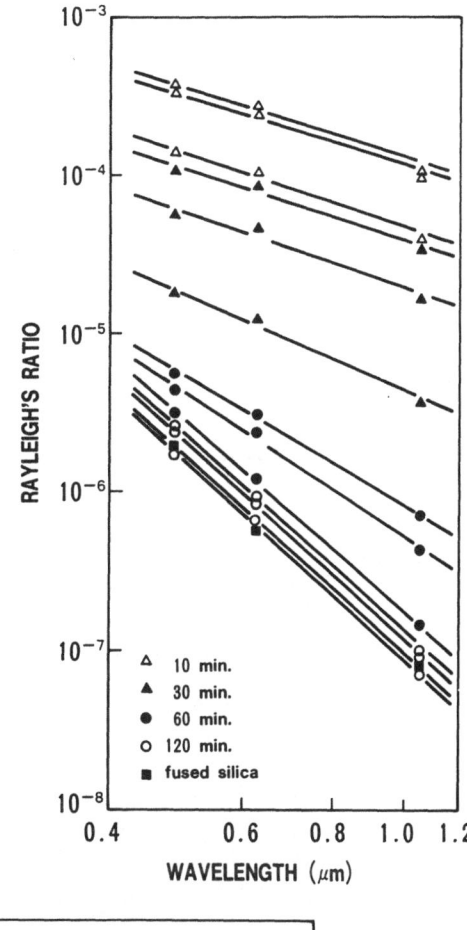

FIGURE 5. Scattering characteristics
for ZBLAN glass (18).

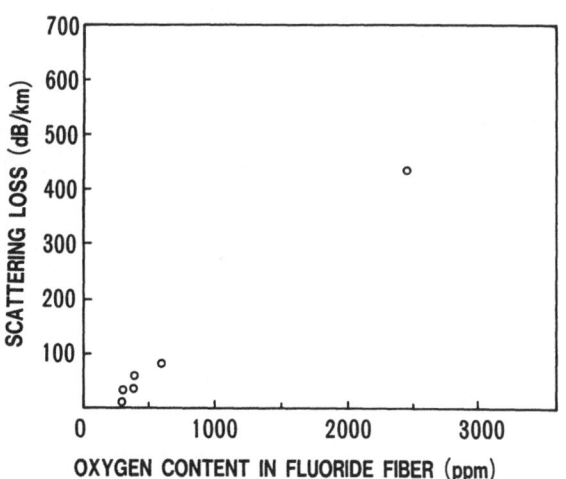

FIGURE 6. Scattering loss due to imperfections vs. oxygen content in fluoride glass
fiber (19).

been obtained at 0.63 μm using a low scattering fiber segment. Its value extrapolated to 2 μm is 0.08 dB/km, which confirms the possibility of achieving a transmission loss below 0.1 dB/km in the 2 μm band.

In measuring the scattered power from fluoride glass fibers using a cubic detector, some peaks at visible to near-infrared wavelengths were observed, which seemed to originate in the photoluminescense effect of rare earth impurities (21).

5. TRANSMISSION LOSS

The lowest reported loss of 3 dB/km was achieved at 2.35 and 2.55 μm by largely eliminating the causes of absorption and scattering in ZBLAYL glass fiber (12, 22). The fiber measured was 30 m long, and along it were observed visible He-Ne laser light scattering points of very small size, less than 1 μm. The spectral loss is dominated by a scattering loss with a λ^{-2} dependence and by transition metal and OH impurity absorption peaks, as illustrated in Fig. 7. The minimum loss has been estimated to consist of 2 dB/km in scattering and 1 dB/km in absorption.

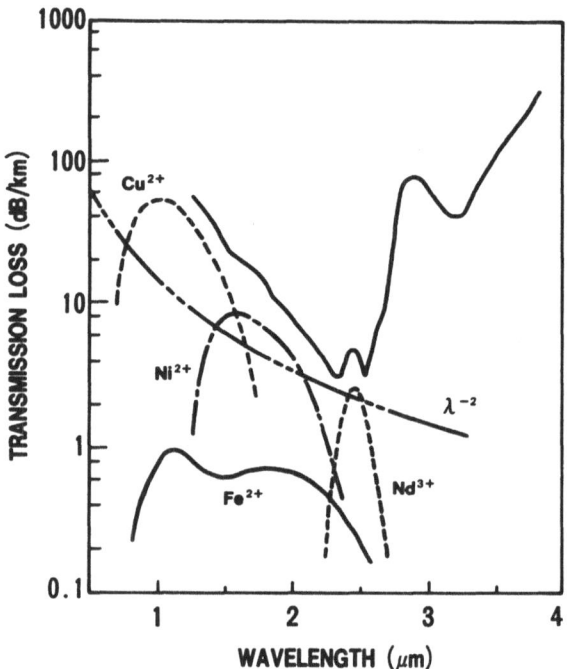

FIGURE 7. Spectral loss of a ZBLAYL fiber (12).

6. SUMMARY AND CONCLUSION

The state of the art of fluoride glass fiber preparation in Japan was reviewed. Steady and persistent studies are being continued to improve the transmission losses and fabrication processes of fluoride glass fibers. These processes are related to glass compositions, purification of raw materials, preform fabrication processes, fiber drawing, analysis of loss causes, and other areas.

Although the minimum loss reported is still far higher than theoretical limits, spectral losses of fibers have been reduced sufficiently to enable use in short haul applications such as infrared optical sensing and spectroscopy. Practical uses of fluoride glass fibers are expected to be important because fabrication processes will progress largely through actual applications. Practical applications will also be helpful in realizing ultra-low-loss properties, which are the final goal of infrared fiber development.

Lowering the minimum loss of fluoride glass fibers far below that of high silica fibers is a difficult yet feasible task. In order to accomplish it, some kind of technical breakthrough seems to be necessary. Moreover, investigations of the optical and mechanical reliability of fibers will be essential. Consequently, persistent research and development must be continued.

REFERENCES

1) Mitachi S., Ohishi Y. and Takahashi S., Jpn. J. Appl. Phys. 23, L 726 (1984).
2) Kanamori T. and Takahashi S., Jpn. J. Appl. Phys. 24, L 758 (1985).
3) Tokiwa H., Mimura Y., Shinbori O. and Nakai T., IEEE J. Lightwave Technol. 3, 574 (1985).
4) Shibata T., Takahashi H., Kimura M., Ijichi T., Takahashi K., Sasaki Y. and Yoshida S., Proc. 3rd Int. Symp. Halide Glasses, (1985).
5) Ohsawa K. and Shibata T., IEEE J. Lightwave Technol. 2, 602 (1984).
6) Ohsawa K., Shibata T., Nakamura K. and Yoshida S., Proc. 7th ECOC, 1.1 (1981).
7) Mitachi S., Terunuma Y., Ohishi Y. and Takahashi S., IEEE J. Lightwave Technol. 2, 587 (1984).
8) Nakai T., Mimura Y., Tokiwa H. and Shinbori O., Electron. Lett. 21, 625 (1985).
9) Nakai T., Mimura Y., Shinbori O. and Tokiwa H., Jpn. J. Appl. Phys. 24, 1658 (1985).
10) Nakai T., Mimura Y., Tokiwa H. and Shinbori O., to be published in IEEE J. Lightwave Technol.
11) Mitachi S., Sakaguchi S. and Takahashi S., to be published in Physics and Chemistry of Glasses.
12) Iwasaki H., Proc. SPIE, Jan. (1986).
13) Tran. D.C., Fisher C.F. and Sigel G.H., Electron. Lett. 18, 657 (1982).
14) Tokiwa H., Mimura Y., Shinbori O. and Nakai T., IEEE J. Lightwave Technol. 3, 574 (1985).
15) Ohishi Y., Mitachi S. and Takahashi S., IEEE J. Lightwave Technol. 2, 593 (1984).
16) Tokiwa H., Mimura Y., Nakai T. and Shinbori O., Electron. Lett. 21, 1131 (1985).
17) Yoshida S., this volume.
18) Nakai T., Mimura Y., Tokiwa H. and Shinbori O., IEEE J. Lightwave Technol. 3, 565 (1985).
19) Mitachi S., Sakaguchi S., Yonezawa H., Shikano K., Shigematsu T. and Takahashi S., Jpn. J. Appl. Phys. 24, L 827 (1985).
20) Ohishi Y., Kanamori T., Mitachi S. and Takahashi S., Appl. Opt. 24, 3227 (1985).
21) Ghosh G., Kachi S., Nakamura K. and Kimura M., to be published in Jpn. J. Appl. Phys..
22) Kanamori T., Sakagushi S. and Takahashi S., (1985) Nat. Conf. Rec. on Sc. Tech., IECEJ, 417.

DISCUSSION

Q: J. Lucas. Could you please comment on the composition of the fiber that was claimed to reach 0.7 dB/Km? In order to avoid the 2.5 µm absorption of Nd^{3+}, do you think that LaF_3, which is always contaminated by Nd, may have been replaced by another rare earth like YF_3?

A: S. Yoshida. I believe there was some residual Nd-induced loss.

Q: J. Lucas. Do you have an aging problem with KRS-5 crystalline fibers?

A: S. Yoshida. We can obtain very stable fibers by annealing them after extrusion.

C: C. Moynihan. The viscosities reported by Tokiwa et al. (KDD, Japan) in their studies of development of scattering losses with time and temperature may be in error (undervalued) in the 10^3-10^6 Poise region. In fact these viscosities were extrapolated from data measured above 10^7 Poise and, just above this temperature range, the η-T curves become very non-Arrhenian, making such extrapolations difficult and prone to error.

Q: A. Sarhangi. What kind of container was used for the NF_3 RAP application?

A: S. Yoshida. The container used by KDD workers consisted of a vitreous carbon or graphite crucible in a Telfon-coated silica reactor.

Q: H. Poignant. What was the length of the 0.7 dB/Km loss fiber and what was the error in this measurement?

A: S. Yoshida. I was told by NTT that the measured fiber was 30 m long and that the error was ±0.6 dB/Km (∿0.02 dB measured for 30 m of fiber).

VAPOR DEPOSITION OF FLUORIDE GLASSES

Ahmad Sarhangi

Applied Physics Dept., Corning Glass Works, Corning, New York 14831

ABSTRACT

The state of the art of fluoride glasses for fabrication of infrared fiber optics is reviewed. Different approaches for making ultra-pure raw fluoride materials and direct fabrication of fluoride preforms by chemical vapor deposition (CVD) are discussed. The results of CVD of metal fluorides for fabrication of ultra-low-loss fiber optics are discussed and recent progress in deposition of fluoride glass is noted.

VAPOR DEPOSITION OF FLUORIDE GLASSES

Non-silica-based infrared fiber materials with extremely low loss have attracted considerable interest. To achieve the ultra-low losses ($\sim 10^{-2}$ dB/km), the following fundamental material requirements are desirable:

1. The optical window should be sufficiently wide.

2. Scattering losses should be minimized.

3. Operation as close as possible to the material zero dispersion point.

4. Impurity and defect absorptions must be very low.

5. The fiber material must form a stable glass.

Great progress has been made towards making ultra-low-loss heavy metal fluoride glasses by conventional melting techniques. Purification of fluoride raw materials has been accomplished by numerous processes, including sublimation, distillation, extraction, crystallization, and chemical vapor deposition (CVD). The required level of transition and rare-earth metal impurities should be less than about 0.1 ppb to achieve the theoretical limit of 10^{-2} to 10^{-3} dB/km loss for these glasses.[1] Refinement of the raw materials and preparation methods has reduced heavy metal fluoride fiber attenuation from 500 dB/km in 1980[2] to 20 dB/km in 1982[3] to 6.5 dB/km in 1984[4] to 4 dB/km in 1985[5] to 0.9 dB/km in 1986.[6] Losses will, no doubt, continue to decrease up to a limit imposed by contamination inherent in the process of purification, handling, mixing, and crucible melting of fluoride raw materials. The development of cleaner methods for fabrication of fluoride fiber optics and bulk glasses is thus essential.

Chemical vapor deposition of fluoride glasses for optical waveguide application is very different from CVD of silica-

based fiber optics. The requirements for fabrication of
fluoride optical waveguides are:

1. Existence of a metal source compound with appreciable vapor
   pressure.

2. The source compound should be stable at ambient temperatures but react with a fluorine source gas to form
   nonvolatile fluoride deposit and volatile by-products at
   the conditions of deposition.

3. The deposition process should be carried out in an inert
   atmosphere. This suggests that CVD of fluorides could be
   an ICVD-type process, but other processes have been used
   (vide infra).

4. The substrate glass tube should have compatible physical
   properties with the fluoride glass and withstand the
   deposition process requirements.

The following summary highlights many different approaches
described elsewhere for making ultra-pure raw fluoride
materials or the direct fabrication of fluoride preforms by
chemical vapor deposition.

M. Bridenne has reported preparation of ultra-pure ZrF_4 raw
material by the CVD method.[7] Zirconium tetraborohydride
vapor, $Zr(BH_4)_4$, was reacted with 10% F_2 or HF to give ZrF_4.
$ZrCl_4$ was used as starting material to synthesize the volatile
$Zr(BH_4)_4$ source material and the product, $Zr(BH_4)_4$, was purified by sublimation under vacuum at 25°C. The borohydrides of
the 3d elements are unstable and their decomposition products
are nonvolatile at this temperature.

$$ZrCl_4 + 4 \ LiBH_4 \longrightarrow Zr(BH_4)_4 + 4 \ LiCl$$

$$Zr(BH_4)_4 + 16 \ F_2 \xrightarrow{0°C} ZrF_4 + 4BF_3 + 16 \ HF$$

$$Zr(BH_4)_4 + 16 \ HF \longrightarrow ZrF_4 + 4 \ BF_3 + 16 \ H_2$$

The analysis of the products showed:

	Ni, ppb	Cu, ppb
$Zr(BH_4)_4$	700	300
$ZrF_4(F_2)$	400	250
$ZrF_4(HF, 40\%)$	2200	350

A manufacturing method for making fluoride compounds by CVD
was also described by N. Mitachi, Japan Patent Application
No. 82-51146.[8] A volatile source compound, e.g. alkyl,
alkoxide, or chloride is packed in Bubbler 4 (Fig. 1).
Hydrogen fluoride is the fluorine source and is generated from
NH_4F, or $NH_4F \cdot HF$, or liquid HF. In an experiment, $ZrCl_4$ was
used as source compound. Argon carrier gas was passed through

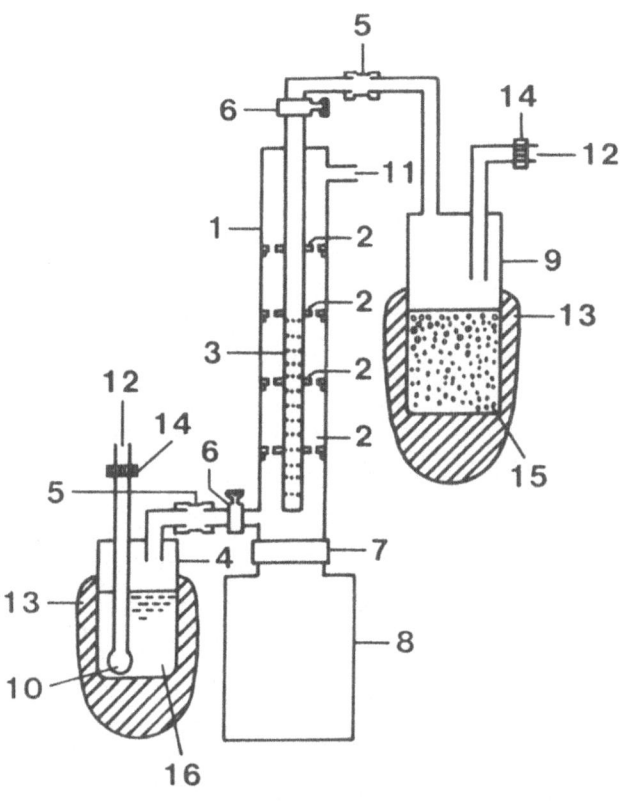

$$ZrCl_4 + 4\,HF \longrightarrow ZrF_4 + HCl$$

Fig. 1. Manufacture of ZrF_4 by CVD.
      (After ref. 8)

Bubbler 4 while it was heated to 340-350°C. The HF from Bubbler 9 passed through the three small holes of the poly-tetrafluoroethylene (PTFE) Pipe 3 to homogeneously diffuse into the PTFE reaction vessel. From reaction of $ZrCl_4$ and HF, ZrF_4 was formed which was collected in the Container 8.

$$ZrCl_4 + 4\ HF \longrightarrow ZrF_4 + 4\ HCl$$

The analysis of the fluorides made by this technique is shown in Table 1.

Mitachi, Japan Patent Application No. 82-175434[9], also described making fluoride preforms from reaction of organo-metallic source compounds with a mixture of $H_2 + F_2$ gas (Fig. 2). Bubbler 14 contains Zr $(OC_4H_9)_4$, bubbler 15 contains $Ba(OC_9H_{19})_2$, and bubbler 16 contains $Al(i-C_4H_9)_3$. Core Burner 11 produces ZrF_4 and BaF_2 from reaction of $Zr(OC_4H_9)_4$ and $Ba(OC_9H_{19})_2$ with HF. HF is formed from reaction of H_2 with F_2. Cladding Burner 12 produces AlF_3 from reaction of $Al(i-C_4H_9)_3$ with HF. The composition of the desired glass can be controlled with these burners and carrier gas flows. The reaction is carried out in an inert atmosphere. Like axial vapor deposition of oxides, the preform is pulled up as it grows. A glass consisting of BaF_2, AlF_3, ZrF_4, 25 mm diameter and 150 mm long was made by this technique which was drawn to a fiber. The fiber had "very low loss" at 2.5 and 3.5 μm.

Beryllium fluoride is a well-known glass former which also has the potential to have an even lower loss than the currently used SiO_2-based glass optical waveguides. The intrinsic optical loss in a medium originates from three sources: Rayleigh scattering, multiphonon IR absorption, and electronic UV absorption. Dumbaugh has reported the UV and IR transmission data for BeF_2 base glasses.[10] He has shown that BeF_2 base glasses absorb at shorter wavelengths than silica-base glasses, including pure SiO_2. Contamination of the melt with crucible material and the presence of transition and rare-earth metals in the raw material was shown to keep BeF_2 glasses from transmitting to their fundamental absorption edges. Oxygen and hydroxyl impurities also lower the ultra-violet transmittance of potential BeF_2 glasses. The infrared spectrum shows the presence of a hydroxyl group at 2.73 μm. Mackenzie demonstrated the effectiveness of the fluorination cycle and the removal of hydroxyl impurities in the raw material.[11] There, the fundamental OH infrared absorption band at 2.73 μm was removed from BeF_2 glass samples made from purified material and optical transmission improved in the region of 0.15 to 4.5 μm.

From material properties, the intrinsic or Rayleigh scattering losses for a single component glass can be calcu-lated. From comparison of physical properties of BeF_2 and SiO_2, it becomes clear that at any wavelength the BeF_2 glass should have lower scattering. Currently used operating wave-lengths for SiO_2 fibers are 0.85, 1.3, and 1.55 μm. In this region, the calculated intrinsic scattering of BeF_2 is an order of magnitude less than SiO_2. At present the lowest attenuation measured on a silica-based single-mode fiber is about .154 dB/km at 1.55 μm. This Rayleigh scattering limited attenuation allows for 100-200 km between repeaters. If a

TABLE 1

PURITY OF VAPOR DEPOSITED FLUORIDES - MITACHI ET AL

Fluoride Compd.	Impurity (weight %)							
	Zn	Al	Fe	Ni	Cu	Si	Ca	Li
ZrF$_4$ by Mitachi et al	0.002	ND	<0.001	0.001	0.0006	ND	ND	ND
BaF$_2$ by Mitachi et al	ND	0.088	ND	ND	ND	0.042	0.024	ND
GdF$_3$ by Mitachi et al	ND	ND	0.002	0.0003	0.0005	ND	ND	<0.05
ZrF$_4$ by conventional method	0.001	0.085	0.091	0.002	0.0007	ND	ND	ND

Note: ND means none detectable (below detection limit)

Reference: Japan Kokai 57-51146 (NTT), published March 25, 1982.

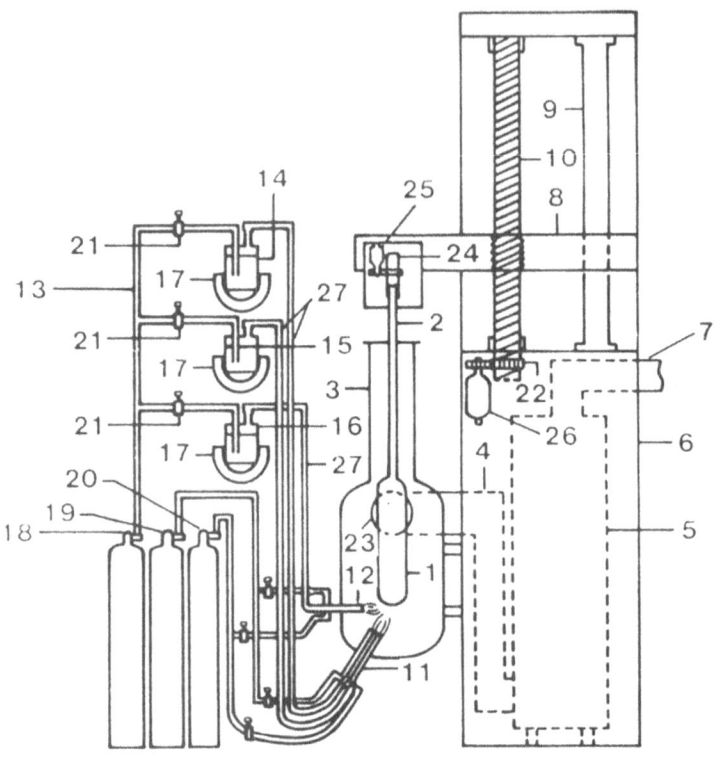

$$H_2 + F_2 \longrightarrow 2\ HF$$
$$Zr\ (O\ C_4H_9)_4 + HF \longrightarrow ZrF_4 + \ \ldots\ldots\ldots$$
$$Ba\ (O\ C_9H_{19})_2 + HF \longrightarrow BaF_2 + \ \ldots\ldots$$
$$Al\ (i-C_4H_9)_3 + HF \longrightarrow AlF_3 + \ \ldots\ldots\ldots\ldots$$

Fig. 2. Manufacture of fluorozirconate glass
preforms by CVD. (After ref. 9)

fiber attenuation of even 10^{-2} dB/km can be achieved, it would mean repeater spacing of ~ 1200 to 3300 km. Such a fiber could be very important for submarine communication systems. Another potential advantage of BeF_2 and other fluoride glasses over SiO_2 glasses is their resistance to radiation damage. Beryllium fluoride has a higher ionization potential ($\simeq$ 18 e.v.) than silica ($\simeq$ 12 e.v.).

In U.S. Patent No. 4,378,987 (1983), S. Miller et al[12], described the apparatus and method for manufacturing fluoride preforms as shown in Fig. 3. Bubbler 64 contains a fluorinating reagent gas, and bubblers 66 and 68 contain organometallic source compounds of beryllium and aluminum. Optical core and cladding ratio and compositions can be adjusted by varying the temperature and carrier gas flow. AlF_3 and BeF_2 may form according to the following reactions using BF_3 as the fluorinating agent:

$$3\ Be(CH_3)_2 + 2\ BF_3 \longrightarrow 3\ BeF_2 + 2\ B(CH_3)_3$$

$$Al(CH_3)_3 + BF_3 \longrightarrow AlF_3 + B(CH_3)_3$$

Investigations at Corning Glass Works into the various available CVD processes have indicated a number of potential difficulties. These include corrosiveness of materials, stability of source materials and collection efficiency. All of these aspects can affect product purity. For example, impurity levels well in excess of 1% have been observed at Corning in AlF_3 powders deposited from the reaction of aluminum organometallics with excess 100% HF at 150°C.

Steady progress has been made in overcoming these difficulties, and Corning scientists have recently succeeded in depositing very pure AlF_3. The level of transition metal impurities was measured in the starting material, organometallic source compound, and the deposited AlF_3 to see how effectively the impurities have been removed in each step.

Sample	Impurities, parts per billion				
	Fe	Cr	Cu	Mr	Ni
AlX_3	7968	1242	725	289	228
Aluminum-organometallic	327	367	193	26	75
AlF_3*	<149	<17	<35	<22	<47

*All values below detection limits noted.

We are continuing our examination of vapor deposition of fluoride materials. These results suggest that CVD offers fluoride purities comparable to today's oxide fibers or electronic grade materials. Many of the required source materials are available commercially to prepare fluoride glasses. Using the methods noted in this paper, conventional fiber processing steps beyond deposition, including collapsing, drawing and coating steps, may be based upon known technology.

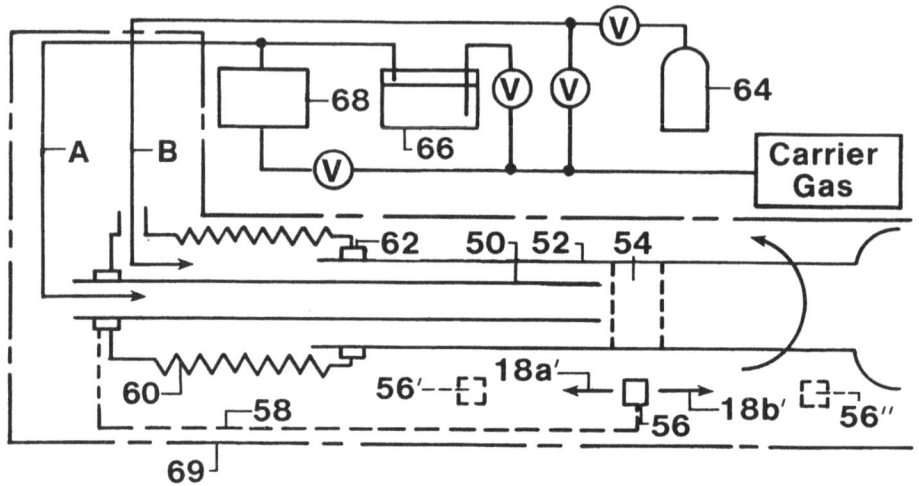

$$3 \text{ Be (CH}_3)_2 + 2\text{BF}_3 \longrightarrow 3 \text{ BeF}_2 + 2 \text{ B(CH}_3)_3$$

$$\text{Al (CH}_3)_3 + \text{BF}_3 \longrightarrow \text{AlF}_3 + \text{B(CH}_3)_3$$

Fig. 3. Manufacture of fluoroberyllate glass
preforms by CVD. (After ref. 12)

REFERENCES

1. P. W. France, S. F. Carter, and J. R. Williams, Proceeding of Third International Symposium on Halide Glasses, Universite de Rennes, France, June 1985.

2. S. Mitachi and T. Manabe, Japanese Journal of Applied Physics, Vol. 19, p. L313 (1980).

3. S. Mitachi and T. Miyashita, Electron. Letters, Vol. 18, p. 170 (1982).

4. D. C. Tran, Proc. NATO Advanced Study Institute, Tenerife (1984).

5. D. C. Tran, M. J. Burk, K. H. Levin, C. F. Fisher, P. Hart, L. Busse, G. Lu, and G. H. Sigel, Jr., Third International Symposium on Halide Glasses, Université de Rennes, France, June 1985.

6. D. C. Tran, Naval Research Laboratory, Washington, DC, Optical Fiber Communication Conference, Technical Digest, Atlanta, Georgia, February 24-26, 1986.

7. M. Bridenne, G. Folcher, H. Marquet-Ellis, Proceeding of Third International Symposium on Halide Glasses, Université de Rennes, France, June 1985.

8. Japan Kokai 57-51146 (NTT), published March 25, 1982.

9. Japan Kokai 57-175743 (NTT), published October 28, 1982.

10. W. H. Dumbaugh and D. W. Morgan, J. Non-crystalline Solids, 38 and 39, p. 211-216 (1980).

11. C. M. Baldwin and J. D. Mackenzie, J. Non-crystalline Solids, 31, p. 441-445 (1979).

12. U.S. Patent No. 4,378,987 issued April 5, 1983.

DISCUSSION

Q: P. Klein. How did you analyze your deposits?

A: A. Sarhangi. By atomic absorption and plasma emission spectroscopies.

Q: P.W. France. Could you comment on the relative merits of thermal CVD versus microwave or R.F. plasma CVD?

A: A. Sharhangi. Each one has advantages and disadvantages. The plasma process has high collection efficiency ($\sim$ 100%), but it needs a vacuum system and some time. Also, it is difficult to generate a stable plasma. The thermal process is simpler, but the collection effciency is less than 100%. Both R.F. and microwave plasma could be used, with the particular kind being defined by the specific process.

Q: J.D. Mackenzie. What was the glass tube on which you deposited BaF_2?

A: A. Sarhangi. K-F is a chloro-fluoro polyethylene which is transparent. Its structure is similar to that of teflon, except that it has both Cl and F in the polymeric chains.

Q: A.C. Wright. What is the glass-forming range in the AlF_3-BeF_2 system?

A: A. Sarhangi. The range of compositions we use is near 92.5 BeF_2-7.5 AlF_3.

Q: D. Tregoat. Have you done a carbon analysis for your AlF_3 deposits? The black color may be due to carbon.

A: A. Sarhangi. A chemical analysis of carbon is not easy to do and the detection limit is high.

WAVEGUIDE SYSTEMS DEVELOPMENT

S.YOSHIDA

THE FURUKAWA ELECTRIC CO., LTD.
9-15, 2-CHOME, FUTABA, SHINAGAWA-KU, TOKYO 142, JAPAN

ABSTRACT
Development in Japan of waveguide systems using halide glass fiberoptics is reviewed. As the optical losses of fluoride glass fibers obtained to date are still far higher than theoretical values, studies related to ultra-low-loss telecommunication systems are limited to basic subjects. On the other hand, short haul fibers are beginning to find practical applications, such as in infrared sensing technology. Examples of systems, as well as light sources, detectors and cabling techniques are described.

1. INTRODUCTION
The ultimate field of application of halide glass fibers is of course in long distance repeater-less communication systems. To realize such systems, various kinds of devices and components, together with ultra-low-loss fiber cables, must be developed. These include light sources, detectors, couplers, dividers, multiplexers and switches, operable in infrared wavelength regions.

However, as the transmission losses of fluoride glass fibers experimentally obtained to date are still far higher than theoretical values, realization of such systems remains as a goal for the future. Therefore, development of devices and components aimed at telecommunication applications has not proceeded, and only basic studies have been performed on limited subjects in this area. These are related to fiber cable properties, such as estimations of dispersion characteristics, microbending losses and splicing losses of single mode fluoride glass fibers.

On the other hand, infrared transmitting losses of fibers have become low enough to enable short haul applications, for instance, remote infrared sensing and laser light transmission. Work on some such systems, comprising devices and fiber cables, is progressing gradually on a trial basis.

2. INVESTIGATIONS RELATED TO TELECOMMUNICATION SYSTEMS
Long distance communication systems using fluoride glass fibers will aim for repeater-less low loss transmission and huge data capacities. Consequently, single mode fiber with low dispersion as well as low loss will be indispensable for such systems.

The dispersion in a single mode fiber is determined by the refractive index spectra of the fiber materials and the waveguide structure. The material dispersion has been calculated for some fluoride glasses, using their measured refractive index spectra. The material dispersion zero is usually located at wavelengths around 1.5 μm, somewhat shorter than the predicted minimum loss wavelengths (1). Nevertheless, slopes of the dispersion curves are far gentler than for silica, and consequently, the dispersion at the minimum loss wavelength is so small that cancellation of the material dispersion by the waveguide dispersion has been thought possible. In designing a dispersion shifted waveguide structure, a smaller core diameter with a large refractive index difference (Δn or $\Delta = \Delta n/n_1$) will be necessary.

Moreover, it was clarified that there exists an optimum design for fluoride glass fibers, in which the total dispersion would be held to around zero over a broad wavelength range (2). An example of an optimum design sets the core diameter and Δn equal to 6 μm and 0.02 respectively, which are less critical than those of a single mode silica fiber.

As the core diameters of single mode fibers are very small, such technology as built-in casting and rotational casting cannot be directly applied to preform fabrication. A preform prepared by built-in casting was inserted into a jacketing glass tube which was also prepared by upsetting the mold cast. The preform-in-tube thus obtained was drawn into single mode fibers (3).

Compositions and refractive indexes of glasses are shown in Table 1, and the refractive index profile of a fiber as measured using an interference microscope is illustrated in Fig. 1. Core, cladding and fiber diameters were 18, 92 and 170 μm respectively, and the refractive index differences at the boundaries were 0.3% and -0.1%.

TABLE 1 Glass composition and refractive index of single mode fiber

	ZrF$_4$	BaF$_2$	GdF$_3$	AℓF$_3$	PbF$_2$	n$_D$
Core	60.2	31.2	3.8	3.8	1	1.519
Cladding	59.2	31.1	3.8	5.7		1.513
Jacketing	60.5	31.7	3.9	3.9		1.516

(mol %)

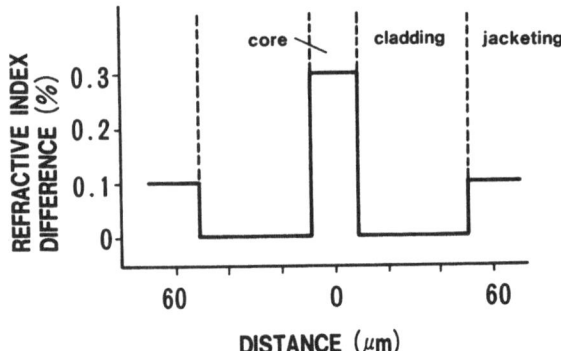

FIGURE 1 Refractive index profile of a fluoride glass single mode fiber

The cutoff wavelength was obtained from measurements of the bending loss spectrum, which is conventional method for silica fibers. The single mode region wavelength was greater than 2.7 μm, corresponding to the cutoff of the LP$_{11}$ mode, as seen in Fig. 2. This cutoff wavelength was in good agreement with the value of 2.77 μm estimated from the core diameter and Δn. The minimum transmission loss of this fiber was reported to be 160 dB/km at 3.28 μm, with an OH absorption peak of 450 dB/km at 2.85 μm.

In long distance transmission systems, micro-bending losses and splicing losses are added to the original fiber loss. As the fluoride glass fiber loss is expected to be

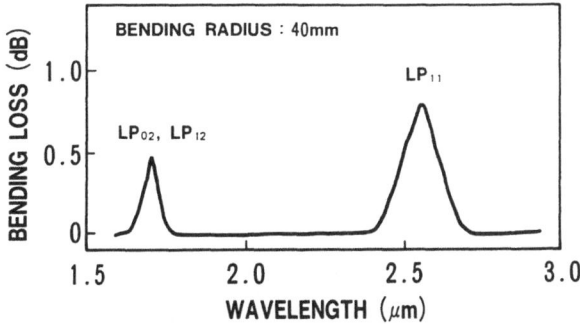

FIGURE 2 Bending loss spectrum of a fluoride glass single mode optical fiber

extremely low, on the order 0.01 to 0.001 dB/km, these additional losses will greatly affect the system design. The microbending losses are caused by cabling, and the splicing losses are determined by the offset and the number of splices per kilometer.

The microbending loss of single mode fibers increases rapidly with increasing wavelength. Its dependence on fiber parameters such as Δn and the core diameter is mainly determined by the correlation function of microbends produced in cabling processes. For silica fibers, the microbending loss spectra were calculated assuming that the random bends followed a Gaussian correlation function of the spatial frequency (4). Spectra of measured loss increases due to cabling proved to be in good agreement with the calculated curve, and the microbending correlation length could be estimated from these results.

Based on the fact that the microbending correlation length directly depends on the fiber stiffness, microbending loss spectra of fluoride glass fibers for various V values and Δn have been calculated, replacing the stiffness of the silica fiber by that of a fluoride glass fiber (5). The results are shown in Fig. 3, where the fiber diameter was assumed to be the same 125 µm as for a conventional silica fiber. In order to reduce microbending losses, higher Δn have been desired, and a loss of 0.001 dB/km was estimated for the entire 2-4 µm region given a Δn of 1%.

Splicing losses have also been estimated, and are shown in Fig. 3. Fusion splicing may be applied for fibers made of stable fluoride glasses. The splicing offset and the number of splices were assumed to be 0.4 µm, which is typical for silica fibers, and one per 10 km interval, respectively. In contrast with microbending losses, splicing losses increase with increasing Δn. At 2.5 µm, the loss has been determined to be 0.003 dB/km for a Δn of 1%, and 0.0015 dB/km for a Δn of 0.5%. The total additional loss is estimated to be 0.003 dB/km at 2.5 µm or longer, for a Δn of 0.7% or larger, as seen in Fig. 4. Fortunately, a higher refractive index difference is also desirable for cancellation of material dispersion by waveguide dispersion.

3. SHORT HAUL APPLICATIONS

Potential applications of infrared fiberoptics in the near future seem to lie in optical sensing systems and laser power transmission. Fluoride glass fibers will be applied in the remote monitoring of temperature, infrared images and radiation, in remote infrared spectroscopy, and in middle infrared laser light transmission.

Remote temperature monitoring systems using high silica fibers, through which radiometric energy is transmitted, are already in practical use in industry. However, they can be applied to higher temperature ranges only, so that expectations are being placed on fluoride glass fibers for measuring lower temperatures, down to around 80°C or so.

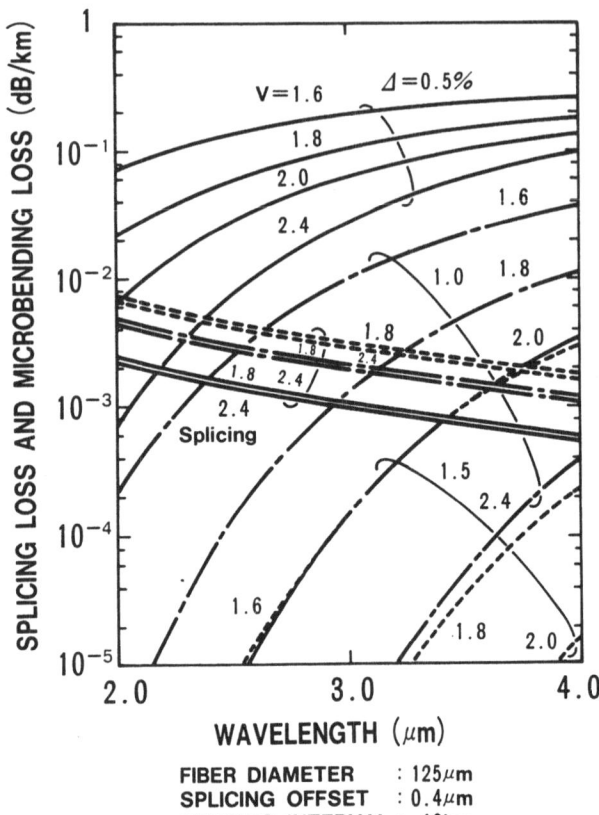

FIGURE 3 Calculated splicing and microbending losses for parameters Δn and V

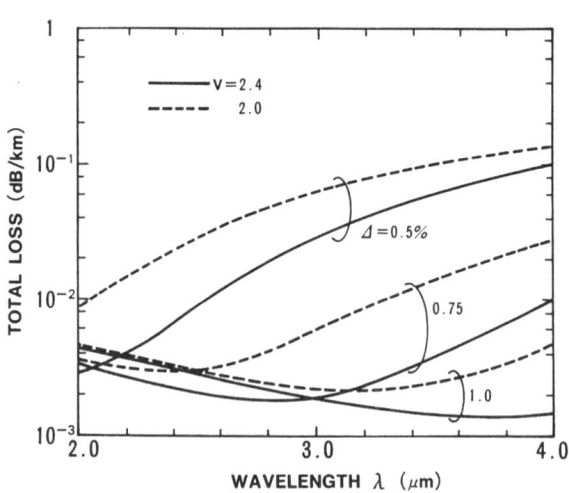

FIGURE 4 Total additional loss of fluoride glass fibers

A radiometric temperature sensing system using infrared fibers was investigated (6). A schematic diagram of the system is shown in Fig. 5. The signal-to-noise ratio at the detector output was calculated and measured for fibers of several materials, using lenses and detectors matched in wavelength range to the fiber. In the case of a

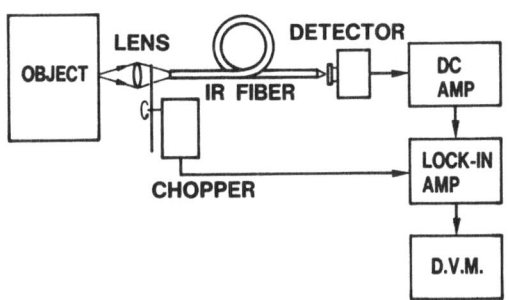

FIGURE 5 Block diagram of temperature sensing system using an infrared optical fiber

fluoride glass fiber, a CaF_2 lens and an InSb detector were selected as suitable. In calculations, the loss spectra of fibers and sensitivities of detectors actually available were used. Results of experimentally measured SN ratios as well as those calculated are shown in Fig. 6. The experiment shows that a fluoride glass fiber of 160 µm core diameter and 28 m long enables temperature measurement above 180°C with an SN ratio larger than 10 dB. In the case of a seven-fiber bundle of the same core diameter and 1.3 m long, the detectable temperature limit at a 10 dB SN ratio went down to 60°C.

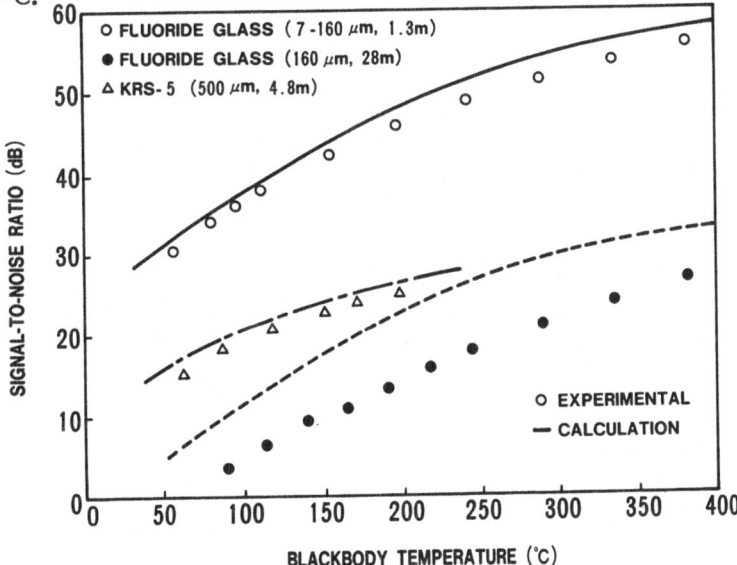

FIGURE 6 Detected signal-to-noise ratio vs. blackbody temperature to be measured

In one steel producer's rolling plant, a fluoride glass fiber system has been used on a trial basis to measure the temperature of steel sheets, and proved to have better sensitivity at around 200°C than that of silica fiber systems. From these studies, it seems that radiometric temperature sensing systems using infrared fibers can be regarded as useful in lower temperature ranges.

A prototype system for remote sensing of the focal plane of an infrared image has been set up using a tape cable composed of 80 parallel fluoride glass fibers, as illustrated in Fig. 7 (7). Each component fiber was 100μm in core diameter, with a 10 μm thick teflon cladding, and 3 m long. Two dimensional images of the object to be

TAPE CABLE COMPOSED OF 80 FLUORIDE GLASS FIBERS

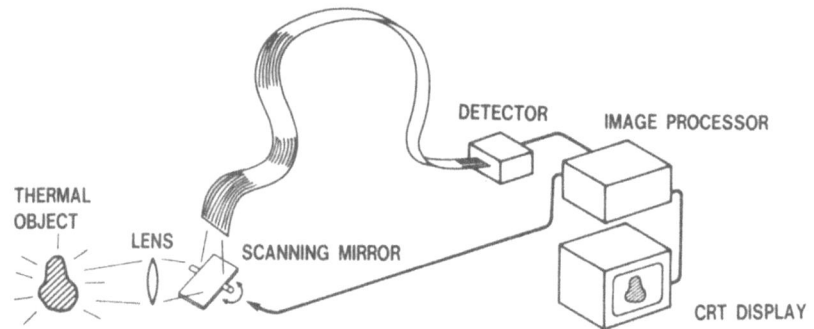

FIGURE 7 Illustration of an infrared imaging system using fluoride glass fiber tapecable

monitored can be focused by a lens on the front end of the fiber array, using a mirror which scans perpendicularly to the array. Each fiber transmits an infrared signal to the detector, and the detected signal for each scanning time slot is digitized, linearized to temperature and stored in a memory. An image processer constructs an infrared image on a CRT screen by processing signals thus stored. The temperature distribution can be displayed as a color image. Fig. 8 is a photograph of the thermal image of seven segment heaters at about 350°C, as displayed on a CRT.

The lowest temperature displayed using this system has been 150°C, and the temperature resolution was 2°C. The minimum time to form one frame image was 4 sec. Although a detector array directly coupled to the output fiber array is desired, an InSb detector with a scanning mirror and a focusing lens, used for convenience, seems to have degraded the efficiency of this system.

As for capacity to transmit laser power, a ZBGA glass fiber was tested using a HF laser (8). The teflon-FEP clad fiber, with a 300μm core diameter and 4 m long, and with a loss of about 0.4 dB/m in the 2.5 - 2.7μm region, was able to transmit 0.6 W of optical power in the 2.7 μm band without fiber damage. As one application, a laser printer was proposed, and studied experimentally. Commercially available thermal printing papers were subjected to the fiber output, and it was found that the most suitable printing power was about 100 - 250 mW for a spot of about 1 mm in diameter. Fluoride glass fibers available at present have been used successfully in such applications.

4. SOURCES AND DETECTORS

Fluoride glass fibers can also be applied in remote infrared spectroscopy, such as in organic gas monitoring systems. Lasers wavelength-matched to fluoride glass fibers may be constructed from an InGaAsSb or InAsPSb active layer grown on either InAs or GaSb substrate. Though development of such lasers has been started, work on PbSnTe lasers operating at somewhat longer wavelengths has made progress in recent years.

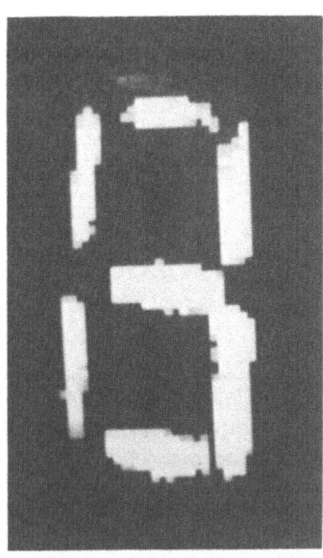

FIGURE 8 Photograph of a thermal image obtained through
the fluoride glass fiber tape cable system

As optical sources for high resolution spectroscopy, PbSnTe lasers are well known in the wavelength range of 6 - 30 μm. Although highly reliable high-power lasers, with output powers of up to 30 mW, have already been developed, further improvements, such as a higher operating temperature, have been desired.

PbSnTe-PbTeSe lattice-matched buried heterostructure single mode lasers having an active layer of cross section 4 x 1μm have been fabricated by LPE techniques (9). These lasers emit light at around 8 μm, and maximum powers of 1.7 mW at 50 K and 300 W at 107 K were obtained in single mode cw operation. They will be used in simple optical systems without gratings.

To raise operation temperatures further so as to enable use in thermoelectric coolers, multiple quantum well lasers have been developed (10). These are lasers with a mesa-strata geometry, in which a PbSnTe-PbTeSe superlattice structure is fabricated on a PbTe substrate by hot wall epitaxy. These lasers worked well in pulsed operation up to 204 K at a 6 μm wavelength, and in cw operation at 130 K and 6.6μm.

Using a similar technique, MQW lasers which have PbEuTeSe barrier layers and PbTeSe well layers have been developed (11). The temperature dependence of the wavelength emitted, shown in Fig. 9, ranges from 4.5 to 6μm. The output power is 1.3 mW at 77 K and the threshold current density is 1 kA/cm^2. The maximum operable temperature has been reported as 133 K for cw and 170 K for pulse operation.

Detector materials matched to fluoride glass fibers in wavelength include InAs and InSb, but no remarkable progress seems to have been reported recently. Meanwhile, high speed photovoltaic HgCdTe diodes were fabricated by planar techniques (12). The bandwidth of the diode, of diameter 150 μm, was 370 MHz at a reverse bias voltage of -400 mV. Direct heterodyne sensitivity measurements indicated a heterodyne NEP at 380 MHz of as low as 6.0 x 10^{-20} W/Hz.

A HgCdTe photoconductive diode array of 200 elements was also developed; this was composed of four chips each with 50 elements (13). Highly uniform sensitivity

310

was achieved as the result of unified alloy compositions through improved zone melting, unified carrier concentrations through In impurity doping, and improved array fabrication processes.

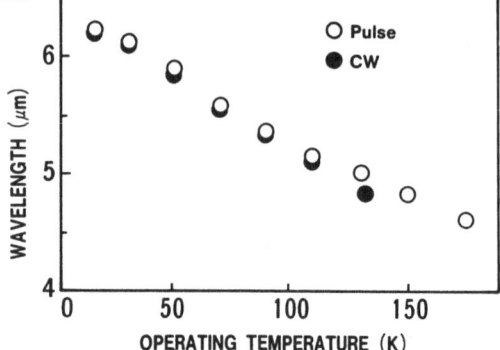

FIGURE 9 Operating temperature vs. wavelength of MQW laser

Superconducting $BaPb_{0.7}Bi_{0.3}O_3$ (BPB) thin films were proposed as detectors which are capable of detecting high-speed optical signals in the 1-10μm wavelength range. The principle of BPB optical detectors is based on the fact that the superconducting energy gap of the BPB thin film is changed by the incident light intensity. Transmission experiments of 3 μm band signals were carried out using a fluoride glass fiber and the superconducting BPB detector. Low level signals of 10^{-9} W up to more than 600 MHz could be detected (15).

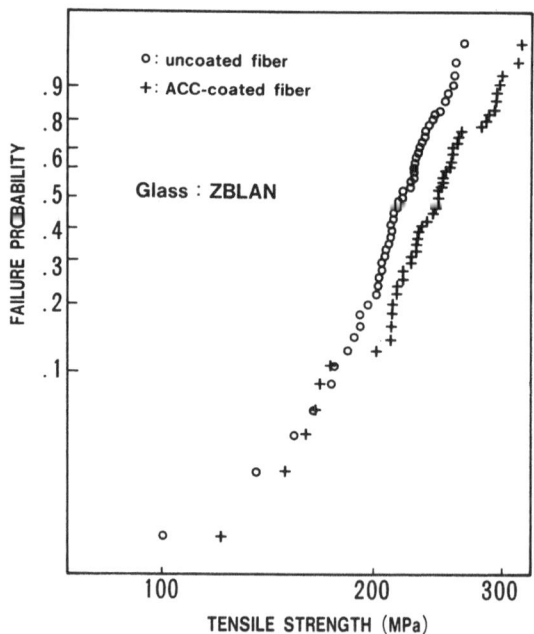

FIGURE 10 Weibull plots of tensile strength for fluoride glass fibers

5. FLUORIDE GLASS FIBER CABLE

Teflon-FEP is known as a primary coating material for protecting fluoride glass fibers, but acrylic composite compounds seem also to be suitable for coating. In the fiber drawing process, ACC plastic coating was applied, with UV curing, immediately after drawing the fiber (12). The tensile strengths of fluoride glass fibers with teflon-FEP cladding were tested, comparing ACC-coated with uncoated fibers. Core, cladding and coating diameters were 230, 260 and 380 μm respectively. Fig. 10 shows Weibull plots of the failure strength of these fibers. The effect of ACC coating has been clarified, and the average tensile strength of ACC-coated fiber was 240 MPa. A low-strength distribution around 150 MPa is believed to be caused by microcrystals and voids in the glass. These extrinsic causes of lowered strengths must of course be eliminated hereafter.

Although ACC coating is effective, it is not sufficient for practical use, so that high-strength protective coating has been desired. A fiberglass reinforced plastic coating was applied to ACC-coated fibers in an off-line process. Loss spectra of the fiber were examined before and after FRP coating, and negligible change was observed. Mechanical properties of the FRP coated fiber and the test method are shown in Table 2. This fiber seems to have a mechanical strength sufficiently high to enable practical application.

TABLE 2 Mechanical strength of FRP coated fiber

Property	Typical values
Crushing strength	180 kg/unit
Bending strength	147 kg/mm^2
Bending limit	80 mmϕ
Tensile strength	36 kg/unit
Elongation	0.5 %

Crushing test

Fiber

Bending test

Fiber

Crushing rate : 0.5mm/min
Width : 50mm

Gauge length : 50mm
Loading rate : 5mm/min

Tensile test
  Gauge length : 250mm
  Loading rate : 5mm/min

6. SUMMARY AND CONCLUSION

Application systems and technologies related to halide glass fiberoptics in Japan were reviewed. As reinforced fluoride glass fiber cables as well as infrared devices become available, actual uses of short haul systems, such as in remote infrared sensing, will be close at hand. By expanding practical applications, technologies for fabrication of stable and reliable fluoride glass fibers should be greatly improved. This will work to promote the reduction of transmission losses as well. Although studies on fluoride glass fiber telecommunication systems are as yet concerned with basics, single mode fibers with higher refractive index differences as well as low losses are desired for use in practical fiber cables with low dispersion.

References

1) Jinguji, K., Horiguchi M., Shibata S., Kanamori T., Mitachi S. and Manabe T., Electron. Lett. 18, 164 (1982)
2) Byron, K.C., Electron. Lett. 18, 673 (1982)
3) Ohishi Y., Mitachi S. and Takahashi S., IEEE J. Lightwave Technol. 2, 593 (1984)
4) Furuya K. and Suematsu Y., Appl. Opt. 19, 1493 (1980)
5) Tokiwa H. and Mimura Y., (1985) Nat. Conf. Rec. on Sc. Tech., IECEJ, 421
6) Shimizu M., Shimoishizaka M. and Yoshida S., Proc. 2nd Int. Conf. Opt. Fiber Sens., 161 (1984)
7) Kimura M. et al., to be submitted.
8) Jinguji K., Horiguchi M., Mitachi S., Kanamori T. and Manabe T., Jap. J. App. Phys. 20, L392 (1981)
9) Nishijima Y., Ebe K., Fukuda H., Shinohara K. and Murase K., Proc. CLEO (1985), 142
10) Shinohara K., Nishijima Y., Ebe H., Ishida A. and Fujiyasu H., Appl. Phys. Lett. 47, 1184 (1985)
11) Ebe H., Nishijima Y., Shinohara K., Ishida A. and Fujiyasu H., Proc. 46th Fall Mtg. Jpn. Soc. Appl. Phys., 199 (1985)
12) Yoshikawa M., Fukuda T. and Akamatsu T., Proc. 1st Sens. Symp., 235 (1981)
13) Itoh M., Takigawa H. and Ueda R., IEEE Trans. ED 27, 150 (1980)
14) Shibata T., Takahashi H., Kimura M., Ijichi T., Takahashi K., Sasaki Y. and Yoshida S., Proc. 3rd Int. Symp. Halide Glass, (1985)
15) Enomoto Y., Ohishi Y., Takahashi S. and Murakami T., Electron. Lett. 21, 219 (1985)

DISCUSSION

Q: H. Poignant. In order to achieve low additional losses due to splicing and microbending, you said that refractive index differences of approximately 0.7% ($\sim$ 0.01) are needed. Don't you think that a large Δn will lead to an increase in the intrinsic Rayleigh scattering of the order of the splicing and microbending losses?

A: S. Yoshida. There is a trade-off here. In order to decrease the refractive index difference with small additional losses, it may be preferable to make the fiber diameter larger.

Q: M. Drexhage. I understand that Furukawa is now offering a fluoride fiber cable for commercial sale. Can you tell us about the optical characteristics and price of such cable?

A: S. Yoshida. We supply a fluoride glass fiber with a loss of about 100 dB/Km. The present cost is rather high.

Q: R. Folweiler. In the Weibull plots for uncoated and coated fluoride fibers, there was surprisingly little difference in strength. Would you care to comment on this?

A: S. Yoshida. The data were obtained on teflon-cladded fibers. That is why the difference was small.

APPLICATIONS OF INFRARED WAVEGUIDES IN REMOTE GAS-SPECTROSCOPY

D. PRUSS

DRÄGERWERK AG
D-2400 LÜBECK 1, POSTFACH 1339, FEDERAL REPUBLIC OF GERMANY

ABSTRACT

The current status of development work on applications of infrared transmitting fibers in the field of gas sensing devices is reviewed with particular emphasis to fluorozirconate materials. Promising performance parameters, which might result by the incorporation of IR-fibers in optical detection schemes are indicated in comparison to conventional techniques. Fibre requirements regarding geometrical data, spectral attenuation, mechanical as well as environmental durability for future applications are summarized. The applicability of frequency modulation gas spectroscopy or evanescent wave gas spectroscopy is briefly outlined. Advantageous features of a methane detection system demonstrate the existence of a new application field of IR-fibers in remote gas spectroscopy.

1. INTRODUCTION

In recent years, progress in fabrication and cabling techniques for low-loss silica fibers has stimulated interest in a wide area of fiberoptic sensor-systems /1-3/. The discovery of glassy materials offering a transmission range up to 5 micron /4/ or even up to 10 micron /5/ has opened a new and promising field of sensor applications. Some of these sensor-concepts rely on parameters, which because of the limited spectral window cannot be addressed or are much more difficult to access by conventional silica fibers. Among those promising applications are optical gas sensor devices. In these systems, a selective and sensitive measure of specific gas-concentrations is required. As IR-active molecules normally exhibit fundamental vibrational excitation states in the region 3 .. 14 micron, this spectral range is particularly suitable in order to achieve high sensitivity. Although the main efforts to develop infrared fibers have been catalyzed by the target to achieve ultra-low-loss properties enabling repeaterless long distance communication systems, such fibers may well be introduced in sensor-systems, which require only moderate attenuation figures depending on the requested distance to bridge between sensor head and supply unit.

The purpose of this paper is to summarize the promising aspects as well as the intrinsic difficulties one has to consider, when using IR-fibers in remote gas spectroscopy.

2. GASSENSOR REQUIREMENTS AND CONCEPTS

The continuous quantitative measurement of gas concentrations at workplace areas, in the field of environmental pollution control and in the case of monitoring of industrial sites exhibiting potential explosion hazards (i.e. mining facilities and offshore installations) gains increasing interest.

The performance requirements for gas detection systems may be summarized as following : Reliable and stable operation, preferably without maintenance should be guaranteed over as long time periods as possible. The system should operate as selective as possible, i.e. it should have a low cross sensitivity regarding response to other gases. The required sensitivity as well as response time strongly depends on the intended application area : Monitoring of doping gases in the environment of semiconductor processing facilities urges for a detection threshold of as low as 5 ppb of for instance arsine or silane. In contrast, surveillance of offshore installations requires only a sensitivity of 500 ppm methane, which equals to 1 % of the lower explosion limit. Whereas both applications ask for a response time of a few seconds, there are industrial processes which require response times in the subsecond range. Finally, depending on wether interest is focused on a portable single sensor system or a stationary sensor-network, cost is also a critical parameter.

Addressing the various requirements, a multitude of different sensor concepts are currently in use. Among these are electrochemical cells, semiconductor sensors, which utilize changes of the surface conductivity as a consequence of gas-adsorption, catalytic combustion sensors and optic sensors based on the principle of differential absorption in the infrared.

3. IR-FIBER-OPTIC GASSENSORS

The introduction of fiber-optics in gas-detection has been first proposed by Inaba et al. /6/ and in the case of methane detection subsequently demonstrated by Chan et al. /7-11/ and Stueflotten et al. /12/. The authors made advantageous use of the fact that various overtone and combination bands of infrared active molecules, especially those involving the C-H bond, are located in the near-infrared region around 1.3 and 1.6 micron. This spectral region may be easily accessed by low-loss silica fibers. A typical arrangement is shown in Fig. 1.

This fiber-optic remote monitoring and detection scheme has a number of useful features. The optical energy is concentrated along the optical fibers instead of dissipated in the open atmosphere. The system is fully optical, hence the detection site is free of any electrical current or voltage. The sensorhead temperature is equal to environmental values and the optical energy density may be kept to a minimum. The system therefore may be considered failsafe. Compared to catalytic combustion sensors presently in use, there is no possibility of permanent or temporary inhibition of response by catalyst poisoning. Moving all optical sensor components like radiation sources, detectors, dispersive elements and various electronic components away from the detections site offers the possibility to access points with space constraints.

Although silica fiber detection schemes already offer promising features, miniature sensor head dimensions become feasable, if fibers having a spectral transmission range to 5 micron are introduced. This can be seen by examining the parameters determining the detection limit. The detection sensitivity of the differential absorption method is limited by three factors : (1) the differential absorption cross section between reference and measuring wavelength; (2) the signal stability of the detection process, which is the sum of the coherently detected signal fluctuations ϵ and the ratio of noise power P_n to signal power P ; and (3) the absorption path length L. The minimum detectable concentration c_{min} is then

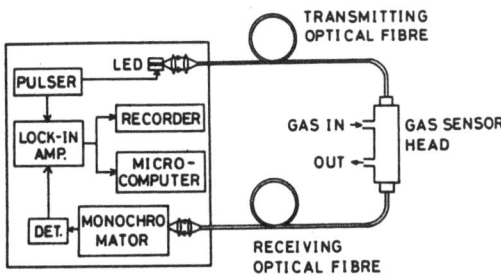

FIGURE 1. Block diagram of the experimental setup for remote absorption measurement of methane gas (after Chan et al. /11/).

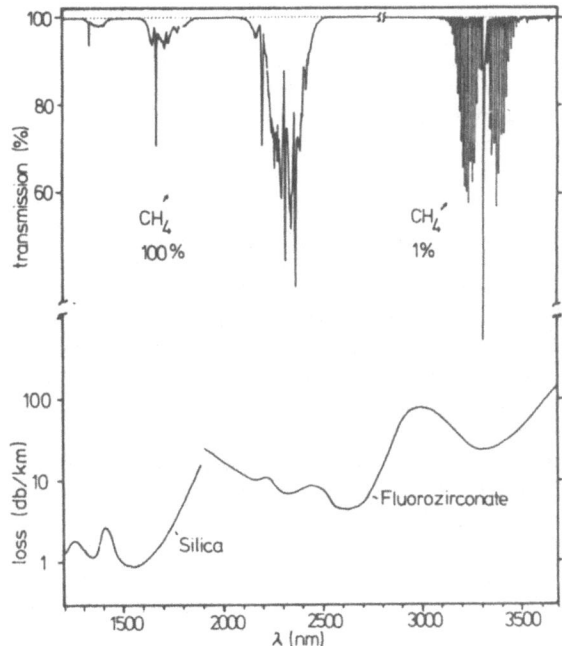

FIGURE 2. Measured transmission spectra of methane in comparison to loss spectra of silica fibers as well as fluorozirconate fibers (Data for ZrF$_4$-fibers taken from France et al. /14/). Note different methane concentrations.

given by /10/ :

$$c_{min} = (\Delta \sigma\ L)^{-1} (P_{n/P} + \epsilon)\qquad\qquad (1)$$

The relevant cross sections of fundamental absorptions are much higher compared to corresponding cross sections of overtone and combination bands. In the case of methane, the measured absorption spectra of both the fundamental and overtone regions (Fig. 2) yields a cross section of $\Delta \sigma$ = $1.5 \cdot 10^{-4}$ (ppm m)$^{-1}$ at 3.3 micron compared to $\Delta \sigma$ = $9 \cdot 10^{-6}$ (ppm m)$^{-1}$ at for instance 1.66 micron. Assuming equal absorption path lengths, the overtone spectral region requires much higher signal stabilities to reach an acceptable detection sensitivitiy, which is easily obtainable in the fundamental region.

4. FIBER REQUIREMENTS

IR-fibers will become widely accepted in sensor applications, if they meet a number of requirements regarding their mechanical and optical properties. Prior to a successful introduction of IR-fibers in communication as well as in sensor applications, the realization of some of these properties is imperative.

The most challenging demand is the quest for durability in aggressive industrial environment. Susceptibility to humid atmospheres, elevated temperatures and mechanical stress should be minimal. For the majority of infrared transmitting materials, this means the necessity to develop protective fiber coatings. Further, the differential attenuation of the fiber material itself should be as stable as possible in the spectral range of interest. The differential absorption characteristic is likely to be a significant problem in remote spectroscopy. Drexhage et al. /13/ have shown experimentally, that multiphonon absorption depends on temperature due to changes in occupancy of phonon states. As this effect might cause different attenuation changes at the reference and measuring wavelength respectively, there is no means to subsequently discriminate against the presence of the medium to be monitored at the detection site.

As already outlined, the required attenuation figures strongly depend on the distance to be bridged. Assuming the sensor head to be located some 100 .. 500 meters away from the support unit, values of 3 .. 20 dB/km would be sufficient in order to secure a reasonable signal power to noise power ratio at the infrared detector. The current status of transmission losses of fluorozirconate fibers as well as estimates of achievable minimum losses in that system have been reviewed by France et al. /14/. According to them, minimum losses in the 3.3 micron region (C-H stretching vibration) should be in the range of 3 dB/km compared to values of 20 .. 50 dB/km currently realized (Fig.2). The ultimate attenuation is mainly influenced by OH^{-} absorption. The intrinsic lower limit, caused by multiphonon absorption should be 1 dB/km , but may only be reached by totally exterminating OH^{-}-radicals.

The core diameter and numerical aperture should be matched to the optical element of lowest etendue. In case of using thermal radiation sources, fat core fibers secure a maximum of optical energy throughput. On the other hand, the applicability of frequency modulation gas spectroscopy with tunable lasers, which would offer far superior detection sensitivity, depends on the availability of single mode fibers, as otherwise modal noise would degrade system performance.

Finally, special fibers with partly removed cladding may be of interest. It has been demonstrated by Tanaka et al. /15/, that methane detection may be carried out by evanescent wave spectroscopy using a silica fiber of reduced cladding diameter and a 3.39 micron HeNe-Laser. Although this technique may offer the possibility of distributed sensing it also may raise problems concerning exposure of the IR-fiber material to aggressive atmospheres.

5. SUMMARY AND CONCLUSION

Further research and development of IR-fiber materials may be stimulated by promising sensor applications. Incorporation of IR-fibers may add unique features to existing sensor concepts. As to gas detection systems, it has been pointed out, that certain fiber-properties have to be specially tailored to sensor requirements. In conclusion, there are application fields which require moderate rather than ultra-low-attenuation figures.

6. REFERENCES

/1/ R.Th. Kersten and R. Kist, Proc. 2nd Int. Conf. on Opt. Fiber Sensors, Stuttgart, 1984 (VDE Verlag, Berlin, Offenbach, 1984).

/2/ P.D.W. Baker, "Fiber Optic Sensors: Development and Design for Industrial Application" (Oyez Scientific and Technical Series, London, 1984).

/3/ Ch.M. Davis, "Fiberoptic Sensors - an Overview", SPIE Proc. Vol. 412 "Fiber-Optic and Lasersensors" (1983) p. 2.

/4/ M. Poulain, M. Poulain and J. Lucas, J. Mater. Res. Bull. 10, 243 (1975)

/5/ C. Le Sergent, "Infrared Glass Optical Fibers for 2 to 11 micron Band" SPIE, Proc. Vol. 320 "Advances in Infrared Fibers II (1982) p. 27.

/6/ H. Inaba, "Laser Radar Studies and Applications in Japan, Sec. III Opt. Fiber Network System for Air-Pollution Monitoring by Differential Absorption Method" in Conf. Abstracts, Ninth Int. Laser Radar Conf., Munich, Germany, p. 61.

/7/ K. Chan, H. Ito and H. Inaba, "Absorption Measurement of Methane at 1.3 micron Using an InGaAsP Light Emitting Diode",Appl. Opt. 22, 3802(1983).

/8/ K. Chan, H. Ito and H. Inaba, "Optical Remote Monitoring of Methane Gas Using Low -Loss Optical Fiber Link and InGaAsP LED in 1.33 micron Region", Appl. Phys. Lett. 43, 634(1983).

/9/ K. Chan, H. Ito an H. Inaba, "An Optical-Fiber-Based Gas Sensor for Remote Absorption Measurement of Low-Level Methane Gas in the Near-Infrared Region", IEEE/OSA J. Lightwave Technol. LT-2, 234(1984).

/10/ K. Chan, H. Ito and H. Inaba, "Remote Sensing System for Near-Infrared Differential Absorption of Methane Gas Using Low-Loss Optical Fiber Link", Appl. Opt. 23, 3415(1984).

/11/ K. Chan, T. Furuya, H. Ito and H. Inaba, Opt. Quantum Electron. 17, 153(1985).

/12/ S. Stueflotten, T.Christensen, S. Iversen, J.O. Hellvik, K. Almas and T. Wien, "An Infrared Fiber-Optic Gas Detection System", in ref /1/, p. 87.

/13/ M.G. Drexhage, B. Bendow, R.N. Brown, P. Banerjee, H. Lipson, G. Fonteneau, J. Lucas and C.T. Moynihan, Appl. Opt. 21, 971(1982)

/14/ P.W. France, S.F. Carter, M.W. Moore and J.R. Williams, "Loss Mechanisms in ZrF_4 Based Fibers" in NATO ARW "Halide Glasses for Infrared Fiber-Optics" ed. R. M. Almeida (1986).

/15/ H. Tanaka, T. Ueki, H. Tai and T. Yoshino, "Fiber-Optic Evanescent Wave Gas Spectroscopy" Post-Deadline Paper at OFS 85, San Diego (1985).

ZIRCONIUM-FREE FLUORIDE GLASSES

Jacques LUCAS

Université de Rennes, Campus de Beaulieu, Laboratoire de Chimie Minérale D, Unité Associée au C.N.R.S. n° 254, 35042 Rennes Cédex (France)

ABSTRACT

Concepts of glass formation in heavy metal fluoride systems are discussed in terms of comparison between fluorozirconates and the other fluoride combina tions. It is concluded that multiplicity in number of cations, coordination numbers, coordination geometries is favorable for stabilizing glassy materials. Examples of quaternary glasses such as BTYbZ based on BaF_2, ThF_4, YbF_3, ZnF_2 or five-component materials such as BIZYT : BaF_2, InF_3, ZnF_2, YF_3, ThF_4 are presented and discussed. Samples of 1 or 2 cm thickness could be prepared in keeping a larger transmission domain ($0.2 \rightarrow 8$ μm) than the fluorozirconate ($0.2 \rightarrow 7$ μm).

1. INTRODUCTION

The initial systems which have been demonstrated to give fluoride glasses are based on zirconium tetrafluoride ZrF_4 and have been discovered in 1974 (1). The most simple, but also the less stable glass, has the composition 2 ZrF_4 – 1 BaF_2, while one of the glass having the lowest tendency to devitrify is the multi-component glass ZBLA : 57 ZrF_4, 34 BaF_2, 5 LaF_3, 4 AlF_3.

Except the BeF_2 based glasses and some very unstable glasses based on AlF_3, ZrF_4 based materials have been considered for many years as the unique candidate for mid-I.R. optical fiber preparation. Some reasons could explain this special situation occupied by the ZrF_4 derived materials. Examination of the crystal structures of different varieties of ZrF_4 and complex fluorides derived from ZrF_4 indicates a very unique and specific situation. The coordination number n and the nature of the ZrF_n polyhedra are extremely diversified lying from n = 6 to n = 8 with about nine ideal types of stereochemistry when F^- ions are surrounding Zr^{4+} cations. See paper 1 "The first ten years" by the same author.

This extreme diversity of choice in a fluorozirconate melt is a kind of application of the so called "confusion principle" and is certainly at the origin of the easy glass formation in these systems. In crossing the liquidus to solidus region, the diffusion kinetics of F^- ions, strongly bonded to Zr^{4+} in many ways, are not fast enough to allow the development of a periodic lattice.

The main key factors for obtaining fluoride glasses are the following :

1) first, it is necessary that the fluoride combination leads to a low liquidus temperature which is not so obvious when the individual constituents are high-melting materials such as $BaF_2 \sim 1500°$ C, $ThF_4 = 1100°$ C... This suggests that the glass-forming region might occur in the vicinity of an eutectic region. In this case, the long range development of the chemical bond in the melt is less opposed by the thermal agitation and consequently the viscosity is higher and the diffusion kinetics of F ions is weak enough to make the competition glass versus crystal going towards glass formation.

2) The second fundamental factor is to create a situation more or less identical to the ZrF_4 chemistry, which means that the key point is to associate in the same fluoride melt the maximum of cations M, M', M"... These cations might be different

in their crystalchemistry and if possible each of them has also to be flexible in its own coordination geometry. In this situation of "confusion", the probability for the formation of a specific individual periodic crystal is weak and the aperiodic 3D polymerization of the melt is maintained when crossing the liquidus → solidus strategic region.

3) The third rule is now the choice of the cations M, M', M"... in order to respond to the factors 1 and 2, which means obtention of low melting materials with a strong chemical association giving a highly viscous melt. It is clear that the size of M must be small and its charge maximum in order to polarise the M-F bond and to give a strong covalency. To increase the viscosity of the melt, in other words to extend the chemical polymerization, it is necessary that the F atoms are in a twofold coordination giving the sequence ..F-M-F-M-F-M... which is not necessarily linear and which corresponds to corner or sometimes edge-sharing polyhedra mechanisms.

However, if the charge of M is too high and the size too small, the long range association is no more possible because of the strong $M^{n+} \to M^{n+}$ repulsions and also because the normal coordination of M ions is satisfied by the formation of a monomeric or dimeric molecule. In this case, it is obvious that the compromise is the formation of individual covalent volatile molecules ; for instance, SiF_4, UF_6, Nb_2F_{10} molecular fumes are formed when Si^{4+}, U^{6+}, Nb^{5+} are introduced in a fluoride melt.

In this chapter devoted to zirconium-free fluoride glasses, we will not discuss the MF_3 based glasses presented in another chapter by JACOBONI, which includes the transition metal FeF_3, CrF_3... and the AlF_3 containing materials.

As discussed later, it is clear that most of the time the addition of AlF_3 to a Zr free fluoride melt increases the viscosity and decreases the devitrification process. However, the formation of the strong Al-F bond has a deleterious effect on the I.R. transparency range of heavy metal fluoride glasses in shifting the I.R. cut-off towards shorter wavelength (2).

2. GLASS FORMATION IN THREE-COMPONENT SYSTEMS

The first combinations which obey the conditions discussed before are three components associations containing for instance the mixtures : BaF_2-ZnF_2-ThF_4 (3), BaF_2-ZnF_2-YbF_3 (4), ThF_4-YbF_3-ZnF_2 (5), CdF_2-ZnF_2-BaF_2 (6).

When BaF_2 is present in the melt, no doubt that Ba^{2+} plays the role of modifier while the cations such as Zn^{2+}, Th^{4+}, Yb^{3+} ... participate to the network formation. If one examines the stereochemistry of these cations, it is clear that they have not the same diversity as Zr^{4+}, but Zn^{2+} with its d^{10} electronic structure is known to be 4- or 6-fold coordinated, Th^{4+} to be 8, 9 fold coordinated with different kinds of polyhedra, and Yb^{3+} to have coordination number 7 or 8 with also various geometries like Zr^{4+}.

Figure 1 which represents the vitreous area in the ternary diagram ThF_4-ZnF_2-YbF_3 is a good illustration of this multiplicity in the coordination possibilities responsible of the confusion in the melt.

The main feature of all these three-component glasses is their very high tendency towards crystallization which needs a very fast quenching of the fluid melt. Figure 2 shows the characteristic temperatures of the ThF_4-ZnF_2-YF_3 glass with a small T_c-T_g = 50° C indicating that this glass is only obtained as chips and has no practical interest.

The melting temperatures for the three constituents are : ZnF_2 : T_m = 872° C ; ThF_4 : T_m = 1110° C ; YF_3 : T_m = 1157° C, and for the glass composition $ThYZnF_9$: Tm = 749° C. This shows clearly that the center of the diagram is in the vicinity of an eutectic region.

The role played by ZnF_2 in these systems is more to decrease the melting

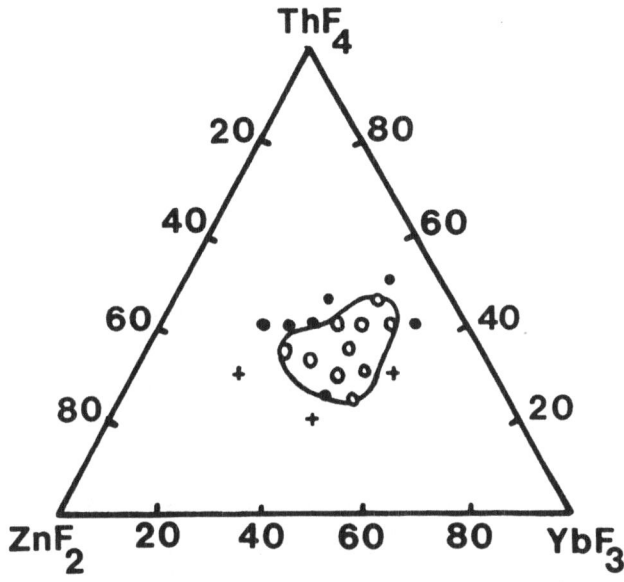

FIGURE 1. Glass formation area in the three component system $ThF_4-YbF_3-ZnF_2$;

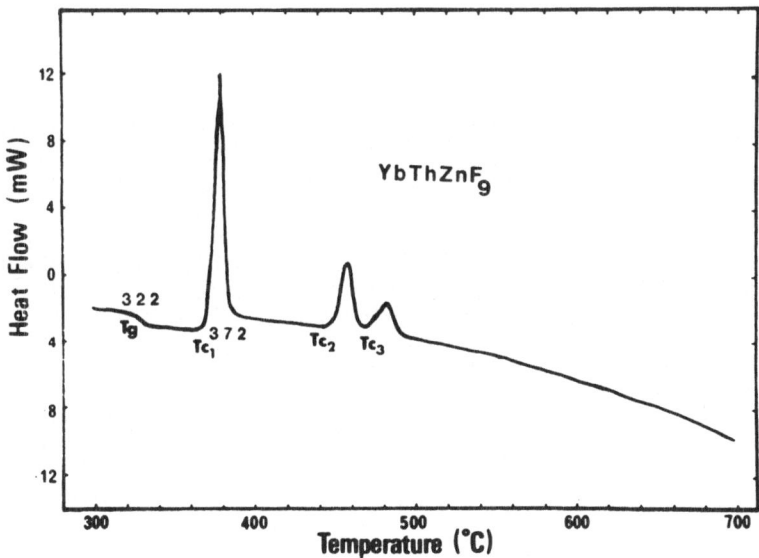

FIGURE 2. D.S.C. curve of the glass $ThZnYbF_9$.

temperature than to act as network former ; in all these glasses, ZnF_2 could be replaced by MnF_2 which also has a low melting temperature.

In the three-component system $CdF_2-ZnF_2-BaF_2$, the vitreous area as indicated on figure 3 is broad. However, very fast quenching is necessary to avoid crystallisation of the small flat chips obtained by fast cooling between two plates of brass.

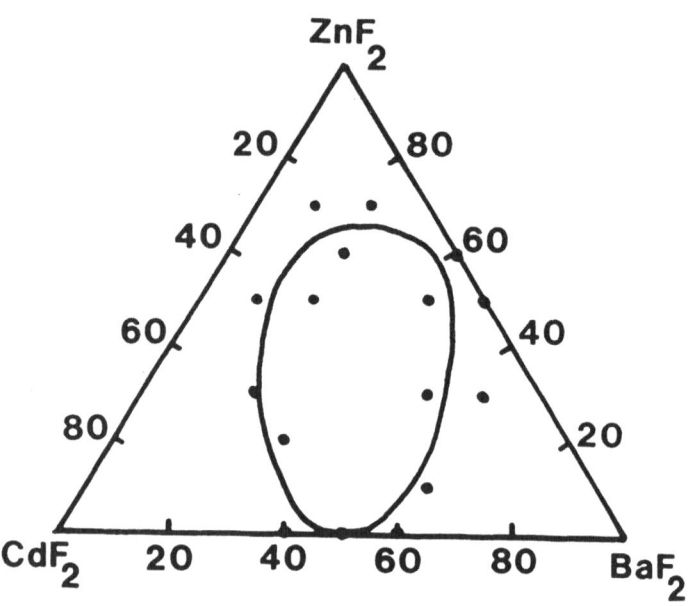

FIGURE 3. Glass forming region in the $ZnF_2-CdF_2-BaF_2$ diagram.

The most stable ternary glasses are based on InF_3 and systematic investigations by BOUAGGAD (7) of three-component systems such as $BaF_2-InF_3-ThF_4$, $BaF_2-InF_3-ZnF_2$; $BaF_2-InF_3-YbF_3$ indicate that the glass formation properties of InF_3 are much more favorable. The three diagrams of the figure 4 show the glass formation area of some three-component systems containing InF_3. A good illustration of this remark is given by the examination of the D.S.C. results on the glass 50 InF_3, 40 BaF_2-10 ThF_4 belonging to the diagram 4a. The value $T_c-T_g = 423 - 330 = 96°$ C is one of the highest observed in ternary systems. The melting temperature is here $T_m = 678°$ C.

The cation In^{3+} is also rather ambiguous regarding its coordination number and geometry. The average ionic radius r $In^{3+} \approx 0.76$ A almost the same as $Zr^{4+} \approx 0,74$ A places it at the frontier between 6-fold coordination like Al and transition metal and 7 and 8-fold environment such as lanthanides and actinides.

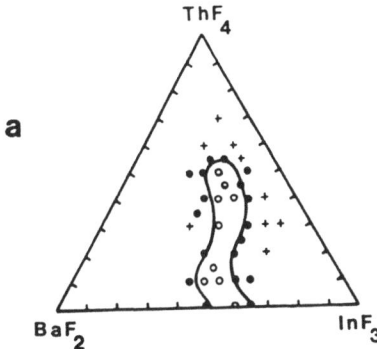

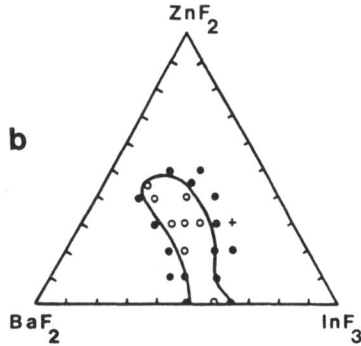

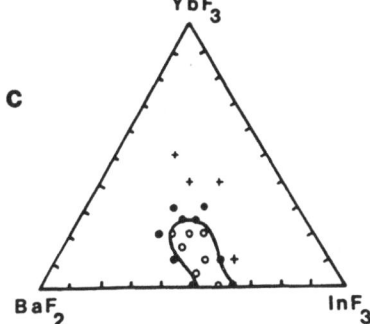

FIGURE 4. Glass forming region in the following diagram :
4a) ThF_4–BaF_2–InF_3
4b) BaF_2–ZnF_2–InF_3
4c) BaF_2–YbF_3–InF_3

3. GLASS FORMATION IN FOUR-COMPONENT SYSTEMS

It appears clearly that the empirical rule to decrease the devitrification pheno-mena in heavy metal fluoride glasses is to multiply the number of cations in the melt. As long as these cations are different in their own coordination spheres, they will multiply the number of particular M–F–M bonds different by the value of M–F–M angles, the bond lengths and so on. This situation is extremely favorable for loosing the periodicity when going from liquidus to solidus.

3.1. The BaF_2–ThF_4 glasses

This rule is verified in many systems and leads to glasses having technological interest regarding their stability towards devitrification.

The systems which have been deeply explored result in the combination of the fluorides: BaF_2–ThF_4–MF_2–LnF_3 with M = Zn or Mn and Ln = Y, Er, Tm, Yb, Lu (8, 9, 10). We will select the diagram BaF_2–ThF_4–YbF_3–ZnF_2 resulting from the systematic addition of BaF_2 to the best glass of the ternary system ThZnYb and giving the glasses called BTYbZ. Figure 5 shows the vitreous volume delimited inside the tetrahedron. As previously mentionned, special attention has been paid to the glass composition lying from $ThZnYbF_9$, center of the basic triangle to the top of the tetrahedron and corresponding to a systematic addition of BaF_2.

Figure 6 indicates the evolution of the temperatures T_g and T_c along this com-position line. It appears that the difference T_c–T_g is maximum for a composition close to 16 BaF_2, 28 ThF_4, 28 ZnF_2, 28 YbF_3. This optimized composition allows,

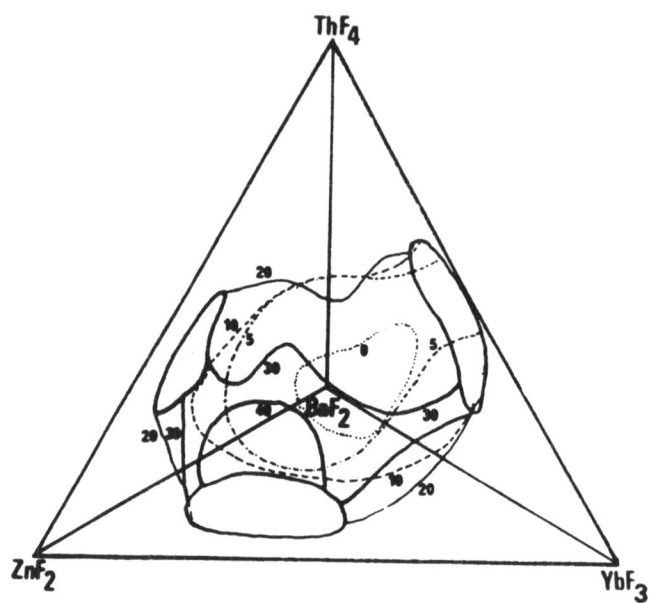

FIGURE 5. Vitreous domain in the quaternary system BaF_2–ThF_4–ZnF_2–YbF_3

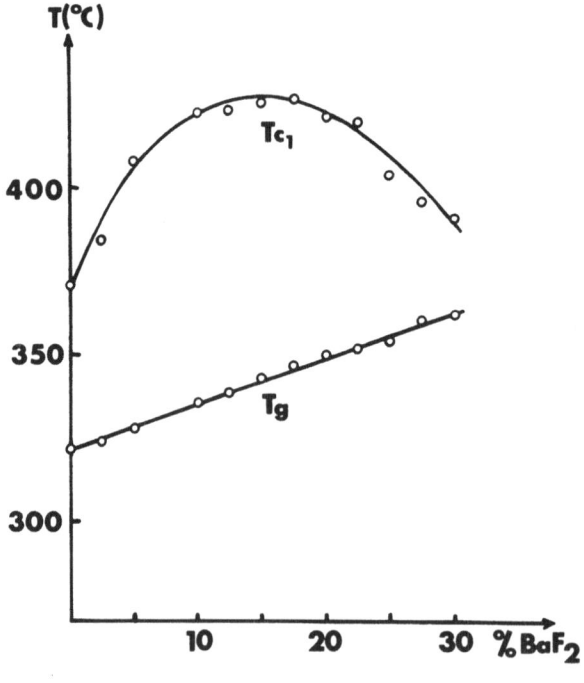

FIGURE 6. Evolution of the temperatures T_g and T_C as a function of BaF_2 ratio (mole %).

as expected, the preparation of samples of about 10 mm thickness.

Exploration of the other systems, for instance with M = Mn and Ln = Y or Lu, indicates that the best glasses are also found around this composition.

3.2. The InF$_3$ glasses

The exploration of InF$_3$ based glasses has been conducted in the four-component system InF$_3$-BaF$_2$-ZnF$_2$-ThF$_4$. The most stable glass regarding the devitrification rate has been found in adding small amount of ThF$_4$ to the ternary glass In-Ba-Zn. For instance, the glass having the highest T_c-T_g corresponds to the composition 30 InF$_3$, 30 BaF$_2$, 30 ZnF$_2$, 10 ThF$_4$. The typical temperatures are : T_g = 305° C, T_c = 413° C, with T_c-T_g = 108° C and T_m = 655° C. In such conditions, it is possible to prepare samples of about 5 to 6 mm thickness.

4. GLASS FORMATION IN FIVE-COMPONENT SYSTEMS

The most significant example illustrating the rule discussed before relative to the need of multication systems is given by the glass BIZYT. This glass has been discovered by BOUAGGAD in a systematic investigation of the diagram BaF$_2$-InF$_3$-ZnF$_2$-YF$_3$ or YbF$_3$-ThF$_4$.

In using the same kind of approach as mentionned before, it has been possible to discover a small region in this multicomponent system close to a deep eutectic region and giving very stable glasses regarding devitrification.

The glass called BIZYT has the composition : 30 BaF$_2$, 30 InF$_3$, 20 ZnF$_2$, 10 YbF$_3$, 10 ThF$_4$. The typical temperatures are T_g = 324° C, T_c = 447° C, T_m = 650° C.

Two comments have to be made concerning this glass :

a) the melting temperatures of the single components are : BaF$_2$: 1355° C, InF$_3$: 1170° C, ZnF$_2$: 872° C, YbF$_3$: 1157° C, ThF$_4$: 1110° C. This indicates that the optimized composition is in the vicinity of a deep eutectic region giving a viscous liquidus in a temperature domain close to 650° C.

b) Assuming that Ba^{2+} plays the role of modifier, the glass network is built up by the cations Zn^{2+}, Yb^{3+}, In^{3+}, Th^{4+} having each of them a very original and specific crystalchemistry and corresponding to coordination number lying from 6 (Zn^{2+}) to 8 (Th^{4+}) and a great variety of polyhedra, almost the same which have been mentionned for the crystalchemistry of Zr^{4+}.

From the composition ZIBYT, it is possible to pour between two brass plates samples of 20 mm thickness.

5. SOME PROPERTIES OF THE HEAVY METAL FLUORIDE GLASSES WITHOUT ZIRCONIUM

5.1. Optical properties

The most interesting feature of the Zr-free fluoride glasses is their large transmission range compared to fluorozirconate glasses. Figure 7 shows the transmission of SiO$_2$ based, ZrF$_4$ based and ThF$_4$-BaF$_2$ based glasses. It is clear that the IR cut-off is shifted towards longer wavelength for the latter family and that they exhibit a continuous high transparency between 0.3 to 7 μm.

The same behaviour is observed for the ZIBYT glasses which have also their multiphonon edge in the region 8-9 μm. Figure 8 shows the variation of absorption coefficient versus wavelength along the IR edge for ZrF$_4$ based and BIZYT glasses. It is obvious that, for a given wavelength, the transparency of this new family of glasses is at least of one order of magnitude better than the fluorozirconate. This suggests that the crossing of the multiphonon edge with the Rayleigh scattering loss will give a minimum in the V shape curve located in the 3-4 μm region with total losses as low as 10^{-3} dB/km.

In taking into account the low devitrification rate of the BIZYT glasses, they

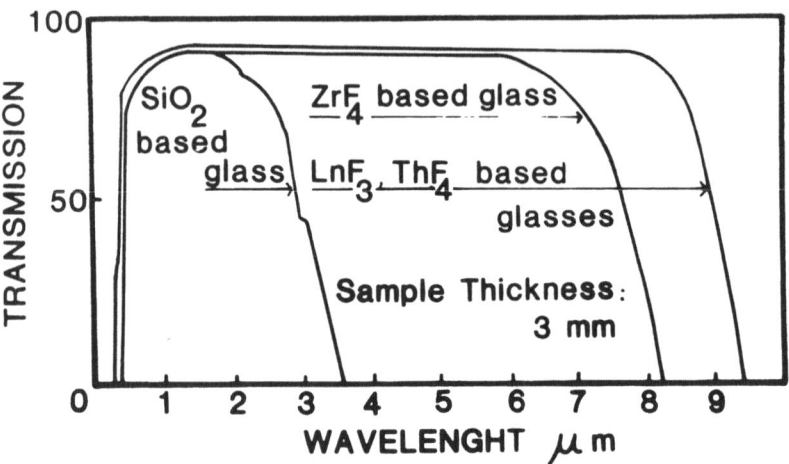

FIGURE 7. Transmission domains of SiO_2 based, ZrF_4 based, ThF_4-BaF_2 and InF_3 based glasses. The thickness of the samples is 3 mm.

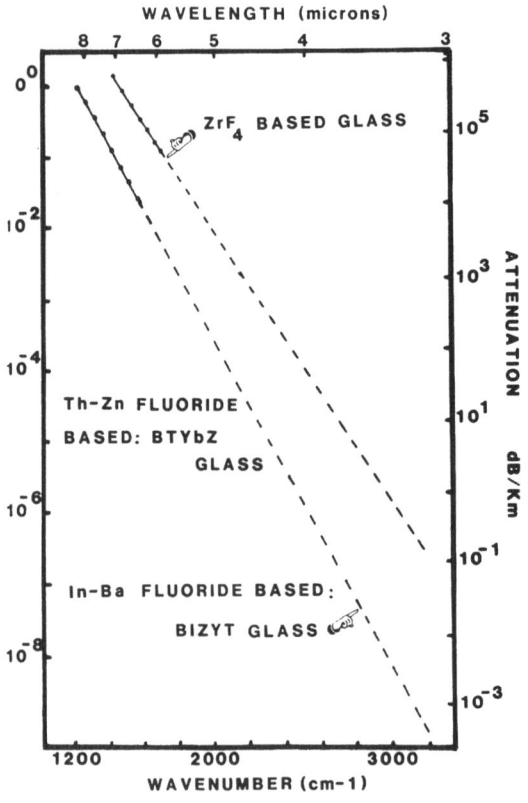

FIGURE 8. Absorption coefficient versus wavelength along the I.R. edge for ZrF_4-based and InF_3-based glasses.

appear as good candidates for low loss and broad transmission range optical fibers.

It must be noted that, when AlF_3 is incorporated in ZrF_4 based glasses, or when it is in combination with ZnF_2 rich materials, it usually stabilizes the glass (11). The same effect is observed with the BaF_2/ThF_4 glasses (2). However, the addition of AlF_3, although beneficial for glass formation, has a deleterious effect on the infra-red transparency range, in shifting the I.R. cut-off towards the 6 µm region.

5.2. Mechanical properties

For many infra-red optical applications, the poor mechanical properties of heavy metal fluoride glasses relative to silicate glasses are of some concern. The Vickers microhardness measured on ZrF_4 based glasses is in the range 250 kg/mm^2, while the same measurements on ThF_4-BaF_2 glasses give values of about 25 % higher (310 kg/mm^2). For the moment, no clear dependence between microhardness and composition is observed.

6. CONCLUSION

These Zr-free fluoride glasses might be considered as alternative solution to fluorozirconate glass for the preparation of ultra-low loss optical fibers. The new discovered family of indium-based glasses with the optimized composition BIZYT, appear to be very promising materials because of their low devitrification rate and their broad transmission domain : $0.2 \rightarrow 8$ µm. Compared to ZrF_4 glasses, the estimated transparency in the mid-I.R. region is about ten times better for a given wavelength. One, of course, needs more informations on the possibility of preform fabrication, purification of the starting material, OH contamination, and so on.

REFERENCES

1. Poulain, M., Poulain, M., Lucas, J. and Brun, P., Mat. Res. Bull., 10, 243 (1975).
2. Bendow, B., Banerjee, P.K., Drexhage, M.G., El Bayoumi, O.H., Mitra, S.S., Moynihan, C.T., Gavin, D., Fonteneau, G. and Lucas, J., J. Am. Ceram. Soc., 66, n° 4, C64 (1983).
3. Fonteneau, G., Lahaie, F. and Lucas, J., Mat. Res. Bull., 15 (8), 1143 (1980).
4. Lucas, J., Slim, H. and Fonteneau, G., J. Non-Cryst. Solids, 44 (1), 31 (1981).
5. Fonteneau, G., El Houari, A. and Lucas, J., Rev. Chim. Min., 20, 321 (1983).
6. Matecki, M., Poulain, M. and Poulain, M., Mat. Res. Bull., 16, 749 (1981).
7. Bouaggad, A., Phil Doct., University of Rennes, 1986, to be published in Mat. Res. Bull., (1986).
8. Fonteneau, G., Slim, H. and Lucas, J., J. Non-Cryst. Solids, 50, 61 (1982).
9. Drexhage, M.G., El Bayoumi, O.H., Lipson, H., Moynihan, C.T., Bruce, A.J., Lucas, J. and Fonteneau, G., J. Non-Cryst. Solids, 56, 51 (1983).
10. Lepage, Y., Fonteneau, G. and Lucas, J., Rev. Chim. Min., 21, 589 (1984).
11. Poulain, M., Poulain, M. and Matecki, M., J. Non-Cryst. Solids, 51, 201 (1982).

DISCUSSION

Q: R. Almeida. How do you put chlorine into a Te-Cl glass?

A: J. Lucas. By using Cl_2(g), $TeCl_4$ or SCl_2.

Q: R. Almeida. How do you explain glass formation in a ThF_4--LiF-NaF-KF glass?

A: J. Lucas. The Li atoms are 4-fold coordinated. Therefore, Li behaves as a network former.

C: G. Frischat. The first model which described glass formation was due to Zachariasen, who stated that the glass former should have a low C.N., e.g. 3 or 4. It is interesting to compare this model to the "principle of maximum confusion" for halide glasses, whose C.N. may vary from 6 to 8 and where different short-range order polyhedra may occur.

R: J. Lucas. The Zachariasen model was established only on the basis of glasses existing in ∿ 1930 and is valid essentially for oxide glasses.

Q: P.W. France. Do you know the refractive index of your new Te_3Cl_2S glasses?

A: J. Lucas. From transmittance data, it can be estimated at around 2.5, about the same as that of other chalcogenides. Measurements are now in progress.

CHEMICAL DURABILITY OF A BaF$_2$-ThF$_4$ BASED GLASS

D. TREGOAT

Université de Rennes, Laboratoire de Chimie Minérale D,
Avenue du Général Leclerc - 35042 RENNES CEDEX (France).
Present address : Laboratoires de Marcoussis, C.G.E. Research Center,
Route de Nozay, 91460 MARCOUSSIS (France).

ABSTRACT
   The behavior of a BaF$_2$-ThF$_4$ based glass when exposed to humid air environments and when immersed in several aqueous solutions has been examined by using infrared transmission spectroscopy, scanning electron microscopy and fluoride ion analysis.
   The $\mathcal{E}$OH values at 2.9 μm are 31 ℓ.mol^{-1}.cm^{-1} for this glass and 19.5 ℓ.mol^{-1}.cm^{-1} for a fluorozirconate based glass. At temperatures higher than 150 °C an hydroxylation process involving a F$^-$/OH$^-$ exchange occurs with an activation energy of 95 kJ.mol^{-1}. At 344 °C the diffusion profile of OH groups into the glass shows a penetration depth of 10 μm after 22.5 hours which corresponds to a diffusion coefficient of $1.8.10^{-12}$ cm^2.s^{-1}.
   Corrosion tests which have been carried out in several aqueous solutions show that the BaF$_2$-ThF$_4$ based glass is not affected in neutral or basic conditions. On the other hand the samples are highly corroded in acidic solutions and the corrosion rates increase with decreasing pH.

1. INTRODUCTION

   In the last ten years increasing interest has been shown towards fluoride glasses. These glasses are theoretically superior to other glasses in their transparency in the mid-infrared region which makes them potential candidates for infrared optical fibers, laser hosts, lenses, etc...

   Besides metallic impurities, water contamination is a limitation to the mid-infrared transparency. Nevertheless only a few studies (1, 2, 3) of the reaction of these materials with atmospheric water have been published. The main conclusions have been that at ambient or moderate temperatures, reaction of the glass with gaseous water is indetectable and negligible, although rapid and severe corrosion could occur if the samples are contacted by liquid water.

   In the present paper, the chemical durability of a BaF$_2$-ThF$_4$ based fluoride glass is examined both under moist air environment at high temperatures and in aqueous media at room temperature.

   The study has been performed on BTYbZ glass samples, the composition and physical properties of which are specified in table 1. All the samples have been melted in air according to the NH$_4$HF$_2$ technique and then treated using reactive atmosphere processing (RAP) with CS$_2$ or helium (4) in order to avoid the 2.9 μm OH absorption peak.

TABLE 1. Physical properties of BTYbZ glass

Composition (mol %)	15 BaF$_2$ - 28.3 ThF$_4$ - 28.3 YbF$_3$ - 28.3 ZnF$_2$
Glass transition temperature, Tg	344 °C
Crystallization temperature, Tc	426 °C
Melting temperature, Tm	665 °C
Refractive index, n	1.536
Optical transmission window	0.25 → 8 µm
Thermal expansion coefficient	16.10^{-6}°C^{-1}
Knoop hardness	310 kg.mm^{-2}
Density	6.43

2. EXPERIMENTAL RESULTS

2.1. OH extinction coefficient at 2.9 µm

Given the contribution of the 2.9 µm OH absorption band on loss increase in the theoretical ultralow loss region (2.5-3.5 µm), the value of the molar extinction coefficient for this extrinsic species is of great interest.

This coefficient has been measured by determining the amount of HF evolved during the heating of the glass under steam and the corresponding intensity of the OH absorption peak.

A recent study (5) of the reaction with gaseous water in and somewhere below the Tg region (212-347 °C) has shown that at constant temperature the total OH absorbance Ln (To/T) increases with the square root of time :

$$Ln\ (To/T) = (kt)^{\frac{1}{2}} \tag{1}$$

Gaseous HF was detected as a product of the reaction. These two observations, along with the absence of any sizeable water bending absorption band at 6.1 µm, indicated that the main corrosion process involves a rapid reaction of the fluoride ion in the glass with gaseous water :

$$F\ (glass) + H_2O\ (g) \longrightarrow HF\ (g) + OH\ (glass) \tag{2}$$

This is accompanied by diffusion of OH groups from the surface into the glass bulk. No decrease of the OH absorption band is observed after heating of the hydroxylated glass samples under dry conditions, ruling out a condensation of the OH groups. The rate constant k of equation [1] is temperature dependent with an activation energy of 95 kJ.mol^{-1}.

According to reaction [2], the amount of HF evolved during the hydroxylation process at temperatures near Tg is equal to the amount of OH introduced into the glass. Based on the assumption that the Lambert-Beer law is valid for the OH stretching vibration peak in the infrared spectrum of fluoride glasses, the hydroxyl concentration is directly proportional to the intensity of this peak. The extinction coefficient in liter.mol^{-1}.cm^{-1} is given by :

$$\varepsilon = A.10^{-3}/2\ Cs \tag{3}$$

where A is the absorbance of the OH peak, Cs is the OH concentration per unit of area and equal to the amount of HF measured using a fluoride ion selective electrode.

On Fig. 1a and b are shown the absorbances of the OH peak as a function of OH concentration per unit of area for two glasses. Calculated values are :

$\mathcal{E}$ = 31 ± 3 ℓ.mol^{-1}.cm^{-1} for BTYbZ glass
$\mathcal{E}$ = 19.5 ± 4 ℓ.mol^{-1}.cm^{-1} for BALLA glass (34 BaF$_2$ - 57 ZrF$_4$ - 4 AlF$_3$ - 5 LaF$_3$)
These values correspond to loss increases of respectively 11690 and 5200 dB/km for 1 ppm OH (6).

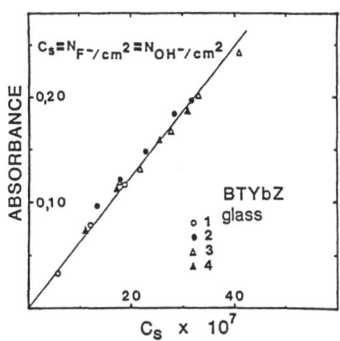
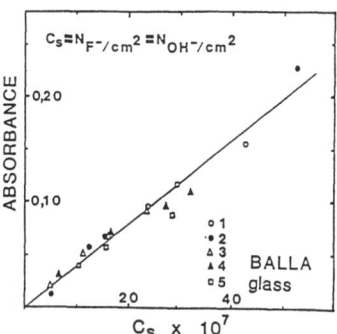

FIGURE 1a and b. Absorbance of the OH band at 2.9 µm versus OH concentration per unit of area for BTYbZ and BALLA glass samples.

2.2. Behavior in humid air environment

A previous study (5) has shown that a reversible water adsorption takes place at temperatures lower than 150 °C and that above this temperature the hydroxylation reaction occurs. The total amount of OH diffused into the glass has been measured but no indication has been given about the penetration depth.

This work has been performed on a CS$_2$ treated BTYbZ disk with no detectable OH absorption band. The OH diffusion profile after reaction of the glass with ambient atmosphere for 22.5 hours at 344 °C has been determined by monitoring the decrease of the 2.9 µm absorption peak as increasing material thicknesses are removed from the surface by polishing.

The infrared spectra in the 2.5-4 µm region are shown in Fig. 2 as a function of the thickness x. There is a small narrow parasitic absorption peak centered at about 2.8 µm due to a small amount of hydroxylated devitrified material as previously described (5). This material seems to be present within a one µm distance from the surface, as indicated by Fig. 2.

The 2.9 µm absorbance A(x) = 0.5 Ln (To/T) for one surface is plotted in Fig. 3 as a function of the thickness x.

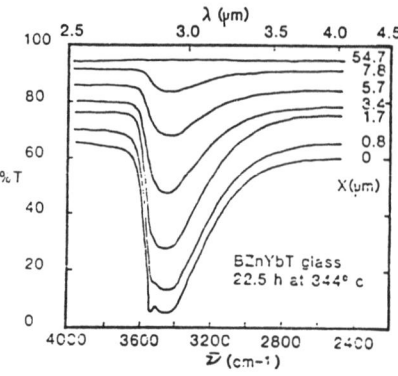

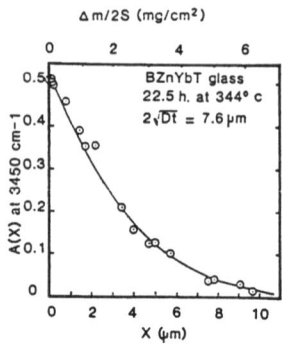

FIGURE 2. IR spectra of BTYbZ glass heated in air for 22.5 hours at 344 °C as a function of thickness x removed from each surface. For clarity each spectrum is displaced downward.

FIGURE 3. Absorbance at 2.9 μm due to OH inside one surface versus thickness x.

The shape of the A(x) versus x plot can be explained assuming OH diffusion into the glass and constant OH surface concentration, C(o) produced by reaction [2].

Calculations (7) according to Fick's second law show that the shape of the A (x) versus x plot should be given by :

$$A(x) = A(o) \left[ierfc \left(x / 2 \sqrt{Dt} \right) / ierfc \left(o \right) \right] \qquad [4]$$

where A(o) is the initial absorbance before polishing, "ierfc" stands for the integral of the complementary error function, D is the interdiffusion coefficient and t is the diffusion time.

The best fit using equation [4] is obtained with a value of $2 \sqrt{Dt}$ = 7.6 μm and is shown in Fig. 3 as a solid line.

There is an excellent agreement between the shape of the experimental plot and that predicted from equation [4]. The corresponding interdiffusion coefficient D at 344 °C is $1.8.10^{-12} cm^2 .s^{-1}$ which leads to a penetration depth for OH radicals of about 10 μm after 22.5 hours.

2.3. Behavior in aqueous solution

The effects of aqueous corrosion at room temperature on a BaF_2-ThF_4 based glass with the composition given above have been investigated using infrared transmission spectroscopy, scanning electron microscopy and fluoride ion analysis.

Corrosion tests have been carried out in 100 ml volumes of several aqueous solutions on samples previously treated under CS_2.

Table 2 contains a summary of the results obtained after immersion of the samples at various pH.

TABLE 2.

Solutions	pH	Leaching time	IR absorption peaks	Remarks
NaOH, 3N		72 h	none	-
NaOH	11.6	72 h	none	-
NH$_4$OH	12.5	72 h	none	-
buffer	8	2 months	none	-
buffer	7	2 months	none	-
distilled water	5.6	15 days	OH, H$_2$O	hydrated crystalline crust (ThF$_4$, n H$_2$O) - pH ↘ 2.8
CH$_3$CO$_2$H	3.5	72 h	OH, H$_2$O	corroded surface
HF (48 %)		72 h	OH, H$_2$O	hydrated and cracked surface. YbF$_3$, n H$_2$O and ThF$_4$, n H$_2$O in excess
HNO$_3$	2	72 h	OH, H$_2$O	corroded surface
HCl	2	72 h	OH, H$_2$O	corroded surface
H$_2$SO$_4$	2	72 h	OH, H$_2$O, SO$_4^{2-}$	hydrated crystalline crust (BaSO$_4$, n H$_2$O and ThF$_4$, n H$_2$O).

For basic and neutral solutions, an examination of the BTYbZ glass surface as well as the infrared transmission spectra show that no attack has taken place after the indicated soak periods.

In acidic solutions, the corrosion rate increases with decreasing pH. Two absorption bands can be seen : the first and stronger one at 2.9 µm is related to OH stretching and the second at 6.1 µm is due to the bending of water molecules.

Examination of the sample surface shows in some cases the formation of crystalline materials. For example after immersing of the glass for 15 days in deionized water, the pH has dropped down from 5.7 to 2.8 and the surface is covered with small filament like crystals of hydrated ThF$_4$.

The surface of the sample immersed during 72 hours in 48 % HF shows a highly cracked and opaque layer. This layer is made up of poorly crystallized hydrated microcrystals of ThF$_4$ and YbF$_3$ remaining after the fast dissolution of ZnF$_2$ and BaF$_2$.

For other types of acidic solutions (HCl, HNO$_3$, CH$_3$CO$_2$H), all the glass components dissolve and no crystalline interface is formed (rough surface). Fig. 4 shows the leach rate of fluoride ion normalized to composition for two BTYbZ glass samples, the first one being previously treated under CS$_2$ and the second one under helium.

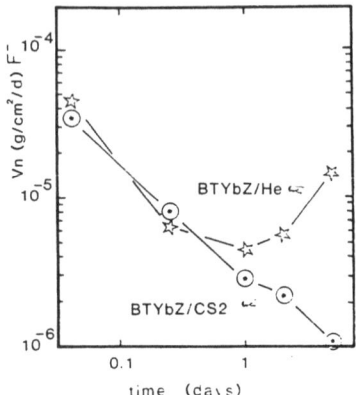

FIGURE 4. Leach rates of fluoride ions, normalized to composition for He and CS$_2$ treated BTYbZ glass samples immersed for 1, 5, 18, 24 and 72 hours successively.

It can be seen from Fig. 4 that the leach rates are similar during the two first immersions. The leach rates continues regularly to decrease for the CS$_2$ treated sample. In contrast a very sharp rate increase is observed for the He treated sample which leads to a partial devitrification of the surface as shown on Fig. 5.

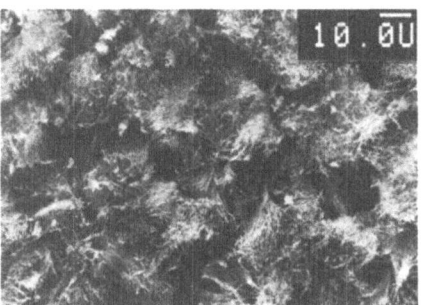

FIGURE 5. He treated BTYbZ glass surface after leaching in deionized water for 72 hours at room temperature.

The small needle-shaped crystals have been identified as hydrated ThF$_4$ and the infrared transmission spectrum shows a strong 2.9 μm OH absorption band and a weaker one at 6.1 μm due to free water molecules.

For the same time conditions, the surface of the sample treated under CS$_2$ shows no evidence of crystal formation and the infrared transmission spectrum is free of any OH or water absorption band.

3. DISCUSSION

The ε values for BTYbZ and BALLA glasses are to be compared with values found for other glasses :

16 ℓ.mol^{-1}.cm^{-1} for 31.7 BaF$_2$ - 60.5 ZrF$_4$ - 3.8 GdF$_3$ - 4 AlF$_3$

37 ℓ.mol^{-1}.cm^{-1} for 25 BaF$_2$ - 16 CaF$_2$ - 16 YF$_3$ - 43 AlF$_3$

measured by a thermo-gravimetric technique [8] and

25 ℓ.mol^{-1}.cm^{-1} for 30 BaF$_2$ - 30 InF$_3$ - 20 ZnF$_2$ - 10 YbF$_3$ - 10 ThF$_4$ [9]

while the value in silica glasses has been reported as 77-181 ℓ.mol^{-1}.cm^{-1} [10]. These different $\mathcal{E}$OH values can be explained as a different hydrogen bond formation between the OH vibrator and the vitreous network depending on the structure of each glass studied.

The diffusional analysis of hydroxyl groups during heating of the glass at ambient atmosphere leads to some interesting remarks :

Using the values of $\mathcal{E}$ and $2\sqrt{Dt}$ given above, we calculate a value of 0.017 mol.cm^{-3} for C (o), the surface concentration of OH groups at 344 °C. It thus appears that about one out of seven F$^-$ ions is replaced with OH$^-$ right at the air-glass interface, without producing any noticeable devitrification.

However, a partial devitrification can be seen after heating such a sample during longer times (some ten hours) even at 344 °C that is far below Tc. This phenomenon induces an increase of the 2.8 μm parasitic absorption peak due to the formation of a devitrified surface layer while the 2.9 μm OH absorption peak increases less than expected leading to the conclusion that at 344 °C the maximum OH content for which the glass keeps its vitreous state is about 0.017 mol.cm^{-3}.

The activation energy of 95 kJ.mol^{-1} found for the hydroxylation process is to be compared with 96 kJ.mol^{-1} found by PANTANO [11] for the reaction of a ZBLA sample in ^{18}O/water environment at temperatures below Tg.

Presuming that the activation energy for OH diffusion is of the same order of magnitude as the hydroxylating activation energy and using the familiar proportionality between the depth of penetration and $\sqrt{Dt}$, it can inferred that the OH diffusional depth for this glass will be negligible at ambient temperature.

A similar study [8] performed on a 31.7 BaF$_2$ - 3.8 GdF$_3$ - 60.5 ZrF$_4$ - 4 AlF$_3$ has shown that the penetration depth after 6 hours at 300 °C (close to Tg) is about 25 μm which is significantly larger than that for BaF$_2$ - ThF$_4$ based glasses.

If the interdiffusion coefficient behaves like other ionic transport properties [12], the temperature dependence of D for the glass above the Tg is likely to be higher. For example, after heating the sample at 350 °C for 22.5 hours, the OH groups are seen to have penetrated about 50 μm into the glass.

Many authors (13 → 20) have suggested various mechanisms relating to aqueous corrosion of fluoride glasses such as matrix dissolution, ionic exchange or water diffusion.

From the present results, it appears that the H$_3$O$^+$ agressivity may play an important role and the first step of the fluoride glass corrosion could be represented as an electrophilic attack of H$_3$O$^+$ establishing a strong chemical association with the fluoride ions in the glass.

This attack along with a preferential dissolution of ZnF$_2$, BaF$_2$ and YbF$_3$ leads to the glass framework corrosion. As observed during the last immersions of the He treated sample (Fig. 4), this corrosion step induces an increase in leach rate.

The next step will probably be governed by diffusion of both molecular water and leaching species through the hydrated crystallized surface layer.

REFERENCES

1. Pantano, C.G., 3rd International Symp. on Halide Glasses, Rennes, France (1985).
2. Fjeldly, T.A., Hordvik, A. and Drexhage, M.G., 3rd International Symp. on Halide Glasses, Rennes, France (1985).
3. Robinson, M. and Drexhage, M.G., Mat. Res. Bull. 18, 1101 (1983).
4. Tregoat, D., Fonteneau, G. and Lucas, J., Mat. Res. Bull. 20, 179 (1985).
5. Tregoat, D., Fonteneau, G. and Lucas, J., Mat. Res. Bull. 20, 57 (1985).

6. Fonteneau, G., Tregoat, D. and Lucas, J., Mat. Res. Bull. 20, 1047 (1985).
7. Tregoat, D., Fonteneau, G., Moynihan, C.T. and Lucas, J., Com. Am. Ceram. Soc. 68, C-171 (1985).
8. Mitachi, S., Sakaguchi, S. and Takahashi, S., to be published in Phys. Chem. Glasses (1986).
9. Fonteneau, G., Mitachi, S., Christensen, P. and Lucas, J., personal communication.
10. Shelby, J.E., Vitko, Jr.J. and Benner, R.E., Com. Am. Ceram. Soc. 4, C-59 (1982).
11. Pantano, C.G., 3rd International Symp. on Halide Glasses, Rennes, France (1985).
12. Doremus, R.H., Glass Science 154, John Wiley, New York, U.S.A. (1973).
13. Mitachi, S., Phys. Chem. Glasses 24, 146 (1983).
14. Simmons, C.J., Sutter, H., Simmons, J.H. and Tran, D.C., Mat. Res. Bull. 17, 1203 (1982).
15. Frischat, G.H. and Overbeck, I., J. Am. Ceram. Soc. 67, C-238 (1984).
16. Doremus, R.H., Bansal, N.P., Bradner, T. and Murphy, D., J. Mat. Sci. Lett. 3, 484 (1984).
17. Barkatt, A. and Boehm, L., Mat. Lett. 3, 1 (1984).
18. Simmons, C.J., Guery, J., Chen, D.G. and Jacoboni, C., 3rd International Symp. on Halide Glasses, Rennes, France (1985).
19. Doremus, R.H., Bansal, N.P. and Murphy, D.M., Bull. Am. Ceram. Soc. 64, 476 (1985).
20. Loehr, S., Bruce, A.J., Mossadegh, R., Doremus, R.H. and Moynihan, C.T., 3rd International Symp. on Halide Glasses, Rennes, France (1985).

DISCUSSION

Q: G. Frischat. In one of the slides, you showed D_{OH^-} and D_{F^-}. How did you determine D_{F^-}? From the IR measurements, all you get is a binary interdiffusion coefficient $\tilde{D}_{OH^-,F^-}$ which is governed by both OH⁻ and F⁻ as the Nernst-Planck diffusion model suggests.

A: D. Tregoat. D_{F^-} can be determined from the anionic condutivity measurements of Ravaine et al. on the same glass composition, based on the Nernst-Einstein equation. As reported by Doremus, $\tilde{D}_{OH^-,F^-}$ is related to the individual coefficients by:

$$\tilde{D}_{OH^-,F^-} = D_{OH^-} \cdot D_{F^-}/(x_{OH} \cdot D_{OH^-} + x_F \cdot D_{F^-})$$

In the present study, the anionic fraction of OH radicals was $\sim 6\%$, so the following approximations could be made: $x_{OH} \ll x_F \sim 1$ and $\tilde{D}_{OH^-,F^-} \sim D_{OH^-}$.

Q: C. Pantano. Could you comment on the fundamental origin of the difference in corrosion behavior between vapor and liquid water attack?

A: D. Tregoat. The fundamental origin could be related to the different physical status of the water molecules in the two corrosion studies. In one case, the ionic form of water must be taken into account (pH=5.6); the aggressive behavior of H_3O^+ ions was responsible for the first corrosion step (electrophilic attack), through the establishment of a strong association with fluoride anions. In the other case, water is in a molecular form and thus the vapor was less corrosive towards the glass than liquid water.

TRANSITION METAL FLUORIDE GLASSES (TMFG), SYNTHESIS,
PROPERTIES, STRUCTURE

C. JACOBONI

Laboratoire des Fluorures - U.A. 449 - Faculté des Sciences
Université du Maine - 72017 LE MANS Cedex - FRANCE

ABSTRACT
    Despite their less stable character, 3d transition metal
fluoride glasses (3d - TMFG) have the same intrinsic properties as HMFG. Spin-glass behaviour is shown for high content
in paramagnetic ions. Extensive structural studies lead to
octahedral network by corner sharing. I.R. applications of
TMFG could be enhanced by their ability for preparation via
vapor deposition.

1. INTRODUCTION
    At the moment, it seems evident that three quite distinct
fluoride glass families exist with respect to the coordination
number of the glass-forming ion:
    - BeF_2 based glasses which look like silicates (C.N.= 4)
    - MF_3 based glasses (M= 3d metal) - C.N.= 6
    - MF_4 based glasses (M= Zr, Hf, Th, U) with C.N. > 6.
3d transition metal fluoride glasses (TMFG) have not been
studied as extensively as HMFG because of their more difficult
preparation during the early period of fluoride glass research
(working in air).

2. CHEMICAL SECTION
2.1 3d fluoride chemistry
    It is not established whether 3d fluorides are more sensi-
tive to moisture hydrolysis than other 3d compounds but the
oxide derivatives are coloured and often magnetic. As the well
known hydrolysis equilibrium suggests, all the operations involving 3d fluorides need very dry atmosphere and, when possible, an excess pressure of gaseous HF, especially when temperature increases. So the different possible chemical stages
are:
    - Dehydration of hydrated chlorides under dry HCl stream.
    - Fluorination of anhydrous chlorides by dry gaseous HF in
      gold or platinum boats inside monel reactor up to 750°C
      for most of 3d elements: Mn, Co, Ni, Cu, Zn, V, Cr, Fe.
    - Dehydration under dry HF stream of hydrated fluorides re-
      sulting from action of aqueous HF on Ga_2O_3 or In_2O_3.
    - Thermal decomposition of well known ammonium fluorometal-
      lates in the case of Al, In, Y or Lanthanides.

2.2 Glass preparation
    All operations (weighing, mixing, melting and quenching)

are performed inside a dry box (< 1ppm H_2O). The mixture is melted in a covered platinum crucible for 10-15 min at 750°C, and the melt is poured in a preheated brass mould.

2.3 Glass systems

Numerous and quite large vitreous domains have been found in PbF_2-$Mt^{II}F_2$-$Mt^{III}F_3$ (PMF-PMG, [1, 2]), BaF_2-UF_4-$Mt^{II}F_2$-$Mt^{III}F_3$ (BaUMt, [3]) systems (Mt^{II}= Mn, Cu, Zn; Mt^{III}= V, Fe, Ga, In) - Figure 1 - and also in the PbF_2-BaF_2-InF_3 (PBI, [4]) system; the addition of small amounts of AlF_3 and/or YF_3 leads to stable glasses up to 8-10 mm thickness. Classical studies on thermal or mechanical properties show that these glasses are similar to HMFG (Table 1). X-ray devitrification studies confirm the tendency to form mainly binary compounds involving Mt^{3+} ions and the crystallization of high temperature metastable phases.

3. MISCELLANEOUS PROPERTIES
3.1 Chemical durability

The resistance of most TMFG is poor especially in acidic medium, however PBI [5] and BaUMt [6] resist very well (Figure 2). It seems that the large solubility of 3d ions leads to dissolution.

3.2 Ionic conductivity

For TMFG, despite the high PbF_2 content (>40 mole%), the F^- conductivity is low - $\sigma_{200} \sim 10^{-7} \Omega^{-1}$ cm^{-1} - (about two orders below that of HMFG); however, BaUMt shows significant differences according to the nature (oxidation number) of Mt ion, which are interpreted in terms of structural variations.

4. OPTICAL PROPERTIES
4.1 Refractive index

Due to the high PbF_2 content, the refractive index is high and strongly dependent on composition; however, the modulation of n can also be achieved by addition of small amounts of $PbCl_2$ or PbS (Figure 3) [5].

4.2 Luminescence

In such a transparent matrix, various rare-earth ions have been studied; the association of Mn^{2+} and Er^{3+} in PMG glass allows energy transfer [7]. For UO_2^{2+}, interesting long emission lifetimes have been found in PBI glass [8].

4.3 Infra red

As for most fluoride glasses. TMFG have very good transparency (excluding Uranium glasses) as far as 7.5 -8.5 μm, depending on the glass composition. Figure 4 shows the variation of absorption coefficient with frequency for some TMFG compared to fluorozirconates prepared under the same conditions; this confirms that I.R. transparency is an intrinsic property of fluoride media.

4.4 Fibering

Very limited trials have been made to draw fibers of PMG

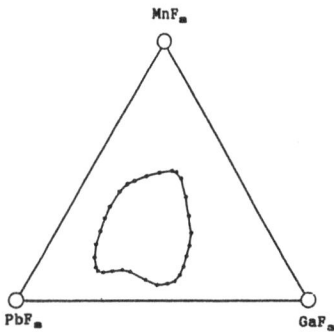
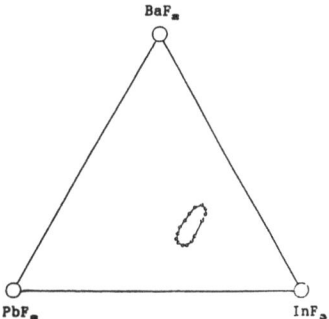

FIGURE 1. Vitreous domains in $PbF_2-MnF_2-GaF_3$, $PbF_2-BaF_2-InF_3$ and $BaF_2-UF_4-MtF_n$ systems (quenched on 230°C brass mould).

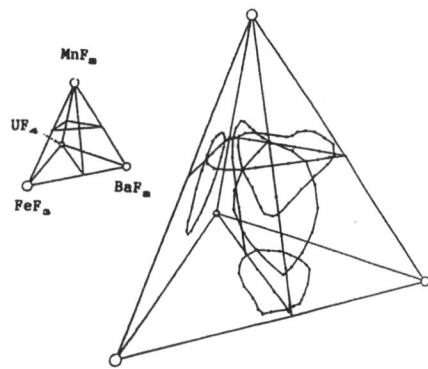

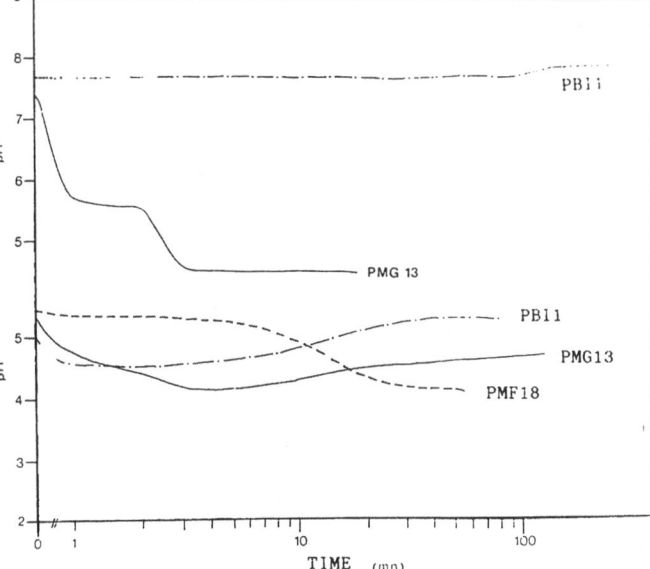

FIGURE 2.

pH variation versus time for various **TMFG**.

TABLE 1. Thermal data ($^{\circ}$C), hardness and optical
characteristics of stabilized TMFG.

Glass composition			Name	Tg	T_x	T_m
$42PbF_2$	$17MnF_2$	$32FeF_3$	PMF18	251	317	520
$3SrF_2$	$5YF_3$	$2AlF_3$				
$36PbF_2$	$24MnF_2$	$35GaF_3$	PMG13	273	336	526
	$5YF_3$	$2AlF_3$				
$19PbF_2$	$23BaF_2$	$47InF_3$	PBI1	250	318	588
$5SrF_2$	$5ZnF_2$	$2AlF_3$				

Expansion coefficient $\sim 170.10^{-7}$ K^{-1}

Hardness Hv_{100} $\sim$ 250-300

Glass	Refractive index	Abbe number	Cut off(μm)
PMG 13	1.5470	44	7.9
PBI0	1.5767	40	8.3

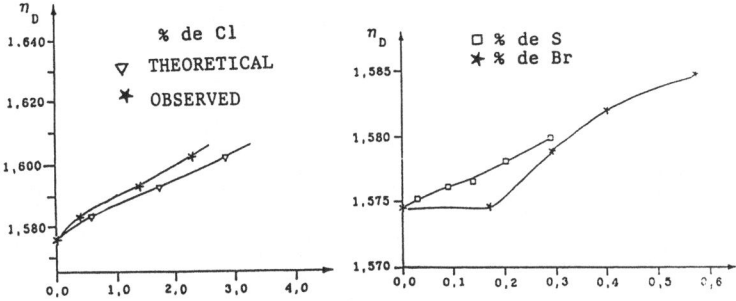

FIGURE 3. Variation of refractive index with Cl, S or Br content.

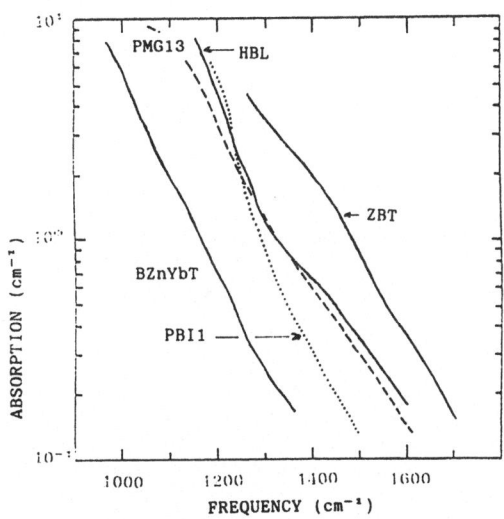

FIGURE 4. Absorption coefficient α (cm⁻¹) or

absorption loss (dB/km) vs. frequency

for TMFG compared to HMFG.

glass and the teflon coated fibers obtained were short and of poor quality because of crystallyzation at the preform stage; the primary technology which has been used prevents definitive conclusions and more serious attempts are needed. A recent S.I.M.S. study of the surface dehydroxylation process for PMG 13 under a reactive atmosphere (N_2 + CS_2) shows that it might be useful to draw fibers in such an atmosphere, since hydrolysis occurs at a temperature as low as the softening point [9].

5. MAGNETIC PROPERTIES

Extensive magnetic studies for high 3d content TMFG corroborate, as expected, the predominance of first neighbour super exchange antiferromagnetic interactions (Figure 5, [2]). In the special case of Mn^{2+} -Fe^{3+} (both $3d^5$ - ^6S ions) glasses, the frustration, induced by the topological disorder leads to spin-glass behaviour (Figure 6, [10]). Magnetic suceptibility measurements of glasses BaF_2-UF_4-MtF_n (n= 2, 3) show that the magnetic behaviour differs with the oxidation number of the 3d ion (Figure 7, [11, 12]); Mt^{2+} ions seems to be more connected to UF_8 polyhedra than trivalent ions (the mean coordination number C.N. -8- for uranium is deduced from the dependence of magnetic susceptibility on temperature).

6. STRUCTURE OF TMFG

The structural knowledge of TMFG comes from various independent ways.

6.1 Coordination number (C.N.) for cations

Visible absorption spectra are consistent with sixfold coordination for major network formers like V^{3+}, Cr^{3+} ions [1]. Extensive E.X.A.F.S. studies confirm octahedral coordination for 3d ions [13], especially by means of very narrow Mt-F bond length distributions in good agreement with classical values observed in well known crystallized compounds (figure 8); other C.N. values are 8 for U^{4+} [14] and about 9 for Pb^{2+} (Figure 9, 10).

6.2 Neutron magnetic diffraction [15]

The "spin glass" behaviour of such glasses as PbF_2, MnF_2, FeF_3 or $2PbF_2$, MnF_2, FeF_3 has given the opportunity to obtain the magnetic correlation functions for Mn^{2+} and Fe^{3+} ions which appear as a unique species (both have ^6S state). Figure 11 shows examples of radial magnetic distribution functions which lead to the following results:
   a) The correlation is antiferromagnetic between first neighbours as already seen by susceptibility measurements.
   b) The average first distance M - M (M= Mn^{2+}, Fe^{3+}) is 3.6 A. This distance implies a corner sharing for MF_6 octahedra.

A lot of neutron magnetic studies on various 3d-TMFG [16] have shown that the vitreous 3d network remains the same, even for a large variation of lead content.

6.3 Neutron diffraction and X-ray diffraction

In pure diffraction methods, such glasses involve the con-

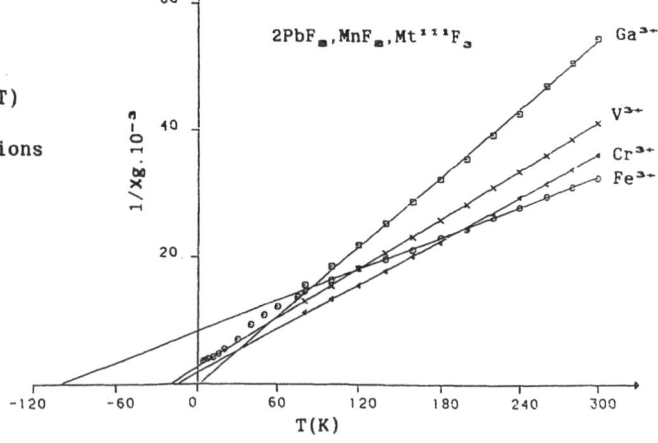

FIGURE 5. $1/Xg = f(T)$ for different Mt^{3+} ions in a glass family.

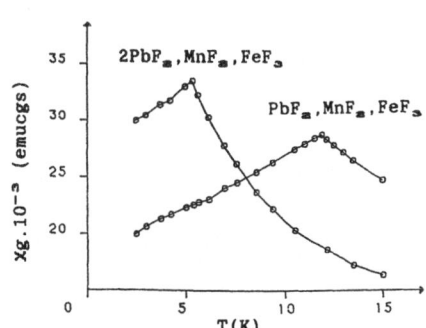

FIGURE 6. Sharp cusps on a.c. susceptibility curves indicate **spin-glass** behaviour.

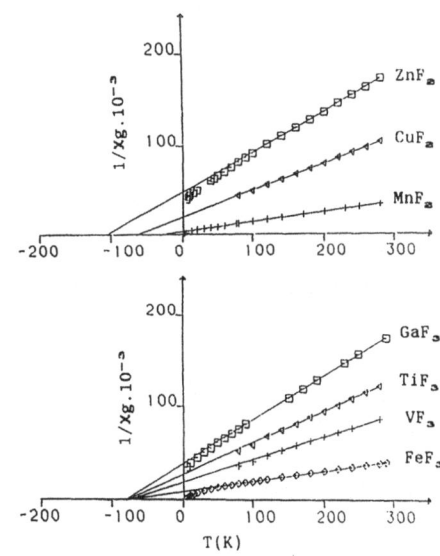

FIGURE 7 $1/Xg$ curves for different oxidation number 3d ions. Note the influence on the paramagnetic Curie temperature.

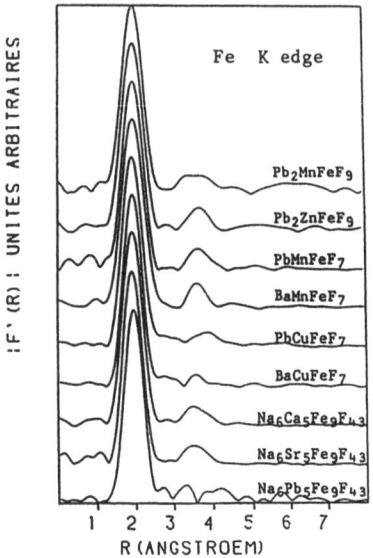

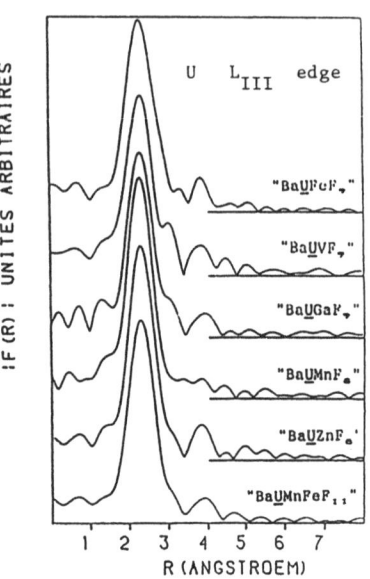

FIGURES 8, 9, 10. Fourier Transform modulus of EXAFS modulations at Iron, Uranium and Lead absorption edge for various **TMFG**.

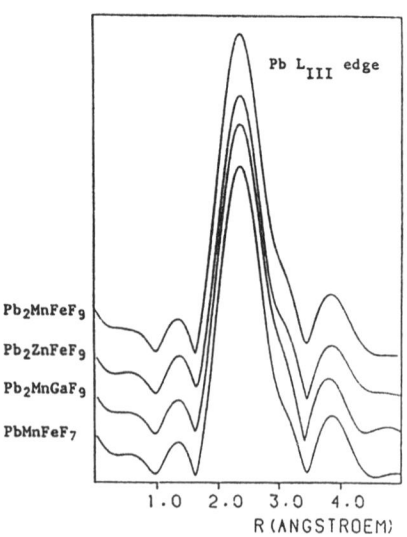

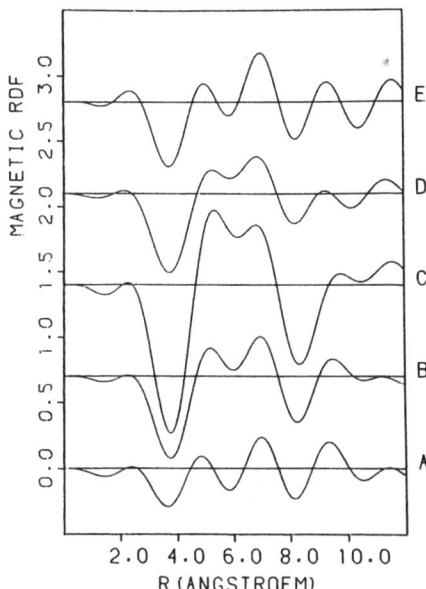

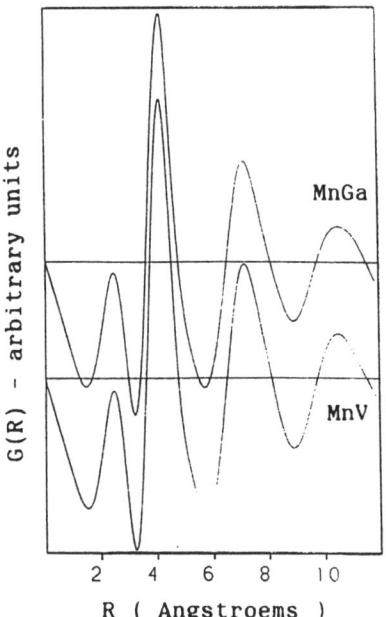

FIGURE 11. $\pi R^2 \rho(R)<S_o.S>$

for $2PbF_2$, $Mt^{11}F_2$, $Mt^{111}F_3$ with

$Mt^{11}Mt^{111}$ = ZnFe (A);

$Mn_{0.4}Zn_{0.6}Fe$ (B); MnFe (C)

$MnFe_{0.4}Ga_{0.6}$ (D); MnGa (E).

FIGURE 12.   X-ray Reduced atomic

distributions for glasses

$2PbF_2$, $Mt^{11}F_2$, $Mt^{111}F_3$ with

$Mt^{11}Mt^{111}$ = MnGa; MnV.

FIGURE 13. Cationic model
which fully fits the neutron
magnetic interference function
for $2PbF_2$, MnF_2, FeF_3 spin-glass.

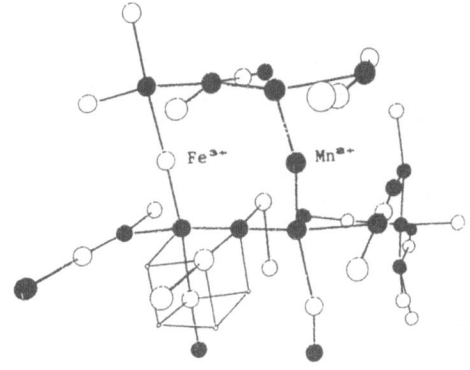

ribution of at least 10 partial pair correlation functions in the RDF. Some special conditions can be stated, tied to the isomorphism of 3d ions and to the relative values of Fermi lengths for Mn and V; this allows one to obtain the partials involving Fe^{3+} for a set of glasses like $2PbF_2$, $Mt^{II}F_2$, $Mt^{III}F_3$ with Mt^{II}= Mn, Zn and Mt^{III}= Fe, V [18]. The M-F distances and C.N. deduced from these neutron data are in well agreement with E.X.A.F.S. ones. The only additional information given by X-ray method concerns the mean Pb-Pb value, but the intense peak near 4.2 A (Figure 12) is masked by the contributions of Pb-Mt and Pb-F.

6.4 Simulation of diffraction data

All the studies performed on TMFG show that the local order is not far from the crystallyzed one. The existence of the crystalline compound $Pb_5Fe_3F_{17}$ [19], located on the edge of the vitreous PbF_2-MnF_2-FeF_3 area, linked to the high level of similarity the glassy and the crystalline behaviour of 3d ions in fluoride media justify the "quasi-crystalline" approach to such glassy state. To support this hypothesis, one can verify the analogy between the $Pb_5Fe_3F_{17}$ structure and the FeF_3 structure (octahedra, corner-shared); this view explains well the first 5 magnetic interactions found in the neutron magnetic study and figure 13 shows a schematic model restricted to 3d (Mn^{2+}, Fe^{3+}) ions which fully fits the neutron magnetic interference function [16]. In fact, this model expresses both the corner linkage of 3d octahedra and the participation of Pb^{2+} in the anionic packing; the good agreement obtained by LE BAIL [20, 21] in the "quasi-crystalline refinement" of $2PbF_2$,$Mt^{II}F_2$,$Mt^{III}F_3$ glasses (Figure 14) with $Pb_5Fe_3F_{17}$ or $KPbCr_2F_9$ as starting models (both crystalline structures imply F^- ions in a near close-packing arrangement and corner sharing of 3d octahedra) is consistent with this concept. Figure 15 shows views of the crystalline starting model ($Pb_5Fe_3F_{17}$ type) and the quasi-crystalline" result, which is the best average local organization in $2PbF_2$,$Mt^{II}F_2$, $Mt^{III}F_3$ glasses; it may be noticed that the distorsion of octahedra seen here is not in good agreement with the E.X.A.F. S. results.

7. CONCLUSIONS AND PERSPECTIVES

TMFG have given the opportunity of various and convergent structural studies which lead to a structure built out of an octahedral network (corner shared, essentially). The magnetic behaviour of most of these glasses (involving Mn^{2+} and Fe^{3+}) is characteristic of spin-glass state.

The optical potential of TMFG is quite similar to that of HMFG but the problem of glass stabilization seems to be more difficult to solve. However, recent studies on vapour deposition of such glasses have given interesting results [22]; nice and regular (10-20 µm thickness) layers of glass can be obtained on ZBLA substrate starting from PMG13 composition. Figure 16 shows the X-ray characterization of the deposit; the chemical analysis is consistent with the composition $45PbF_2$-$55GaF_3$, (MnF_2, YF_3, AlF_3 <1%). Refractive indices are about

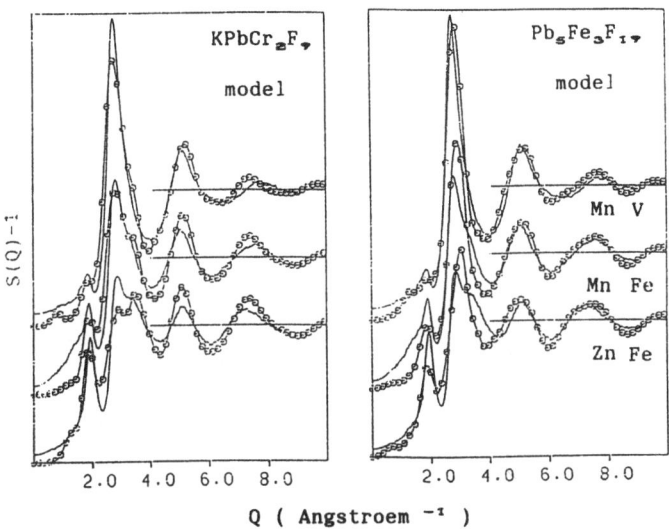

FIGURE 14. Experimental (————) and simulated (ooooo) interference

functions with starting crystalline models $KPbCr_2F_7$ and $Pb_5Fe_3F_{17}$

for glasses $2PbF_2$, $Mt^{11}F_2$, $Mt^{111}F_3$.

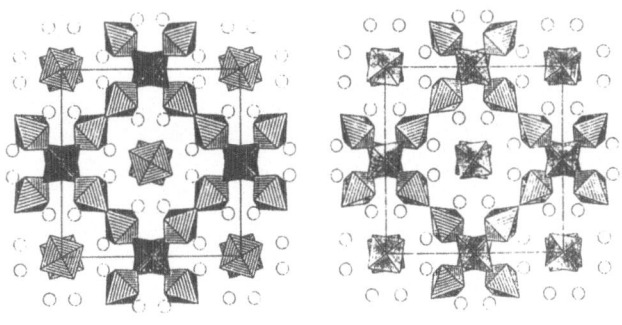

Crystalline model   Quasi-crystalline glass

FIGURE 15. Projection of starting crystalline $Pb_5Fe_3F_{17}$ model, and local

quasi-crystalline arrangment for $2PbF_2$,MnF_2,FeF_3 (from data of FIGURE 14).

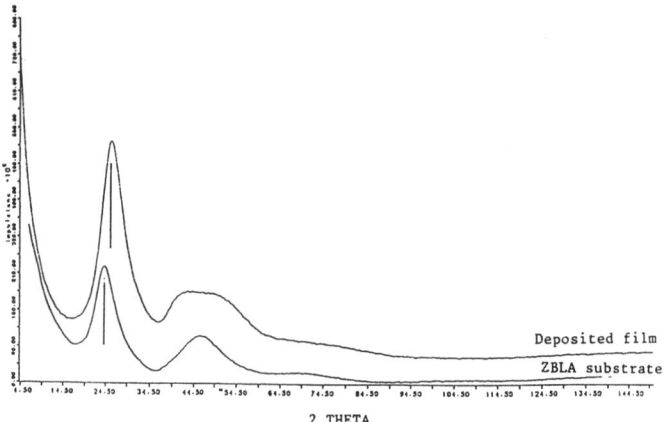

2 THETA

FIGURE 16. X-ray diffraction spectra of **PMG** deposited film
and of the **ZBLA** substrate (reverse side of the slice).
(Cu Kα, step scan 0.2 ° θ, counting time per point 140s).

1.55. Such possibilities for vapour deposition of a glassy mix̱ture could be of interest to increase the refractive index of the core of the preform without introducing heavy anions like Cl^- or S^{2-}; the chemical compatibility of HMFG and TMFG is high enough to obtain a good interface.

REFERENCES

1. Miranday J.P., Jacoboni C. and De Pape R.: J. of Non-Cryst. Sol., 43, 393-401 (1981).
2. Jacoboni C., Le Bail A. and De Pape R.: Glass Technology, 24, 3, 164-167 (1983).
3. Guery J., Courbion G., Jacobini C. and De Pape R.: Materials Chemistry, 7, 715-722 (1982).
4. Auriault N., Guery J., Mercier A.M., Jacoboni C. and De Pape R.: Mat. Res. Bull., 20, 313-316 (1985).
5. Auriault N.: Thesis, LE MANS, France (1986).
6. Simmons C.J., Guery J., Chen D.G. and Jacoboni C.: Third International Symposium on Halide Glasses - RENNES: Materials Science Forum - Volume 5, 329-334 (1985).
7. Reisfeld R., Greenberg E., Jacoboni C., De Pape and Jorgensen C.K.: J. of Solid State Chem., 53, 236-245 (1984).
8. Reisfeld R. and Jacoboni C.: Private Communication, (1985).
9. Auriault N., Jacoboni C., De Pape R., Cottrant J.F. and Chopinet M.H.: Third International Symposium on Halide Glasses - RENNES: Materials Science Forum - Volume 5, 305--309 (1985).
10. Renard J.O., Miranday J.P. and Varret F.: Solid State Com., 35, 41 (1980).
11. Guery J., Courbion G. and Jacoboni C.: Revue de Chimie Minérale, t21, 784-794 (1984).
12. Guery J., Courbion G., Jacoboni C. and De Pape R.: Mat. Res. Bull., 19, 1437-1442 (1984).
13. Le Bail A., Jacoboni C. and De Pape R.: J. of Solid State Chem., 52, 32-44 (1984).
14. Courbion G., Guery J., Le Bail A. and Jacoboni C.: Third International Symposium on Halide Glasses - RENNES: Materials Science Forum - Volume 6, 739-742 (1985).
15. Le Bail A., Jacoboni C. and De Pape R.: J. of Solid State Chem., 48, 168-175 (1983).
16. Le Bail A., Jacoboni C. and De Pape R.: J. of Non-Cryst. Solids, 74, 205-212 (1985).
17. Le Bail A.: Doctorat Thesis, LE MANS, France (1985).
18. Le Bail A., Sacoboni C. and De Pape R.: J. of Non-Cryst. Solids, 74, 213-221 (1985).
19. JACOBONI C., Le Bail A., De Pape R. and Renard J.P.: Solid State Chem. "Studies in Inorg. Chem." Vol. 3: Proceed. of 2nd European Conference. VELDHOVEN (1983).
20. Le Bail A., Jacoboni C. and De Pape R.: Third International Symposium on Halide Glasses - RENNES: Materials Science Forum - Vol. 6, 441-447 (1985).
21. Le Bail A., Jacoboni C. and De Pape R.: International Conf. on the Structure of Non-Cryst. Solids Proceedings, A19, GRENOBLE (1985).

22. Poignant H., Monerie M., Baniel P., Jacoboni C. and Mercier A.M.: To be published.

DISCUSSION

Q: M. Robinson. Was the composition of the crystallized and vitreous deposits the same?

A: C. Jacoboni. I presume so.

Q: R. Almeida. What is the main application of transition metal fluoride glasses?

A: C. Jacoboni. The same as that of fluorozirconate glasses.

Q: R. Almeida. Has the fundamental IR spectrum ever been taken for the transition metal fluoride glasses?

A: C. Jacoboni. No, not yet.

Q: M. Drexhage. Could you clarify some details of the melt evaporation experiments, especially the need for AlF_3 and YF_3 in the melt in order to obtain a vitreous deposit?

A: C. Jacoboni. It appears that the stabilized composition (PMG 13) allows a lower melting temperature and a larger deviation in composition due to selective evaporation.

Q: J.D. Mackenzie. Would you expect the optical properties of your glass to be affected by a magnetic field?

A: C. Jacoboni. Yes, particularly for high contents of 3d para magnetic species, as in glass PMF 18 (PbF_2, MnF_2, FeF_3) which has more than 50% of paramagnetic cations and has a IR cut-off near 7-8 µm.

CHLORIDE, BROMIDE AND IODIDE GLASSES

J.D. MACKENZIE

Department of Materials Science and Engineering
School of Engineering and Applied Science
University of California
Los Angeles, CA 90024, USA

1. INTRODUCTION

Since the discovery of the ease of glass-formation of the fluorozirconates in 1975(1), there has been wide-spread of research on fluorides. Prior to that time, the only glass-forming halides which had received a great deal of attention were those based on BeF_2 as the glass-former. The only non-fluoride halide glass which was most widely known was probably $ZnCl_2$. A critical review on these halide glasses up to 1981 has been published (2). A few other glass-forming halide systems known up to that time include those based on AgCl, AgBr, AgI and $PbBr_2$ (3), $PbCl_2$-$BaCl_2$(4), $SnCl_3$-PbI_2(5), and TlCl (6).

The reasons behind the wide-spread interests in fluorozirconates are two-fold. Firstly, the structures of these glasses are very different from those of oxides. Secondly, they appear to hold promise as candidate materials for infrared transmitting fibers. For the same reasons, there has been increased interest in non-fluoride halide glasses in the past five years although research results reported in the literature for those glasses are a small fraction of those for the fluorozirconates. In this report, some important properties of known glass-forming chloride, bromide and iodide systems are reviewed and projections for the future development presented.

2. SOME KNOWN NON-FLUORIDE HALIDE GLASSES AND THEIR PROPERTIES

Some typical glass-forming systems are shown in Table 1. A few others have been reported by Yamane et al (7) and by Poulin (8). Compared to the fluorozirconates, there is relatively little information on the structures and properties of these glasses beside the fact that many do form glasses fairly readily. The most widely studied property is IR transmission.

(a) Optical Properties

From a simple consideration of the masses of anions associated with the same cations alone, chloride, bromide and iodide glasses would be expected to be more transparent in the IR than fluoride glasses. Limited information available does confirm such a hypothesis. Figure 1 shows the optical transmission of some non-fluoride glasses compared

Table 1. Some glass-forming chloride, bromide and iodide glasses.

$ZnCl_2$	$ZnBr_2$
$ZnCl_2 - KCl$	$ZnBr_2 - KCl$
$ZnCl_2 - KBr - PbBr_2$	$ZnBr_2 - KBr$
$BiCl_2 - KCl$	$ZnBr_2 - KI$
$PbCl_2 - BiCl_3$	$ZnBr_2 - TlCl$
$PbCl_2 - AgCl_3 - CdCl_2$	$ZnBr_2 - TlBr$
$ThCl_4 - KCl$	$ZnBr_2 - TlI$
$ThCl_4 - NaCl$	$ZnBr_2 - KBr - TlBr$
$ThCl_4 - KCl - NaCl$	$ZnBr_2 - KBr - TlI$
$CdCl_2 - BaCl_2$	$ZnBr_2 - KI - KBr - TlBr$
$PbCl_2 - AgCl_2 - CsCl$	$ZnBr_2 - TlBr - TlI$
$PbCl_2 - CdBr_2 - TlI$	$ZnBr_2 - TlBr - GaBr_3$
$PbCl_2 - PbBr_2 - CdCl_2 - TlI$	$ZnBr_2 - RbBr$
$PbBr_2 - AbGr_2 - CsBr$	$ZnBr_2 - KBr - TlBr - BaBr_3$
$PbBr_2 - AgBr - CsBr - CdBr$	
$BiBr_3 - TlBr - PbCl_2$	
$ZnBr_2 - PbBr_2 - TlBr$	
$ZnBr_2 - PbBr_2 - CsBr$	
$ZnBr_2 - PbBr_2 - KBr$	
$GaBr_3 - NaBr$	
$GaBr_3 - TlBr$	
$AgI - CsI$	
$CdI_2 - KI$	
$CdI_2 - KI - CsI$	
$ZnI_2 - KI$	
$ZnI_2 - TlI$	
$ZnI_2 - ZnBr_2 - KI$	

with that of a fluorozirconate. The chloride, bromide and iodide glasses shown in Fig. 1 are, however, extremely hygroscopic. Practical applications of the glasses are obviously not simple. Angell and co-workers (9) have reported that new glasses based on $CuCl-PbCl_2-RbCl$, $PbCl_2-PbI_2-CdCl_2$ and $AgCl-AgI-CsCl$ are not only highly transmitting to 20 microns in the IR, but are also very durable in the ambient atmosphere. One undesirable feature, however, is the relatively low T_g of about 60°C. The use of $ZnCl_2$ glass as a low loss IR fiber was proposed by Van Uitert and Wemple in 1979 (10). Presumably because of the high hygroscopicity of $ZnCl_2$, the proposal apparently has not been exploited since that time. Most recently, Yamane et al. have prepared glasses based on $ZnCl_2$ (11). A glass of the molar composition $48ZnCl_2-48KBr-4 PbBr_2$ apparently was better than $ZnCl_2$ in chemical durability and transmit to approximately 15 microns. After removal of water by reactive atmosphere processing (RAP) using CCl_4 and CBr_4, the absorption loss for a CO_2 laser was found to be 20 dB/m by laser calorimetry. This relatively high loss was attributed to the presence of fine particles of carbon. Presumably, the formation of carbon can be prevented and a lower loss attainable. As for the glasses reported by Angell et al (9) these new glasses also have very low T_g (45°C) and are still hygroscopic. Unless the chemical durability is much improved and T_g significantly raised, it is unlikely that these non-fluoride halide

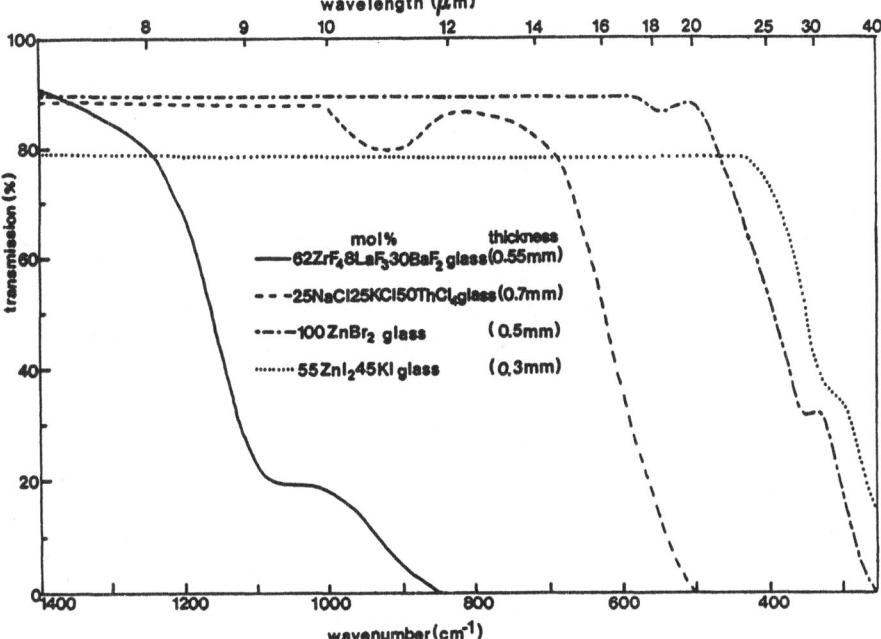

Fig. 1   Infrared transmission spectra of some halide glasses
         (12).

glasses will become strong candidates as IR-transmitting fibers.

One other interesting optical property studied was the photochromism of glasses based on AgI-AgBr-PbBr$_2$-CsBr-CdBr$_2$ (7). The detailed mechanisms of darkening and bleaching are at present unknown and no comparison is available between these new glasses and the well-known photochromic oxide glasses containing silver halide crystals.

(b) Structure

Presumably because of the poor chemical durability, there are few structural studies of chloride, bromide and iodide glasses. The structures of ZnCl$_2$ and ZnBr$_4$ glasses, based on Raman spectra, are postulated to have a three-dimensional random network built up of [ZnCl$_4$] and [ZnBr]$_4$ tetrahedra (13), somewhat similar to the three-dimensional linking of SiO$_4$ tetrahedra in silica glass. There is an alternative suggestion, however, that the zinc ion is 6-coordinated (2). Binary ZnBr$_2$-KI glasses have been studied by Raman scattering (13). The [ZnBr]$_4$ tetrahedra are considered to have become distorted [ZnBr$_3$I] tetrahedra, containing two bridging Br ions, one non-bridging Br ion and one non-bridging I ion with cross-linking K-Br and K-I bonds.

(c) Viscosity and Glass-Forming Tendency

The viscosity of halide melts and glasses is treated elsewhere in this Proceedings (14). Figure 2 shows a plot of log viscosity vs. 1/T for some molten halides. In general, the addition of Group I and Group II halides to ZnCl$_2$ and ZnBr$_2$ tends to decrease the viscosity and lower the activation energy for viscous flow. The glass-forming tendency usually decreases as well because of the lower viscosity at the liquidus temperature.

For oxide glasses, glass-forming tendencies have been correlated with bond energies (15). Poulin has applied the concept of ionic field strengths to glass-formation in fluorides (16). These concepts were modified recently and applied to chlorides, bromides and iodides (17). They both appeared to be compatible with known glass-forming systems. However, the use of these concepts for the prediction of glass-formation have not been reported. Frequently, the appropriate phase diagrams have not been constructed and this renders the prediction of one-liquid phase boundaries impossible.

(d) Chemical Durability, Glass Transition Temperature and Expansion Coefficient

All things being equal, for ionic solids, low glass transition temperature, T_g and high expansion coefficient are associated with poor chemical durability towards water. There are of course some exceptions but this hypothesis is generally sound. The melting points of halides are, as a first approximation, fair indications of the "attractiveness" of a halide for glass preparation. Table 2 gives a comparison of the melting temperatures of halides. Fluorides appear to have higher melting temperatures than the corresponding chlorides, bromides and iodides because of the higher single bond energies between the ions. If the concept that $T_g = 2/3 \ T_m$ is applicable (18), then, in general, fluoride glasses should have the highest T_g among all halide glasses. This appears to be true although data for corresponding halide systems for comparison are scarce. There are many fluorozirconate glasses with T_g in excess of 300°C but few chloride, bromide and iodide glasses have T_g in excess of 150°C.

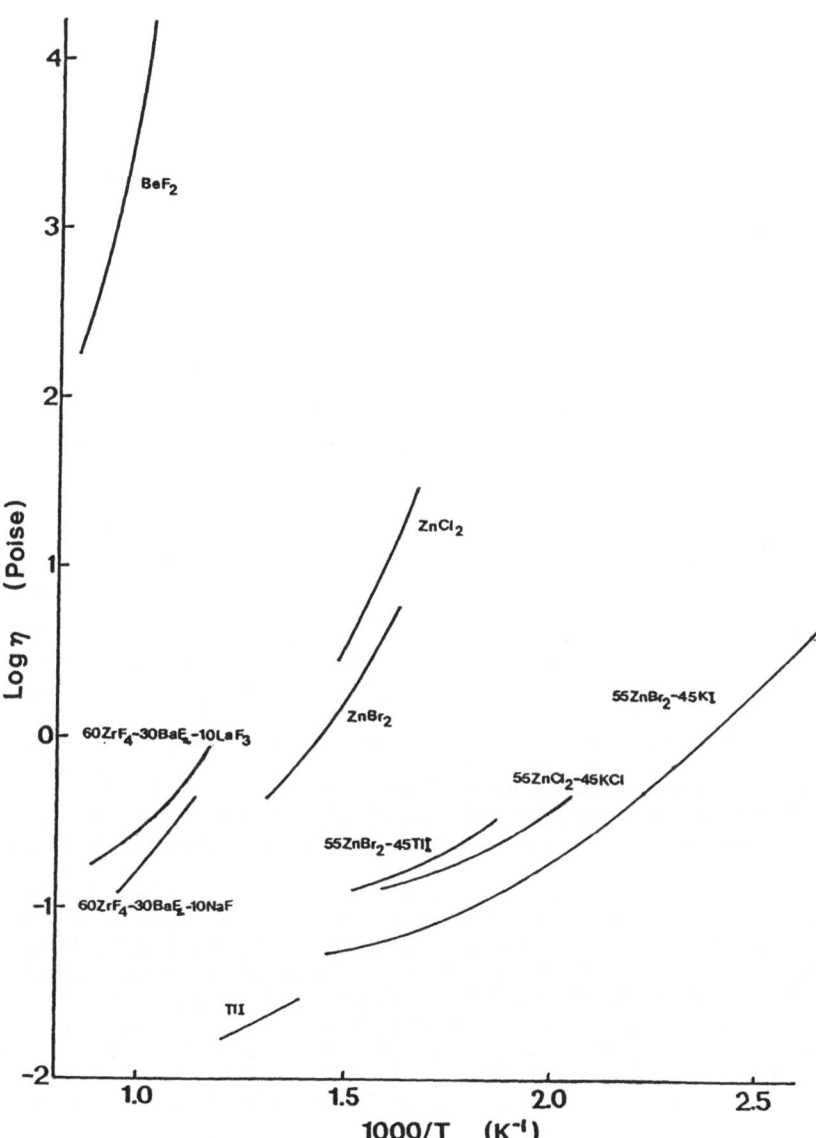

Fig. 2 Viscosity-temperature relation for some halide glasses.

Table 2 Melting Temperatures of Alkali Halides.

Salt	$(2/3)T_m$ °K	T_m °K
LiF	745	1118
LiCl	585	878
LiBr	549	823
LiI	481	722
NaF	844	1266
NaCl	716	1074
NaBr	680	1020
NaI	623	934
KF	754	1131
KCl	695	1043
KBr	671	1007
KI	636	954
RBF	712	1068
RbCl	661	991
RbBr	644	966
RbI	613	920

Similarly, one would expect the thermal expansion coefficient of fluoride glasses to be lowest of all halide glasses. The expansion coefficient of fluorozirconates, for instance, are generally less than 20×10^{-6} per degrees C. A $ZnCl_2$-based bromide glass was reported to have an expansion coefficient of 54×10^{-6} per degree C (11) and a mixed iodide-bromide glass had a value of 34×10^{-7} (7).

There are no quantitative chemical durability studies on chloride, bromide and iodide glasses. Qualitative observations suggest that they are highly unstable in aqueous solutions, and, indeed, even in moist air. There are no known theories for the prediction of chemical durability of halide glasses from their chemical compositions. As a first approximation, it may be assumed that all things being equal,the solubility of the individual components of a glass in water can have an effect on the ultimate chemical durability of the resultant glass. Thus, AgI, AgBr, AgCl, PbI_2, $PbBr_2$, $PbCl_2$, HgI, TlI, TlBr and TlCl, all having low solubility in water, should be desirable components in a halide glass. Such an approach towards the design of a halide glass with good chemical durability still awaits experimental verification.

Figure 3 is a simplified representation of the current status of chloride, bromide and iodide glasses. One would hope to prepare "ideal" glasses with high T_g, with high viscosity at the liquidus temperature but yet with low expansion and better chemical durability. The prospects are not encouraging although science is frequently confronted with the unexpected.

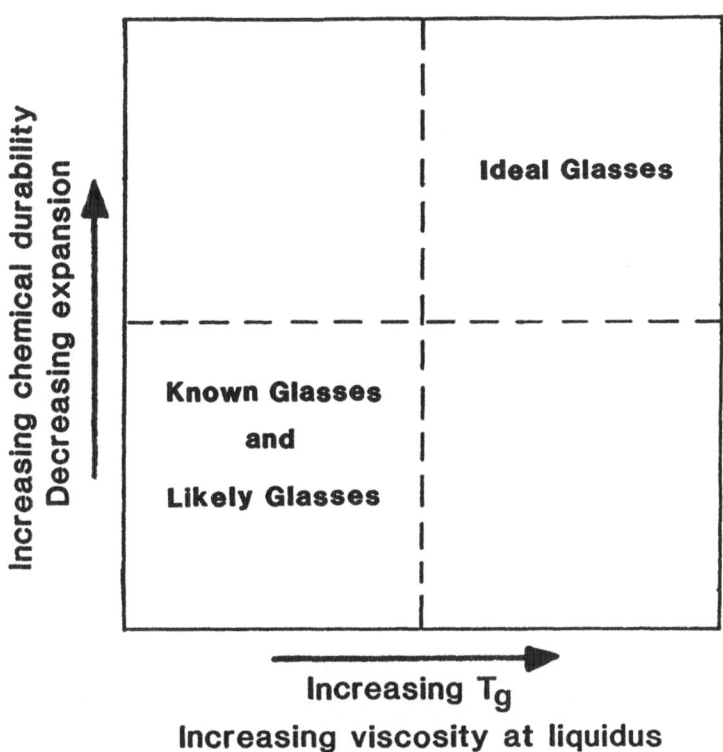

Fig. 3 Properties of known halide glasses versus ideal glasses.

3. CONCLUSIONS

Chloride, bromide and iodide glasses appear to have lower chemical durability, lower T_g and higher thermal expansion coefficients than the corresponding fluoride glasses. The single significant advantage they have is probably the superior transmission in the IR. It is likely that many more non-fluoride halide glasses will be developed in the near future. Structural investigations have been lacking and will be a fruitful field for scientific research.

ACKNOWLEDGEMENTS

The support of the Directorate of Chemical and Atmospheric Sciences of the AFOSR is gratefully acknowledged.

REFERENCES

1. Poulin, M and Lucas J: Mat Res. Bull. **10**, 243 1975.
2. Baldwin CM, Almeida RM and Mackenzie JD: J. Non-Cryst. Solids **43**, 309, 1981.
3. Sun KH: Glass Ind. **27**, 552, 1946.
4. Mellor JE: Comprehensive Treatise on Inorganic and Theoretical Chemistry, Vols. VIII and IX. London: Longmans Green & Co., 1929.
5. Winter A: J. Am. Ceram. Soc. **40**, 54, 1957.
6. Moynihan CT: in Ionic Interactions, Petrucci S(ed): p. 261. New York: Academic Press, 1971.
7. Yamane M et al: J. Non-Cryst Solids **56**, 87, 1983.
8. Poulin M: J. Non-Cryst. Solids, **56**, 1, 1983.
9. Angell CA, Liu C and Sundar HGK: Materials Science Forum, **5**, 189, 1985.
10. Van Uitert L.G and Wemple SH: Appl. Phys. Lett. **33**, 57, 1978.
11. Yamane M et al: Materials Science Forum **6**, 489, 1985.
12. Nasu H et al: Materials Science Forum **5**, 121, 1985.
13. Almeida RM: Materials Science Forum **6**, 427, 1985.
14. Mackenzie JD, Nasu H and J. Sanghera: this Proceedings.
15. Sun KH: J. Am. Ceram. Soc. **30**, 277, 1947.
16. Poulin M: Nature **293**, 279, 1981.
17. Ma, FD, Lau L and Mackenzie JD: J. NonCryst. Solids, in press.
18. Sakka S and Mackenzie JD: J. Non-Cryst. Solids **6**, 145, 1971.

DISCUSSION

Q: P. Klein. Are there any good cyanide-glasses, since cyanogen ($(CN)_2$) is known as a pseudo-halogen?

A: J.D. Mackenzie. Most cyanide glasses, if there are any, are likely to have been investigated by Ubbelhode. They are either water-soluble or have a low T_g.

Q: G. de Leede. Can you explain the low viscosities found in $ZnCl_2$, $ZnBr_2$, etc., at their melting points, as compared to those of oxide glass formers? These halides have rather large ratios of bond strength/$T_{liquidus}$ (of the same order or even larger than in SiO_2) and the structural units are not very different from the oxides.

A: J.D. Mackenzie. I believe that $ZnCl_2$ and $ZnBr_2$ are different from other group II halides, presumably because they have 4-coordinated cations and form melts with a polymeric character.

Q: J. Lucas. Could you comment on the glass formation mechanism in glasses containing halides like AgI, CuCl, etc.? It seems that the "confusion principle" could also be applied to those mixtures which always contain several halides with different structural types.

A: J. D. Mackenzie. I think that the principle must apply to some extent here, not only the cations but also the anions being affected.

C: J. Lucas. There is a correlation between SiO_2, BeF_2 and $ZnCl_2$ glasses in the sense that the basic unit is always a tetrahedron.

Q: G. Frischat. If you take a Na_2O-CaO-SiO_2 glass and substitute 3 weight % of oxygen by nitrogen, T_g values are increased by 100 K and the viscosity increases by at least one order of magnitude. If this is also possible for halide glasses, their properties could also be considerably stabilized.

A: J.D. Mackenzie. Something must get worse. If nitrogen is incorporated into the glass and forms 3-coordinated units, then the IR cut-off will suffer.

Q: M. Drexhage. As a matter of historical interest, could you comment or clarify previous claims for glass formation in HF-based systems?

A: J.D. Mackenzie. I believe that the work reported was not correct, in the sense that other halides such as BeF_2 were also present. I do not know of any truly glass-forming systems containing only HF and alkaline-earth fluorides such as MgF_2.

HALIDE GLASSES AND CHALCOGENIDE GLASSES FOR ULTRA LOW LOSS FIBRE
APPLICATIONS - A COMPARISON

J A SAVAGE

Royal Signals and Radar Establishment, Malvern, UK

ABSTRACT

Mid infrared transmitting glasses have the potential of becoming the next
generation of optical communications fibres. These ultra low loss fibres
will need to demonstrate losses of the order of 10^{-2} to 10^{-3} dB/km at some
particular wavelength between 2.5 and 4.5 μm compared to the present day
loss of 0.2 dB/km at 1.55 μm shown by silicate fibres. Candidate glass
forming systems are fluoride, fluoride-chloride, chloride, chalcogenide and
oxide. The fluoride and fluoride-chloride glasses are leading candidates
at the present time largely because major effort has concentrated upon
them. The recently reported heavy metal oxide glasses based on bismuth and
gallium oxides offer equivalent transmittance range to the fluorides.
These oxide glasses are more resistant to devitrification and are likely to
be synthesised by the vapour route thus offering a potentially serious
challenge to the fluoride glasses for ultra low loss applications if the
scatter losses are shown to be acceptably low. The chalcogenide glasses
are only likely to be seriously considered for these latter applications if
the present drive based on the halide and possibly oxide glasses fails to
yield technically acceptable fibres. Thus the chalcogenides are most
likely to find use as short length fibres for applications in the far
infrared.

1.0 INTRODUCTION

Glass clad glass fibres were first made by Kapany (1959a) and Hirschowitz
et al (1958) and Kapany (1967) applied the name "fibre optics" to this
field and defined it as "the art of active and passive guidance of light
rays in the UV, visible and infrared regions of the spectrum along
transparent fibres through predetermined paths. The technique of making
"multiple fibres" (Kapany 1959b) allowed the fabrication of high resolution
fused fibre optic face plates for cathode ray tubes, and coupling plates
for image intensifiers (Hicks and Kiritsy 1961). Kapany and Mergerian
(1960) investigated materials such as As_2S_3 and AgCl for the mid and far
infrared and Kapany and Simms (1965) studied As-Se-Te and Ge-As-Te glasses
for the far infrared. The work described by Kapany (1967) was done with
high loss materials and was far ahead of its time.

It was recognised by Kao and Hockham (1966) that the high loss exhibited
by commercial oxide glass is not an intrinsic property but is caused by
extrinsic impurity absorption. Later Kao and Davis (1968), Jones and Kao
(1969) and Kao et al (1970) showed that fused silica was capable of
exhibiting a loss as low as 80 dB/km. By 1970 Kapron et al (1970) had
shown that the loss could be reduced to 20 dB/km in a fibre at 0.6328 μm
(HeNe). Fibre fabrication technology was rapidly developed during the
1970s allowing losses of a fraction of a dB/km to be achieved at infrared

wavelengths. In the 1980s, optical communications have become possible using multimode silicate glass fibres in the near IR operating at 0.8-0.9 μm, at 1.3 μm and at 1.55 μm and using monomode fibres operating at 1.3 and 1.55 μm. The loss in these fibres at 1.55 μm is about 0.2 dB/km which is very nearly the intrinsic loss limit of GeO_2-SiO_2 glass.

In the infrared spectral region attenuation in a material is dominated by Rayleigh scattering and multiphonon absorption. To achieve a lower loss medium it is necessary to examine glasses exhibiting a longer infrared wavelength cut off such as the fluorides. If the transmittance window of the glass can be extended farther into the infrared in this way by utilising constituent atoms of higher atomic mass and it can be drawn into fibre form with suitable strength and chemical durability properties then potentially a much lower loss can be achieved. This is due to the fact that the Rayleigh scatter, which has a λ^{-4} dependence upon the propagation wavelength, would be much less for instance by a factor of 16 if the operational wavelength were doubled. Thus it may be possible to achieve of the order of 10^{-2} dB/km loss in a fibre at 2.55 μm and perhaps lower losses at longer wavelengths of 3 to 4.5 μm. For these reasons researchers are now addressing the problems of achieving suitable fibres for use in the mid IR from amongst the fluorozirconate glasses (Poulain et al 1977) and the fluorohafnate glasses (Drexhage et al 1980). There is also interest in far infrared fibres (Klocek 1982) for image transfer and CO_2 laser power transfer for military and medical purposes and in spite of the fact that only short lengths of a few metres of fibre are required in these applications there are problems in finding sufficiently low loss materials. Crystalline halide materials are being examined for high power transfer applications and chalcogenide glasses are being researched for low power applications. Thus leading research aimed at finding suitable fibre materials for the mid and far infrared is being carried out amongst the halide and chalcogenide glasses so that discussion of their relative properties is of current interest.

2.0 GENERAL PROPERTIES OF HALIDES AND CHALCOGENIDES

Before going on to discuss glasses in detail it is instructive to consider some of the relative physical and chemical properties of halides and chalcogenides. The alkali and alkaline earth halides are the most versatile of all of the crystalline bulk infrared optical materials in terms of their optical properties. The alkaline earth fluorides such as MgF_2, CaF_2 and BaF_2 offer the best combination of physical properties; in particular BaF_2 has transmittance capability to about 10 μm. Similarly the heavy metal fluoride glasses amongst the halide glasses offer the best thermal and mechanical properties and a major programme amongst the scientific community is underway to fabricate low loss fibres from them. Chloride, bromide and iodide materials such as KCl, CsI, KBr and AgCl possess very good optical properties and find use where weak mechanical properties and poor chemical durability are not a problem. In general glasses based on these latter three anions are very low melting and offer less attractive physical properties than crystalline materials. Because of these problems many workers have explored making fibre for 8-12 μm applications by extrusion from bulk crystalline materials such as KRS5 and AgCl/AgBr. Crystalline chalcogenide materials have been researched for a number of years and have recently come into some prominence as bulk 8-12 μm optical materials. The most well known of these materials are ZnS, ZnSe and CdTe and they possess moderately low absorption coefficients and average thermal and mechanical properties. Chalcogenide glasses have been produced commercially and used for correcting chromatic aberration in infrared lens systems and are now being researched as fibre materials but the relatively high losses so far obtained in these fibres restricts their use to very short range sensor applications such as image transfer or signal transfer.

Thus having identified families of materials with suitable intrinsic optical properties for future optical fibre applications it is necessary to choose candidate materials for development depending upon the required optical loss and wavelength of operation. Figure 1 shows the theoretical attenuation (Klocek 1982) of halide and chalcogenide glasses compared to silicate glass plotted as a function of wavelength. From this an initial choice of material family can be made for instance fluoride glass for ultra low loss communications applications and chalcogenide or chloride/bromide/ iodide glass for high loss sensor applications in the far IR. The remaining problem is then to choose individual candidates which are amenable to fabrication into a fibre exhibiting sufficient chemical stability, thermal stability and strength. To do this it is necessary to be aware of the physical properties of the various candidate glass materials.

3.0 HALIDE GLASSES
3.1 Bulk Fluorides
It should be noted that all glass compositions quoted in this review are in mole or atomic %. BeF_2 based glasses were researched by Sun (1949, 1979) and Sun and Huggins (1950) who showed that these glasses possessed low refractive indices (BeF_2 n_d = 1.2747) and optical dispersions (BeF_2 V_d = 106.8). This early work has been reviewed by Rawson (1967). Materials of this type are of current interest as active and passive components for high energy lasers (Weber et al 1976). New interest in fluoride glasses as mid infrared fibre optic materials has recently been generated by the discovery of fluorozirconate glasses by Poulain et al (1977) and Lucas et al (1978) and the fluorohafnate glasses by Drexhage et al (1980) transmitting to ~5.5 μm. These glasses contain approximately 50-60% ZrF_4 or HfF_4 together with 30 to 40% BaF_2 and a number of alkali, alkaline earth, transition metal and rare earth fluorides primarily to improve glass stability. Alternative fluoride glasses exhibiting an extended transmittance range to ~7.0 μ m have been reported by Fonteneau et al (1982) and Drexhage et al (1983). These materials are based on ThF_4 and BaF_2 together with rare earth trivalent fluorides (Yb,Y or Tm) and divalent fluorides (Zn or Mg) but are reported to be relatively unstable requiring more rapid quenching than the fluorozirconate glasses. A major problem with all of these glasses has been the relatively narrow glass forming regions and relative thermal instability, resulting in devitrification, and in some glasses, the presence of small crystallites (10-50 μm) (Bendow and Drexhage 1982). Detailed studies of the crystallisation of fluorozirconate glasses has been reported by Weinberg et al (1983) and Parker et al (1985).

Thus emphasis has been placed on complex glass compositions containing four to six fluoride compounds in order to enhance glass stability and also to adjust viscosity values to enable fibre fabrication to be carried out. The background on these materials is contained in a number of reviews (Poulain and Lucas 1978, Bendow and Drexhage 1981, Lucas 1982, Drexhage

370

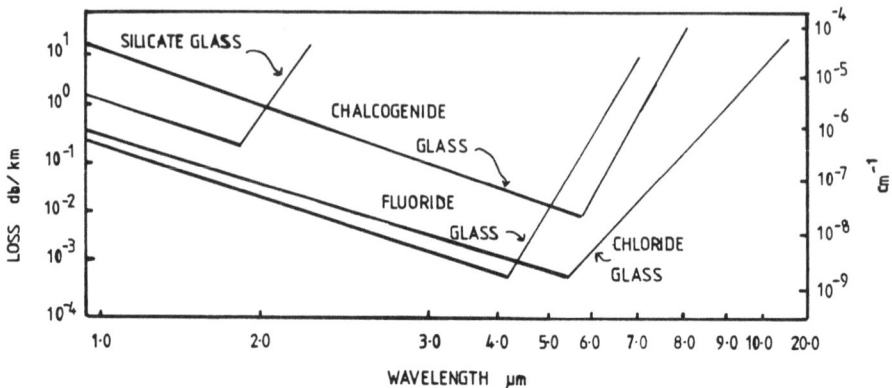

FIGURE 1.
Theoretical V curves for infrared fibre glasses Klocek (1982).

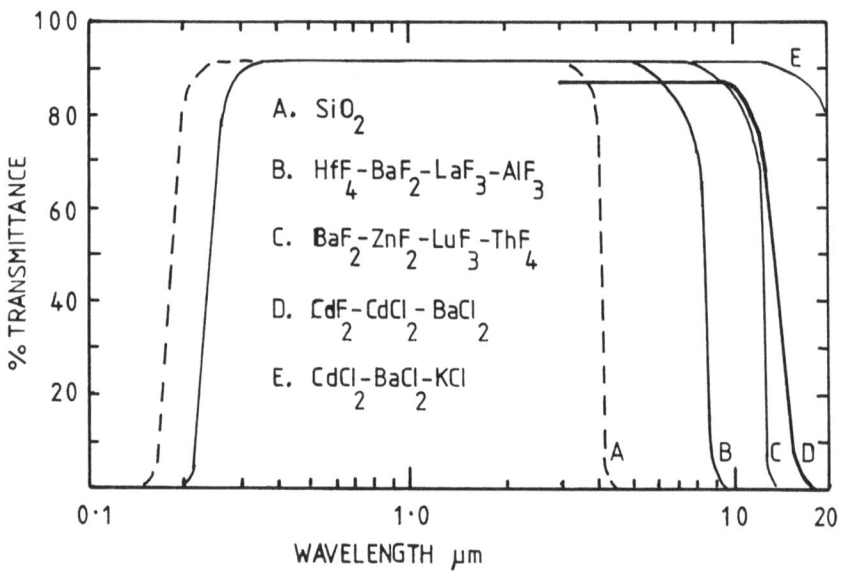

FIGURE 2.
A comparison of the transmittance ranges of halide glasses and silica glass, A, B and C (Drexhage 1985), C and D (Matecki et al 1983).

A SiO_2 5 mm thick
B HfF_4 57 BaF_2 36 LaF_3 3 AlF_3 4   mole %   5 mm thick
C BaF_2 19 ZnF_2 27 LuF_3 27 ThF_4 27   mole %   5 mm thick
D CdF_2 65 $CdCl_2$ 2 $BaCl_2$ 33   mole %   1 mm thick
E $CdCl_2$ 50 $BaCl_2$ 40 KCl 10   mole %   0.4 mm thick.

1985, Sigel and Tran 1984 and Tran et al 1984) while their transmittance
range is shown in figure 2. Basic physical property data on bulk glasses
is relatively sparse since most workers are concerned with identifying
suitable compositions for fibre drawing rather than providing systematic
information on a range of physical properties within particular glass
forming systems. However, enough data exists to enable a general
comparison of physical property information for fluorozirconate,
fluorohafnate, chalcogenide and silica glasses as listed in table 1.
Fluoride glasses offer a broad transparency range from the near UV ~0.3 μm
to the mid-far IR ~6 to 9 μm. Intrinsic Rayleigh scattering varies as λ^{-4}
in silicate glass and Tran et al (1982) have demonstrated this behaviour in
fluoride glasses. A scatter loss of 10^{-3} dB/km has been predicted for
these materials at ~4.0 μm. They also show little susceptibility to high
doses of nuclear radiation in the mid IR (> 3 μm) indicating an improved
performance over silicate glasses where this might be required (Rosiewicz
and Gannon 1981). Refractive indices vary from 1.47 to 1.53 (visible) and
densities from 4 to 6 gm/cc. Glass transition temperatures, T_g, of these
glasses vary from 300° to 450°C, thermal expansion coefficients are in the
range 17–20 x 10^{-6}/°C and the viscosities are low eg. 4 Poise at 490°C with
a very short working range indicating problems for fibre drawing. Hardness
and fracture toughness values are lower than those of silicate glasses but
higher than those of chalcogenide glasses being respectively 225 to 325
Kg/mm^2 and 0.27 to 0.38 MNm$^{-3/2}$. The fluorozirconate glasses are fluorine
ion conductors offering conductivities of the order of 10^{-5} to
10^{-6} ohm^{-1}cm^{-1}. Heavy metal fluoride glasses can be made in a large number
of systems and are likely to offer a wide range of properties. Only the
fluorozirconate, fluorohafnate and rare earth fluoride glasses have been
studied in any detail. Work on other systems is ongoing and may well lead
to improved compositions for fibre drawing. Recent developments (Lucas et
al 1984) on barium, thorium, yttrium, zinc, aluminium fluoride glasses by
substituting indium fluoride for aluminium fluoride have a beneficial
effect on glass stability as well as improving the infrared transmittance
range to ~7 μm. Similarly glasses based on CdF$_2$ reported by Tick (1985)
and Matecki et al (1983) offer alternative glasses with extended IR
transmittance range. Further compositional diversity is afforded by mixed
fluoride-chloride glasses such as those based on CdF$_2$-CdCl$_2$-BaCl$_2$
exhibiting extended IR transmittance range to 9 μm as seen in figure 2.

3.2 Fluoride Fibres

Fluoride fibres have emerged as leading candidates for ultra low loss
optical fibre applications. Such materials are required to possess
attenuation values of the order of 10^{-2} to 10^{-3} dB/km and be capable of
being drawn into fibres exhibiting sufficient mechanical strength and
chemical durability to survive cabling and use in a field environment.
Table 2 (Tran et al 1984) lists typical glass composition types being
utilised in fibre fabrication research. These are of the fluorozirconate-
hafnate types since insufficient data is available on the attractive but
undeveloped alternative glass compositions discussed above to warrant a
major fibre drawing activity. It has been established that the fluoro-
zirconate-hafnate materials potentially offer a low loss window of
<10^{-2} dB/km at 2.55 μm and at present fibres fabricated by the melt process
have yielded losses of ~6.8 dB/km at this wavelength. Major effort is now
being devoted to purification of the raw materials and refinement of glass
compositions and glass and fibre fabrication techniques to further reduce
absorption and scatter with the aim of reaching <1 dB/km. The strength of

TABLE 1

Composition	Density gm/cc	Refractive Index n_d	T_g °C	Thermal Expansion Coefficient x 10^{-6}	Vickers Hardness Kg/mm^2	Poisson's Ratio	Young's Modulus GPa	K_{IC} MNm$^{-3/2}$
58ZrF_4 34BaF_2 8ThF_4 mole %	4.80	1.523	320	–	–	0.28	59.7	–
57ZrF_4 36BaF_2 3LaF_3 4AlF_3 mole %	4.61	1.516	310	18.7	267	0.30	56.0	0.31
57HfF_4 36BaF_2 3LaF_3 4AlF_3 mole %	5.88	1.504	312	17.3	271	0.30	55.0	0.31
62ZrF_4 33BaF_2 5LaF_3 mole %	4.78	1.523	306	18.8	228	–	–	0.25
62HfF_4 33BaF_2 5LaF_3 mole %	5.78	1.514	312	17.7	228	–	–	0.27
22BaF_2 16YF_3 40AlF_3 22CaF_2 mole %	4.00	1.440	430	–	360	0.30	63.8	0.38
Ge28 Sb12 Se60 atomic %	4.40	–	277	13.0	–	–	21.8	–
Ge30 As10 Se60 atomic %	4.36	–	345	13.7	236	–	18.6	–
SiO_2	2.20	–	1100	0.54	635	0.16	73.0	0.74

fluorozirconate fibres has been predicted by Mechalsky et al (1983) to be
as high as 3.8 GPa but an actual breaking strength of 0.46 GPa has been
reported for fibre by France et al (1983). The discrepancy between
predicted and measured strength values may in part be due to the flawed
state of current fibres and the measured strength should increase as the
overall quality of the glass fibres increases. These fibres are also
likely to be more sensitive than the silicate ones to environmental
moisture since normalised leach rates of 10^{-2} to 10^{-3} g/cm^2 day have been
measured by Simmons et al (1983) compared to 10^{-7} to 10^{-8} g/cm^2 day for
pyrex glass. This information indicates that hermetic coatings will be a
necessity to maintain intrinsic fibre strength during field use.

3.3 Bulk Chlorides, Bromides and Iodides

It has been shown that fluoride glasses are suitable for the mid IR but
their transparency is insufficient to cover the far infrared. Thus if
extended mid IR transmittance or far IR transmittance is required then
chloride, bromide and iodide glasses are possible alternative materials to
the heavy metal fluorides or the chalcogenide glasses as can be observed
from figure 1. Until recently this class of halide glass was only of
academic interest since the properties of the known glasses was poor in
terms of chemical durability. However, a number of workers have taken up
this challenge and are investigating the potential of new materials in this
area. Van Uitert and Wemple (1978) seriously considered that ZnCl$_2$ glass
offered potential as an optical fibre material, but the extreme moisture
sensitivity of the known ZnCl$_2$ glasses (Gorre and Pastor 1983) is such a
problem that unless overcome is likely to exclude them as serious
candidates. Shultz (1957) reported a number of binary glasses based on
ZnCl$_2$ containing about 50% KCl, KBr or KI with the ZnCl$_2$-KI system being
reported as the most stable. Savage (1982) also reported binary glasses
with ZnCl$_2$ and up to 25% PdCl$_2$, up to 30% CdI$_2$ and up to 7% CdBr$_2$ with T$_g$
in the range 60°-122°C. A glass of composition ZnCl$_2$ 48% KBr 48% PbBr 4%
was reported by Yamane et al (1985) having T$_g$ of 45-46°C, thermal expansion
coefficient 57 x 10^{-6}/°C and refractive index of 1.63. Hu et al (1983)
have reported that ZnBr$_2$ is vitreous and transparent to about 20 μm but
exhibits poor chemical durability. Glasses based on BiCl$_3$ have been
reported by Angell and Ziegler (1982) in the system BiCl$_3$-KCl and PbCl$_2$-
TlCl-BiBr$_3$ but the T$_g$ values of 25°-45°C are too low for fibre
applications. Cooper and Angell (1983) have reported glasses based on CdI$_2$
such as CdI$_2$-CsI-KI but these have shown poor moisture resistance and low
T$_g$ ~ 10°-35°C. Hu and Mackenzie (1982) have also reported new glasses in
the ThCl$_4$-NaCl-KCl system transparent to around 14 μm. The T$_g$ of one of
these glasses NaCl 30%, KCl 30%, ThCl$_4$ 40% is 130°C and Matecki et al
(1983) reported a T$_g$ of 170°C for a glass of composition CdCl$_2$ 50% BaCl$_2$
40% KCl 10% both offering some promise for fibre applications. Reviews of
halide glass formation are given by Baldwin et al (1981) and Mackenzie
(1983) and an indication of the transmittance range of non fluoride halide
glasses is given in figure 2. It remains to be seen whether further
progress in formulating chloride, bromide and iodide glasses leads to
sufficiently robust materials able to be considered for fibre applications.

4.0 CHALCOGENIDE GLASSES

Since the early 1960s chalcogenide glasses have been researched as
optical component materials for the 3-5 μm and 8-12 μm thermal bands, as
active electronic components in photocopying and for switching applications
and recently as infrared optical fibres. These glasses consist of one of

TABLE 2

Fluoro-zirconate-hafnate glasses used in fibre fabrication research.
(Tran et al 1984)

Core Composition Mol %

$61ZrF_4$ $32BaF_2$ $3.9GdF_3$ $3.1AlF_3$
$51ZrF_4$ $16BaF_2$ $5LaF_3$ $3LaF_3$ $20LiF$ $5PbF_2$
$53ZrF_4$ $19BaF_2$ $5LaF_3$ $3AlF_3$ $20LiF$ (+ Cl BrI dopants)
$56ZrF_4$ $30BaF_2$ $5LaF_3$ $4ThF_4$ $5AlF_3$
$60ZrF_4$ $19BaF_2$ $6LaF_3$ $15NaF$
$27ZrF_4$ $27HfF_4$ $23BaF_2$ $8ThF_4$ $4LaF_3$ $2AlF_3$ $3LiF$ $3NaF$ $3PbF_2$
$53ZrF_4$ $20BaF_2$ $4LaF_3$ $3AlF_3$ $20NaF$
$43AlF_3$ $20BaF_2$ $20CaF_2$ $17YF_3$

TABLE 3   OPTICAL PROPERTIES OF CHALCOGENIDE GLASSES

Glass atomic %	n_2	n_3	n_4	n_5	V_{3-5} [†]	n_8	n_{10}	n_{12}	V_{8-12} [*]	dn/dT °C x 10^{-5}
As40 S60 type B	-	2.395	2.390	2.386	154	-	-	-	-	-1
Ge15 As25 S60	2.30	-	-	-	-	-	-	-	-	-
Ge25 As15 S60	2.22	-	-	-	-	-	-	-	-	-
Ge30 As15 S55	2.25	-	-	-	-	-	-	-	-	-
Ge40 As15 S45	2.30	-	-	-	-	-	-	-	-	-
As40 Se60	-	-	-	-	-	2.7789	2.7789	2.7728	159	-
Ge20 Se80	-	-	-	-	-	2.4071	2.4027	2.3973	143	-
Ge10 As20 Se70	-	-	-	-	-	2.4649	2.4594	2.4526	119	-
Ge10 As30 Se60	-	-	-	-	-	2.6254	2.6201	2.6135	135	-
Ge10 As40 Se50	-	-	-	-	-	2.6108	2.6067	2.6016	176	-
Ge20 As10 Se70	-	-	-	-	-	2.5628	2.5583	2.5528	156	-
Ge30 As10 Se60	-	-	-	-	-	2.4408	2.4347	2.4271	104	-
Ge30 As15 Se55	-	-	-	-	-	2.4972	2.4914	2.4840	113	-
Ge30 As20 Se50	-	-	-	-	-	2.5690	2.5633	2.5560	120	-
Ge33 As12 Se55	-	-	-	-	-	2.5002	2.4942	2.4867	111	-
Ge28 Sb12 Se60	-	2.6263	2.6200	2.6165	165	2.6083	2.6002	11.0 µm 2.5962	99	+8
Ge30 As13 Se57	-	2.4936	2.4887	2.4859	193	2.4784	2.4724	2.4650	110	+7
Ge30 As13 Se47 Te10	-	2.6118	2.6057	2.6024	171	2.5952	2.5897	2.5829	129	+7
Ge30 As13 Se37 Te20	-	2.7412	2.7342	2.7305	162	2.7229	2.7178	2.7117	154	+11
Ge30 As13 Se27 Te30	-	2.8818	2.8732	2.8688	144	2.8610	2.8563	2.8509	185	+15
Si25 As25 Te50	-	-	-	2.93	-	-	-	-	-	+1
Ge10 As20 Te70	-	-	-	3.55	-	-	-	-	-	-
Si15 Ge10 As25 Te50	-	-	-	3.06	-	-	-	-	-	+17

$$\text{† } V_{3-5} = \frac{n_4 - 1}{n_3 - n_5} \qquad \text{* } V_{8-12} = \frac{n_{10} - 1}{n_8 - n_{12}}$$

more of the elements S, Se and Te together with one or more of the elements Ge, Si, As, Sb and some others. They are mainly covalently bonded materials exhibiting room temperature resistivities between 10^3 and 10^{13} ohm cm depending upon the chemical composition. Some of the sulphides offer visible transmittance in the yellow to red part of the spectrum but selenides and tellurides are opaque in the visible spectral region. For a typical window thickness of a few millimetres, sulphides offer transmittance from ~0.7 to 12.0 μm, selenides from ~1.0 to 15 μm and tellurides from ~2.0 to 20 μm as shown in figure 3. Extrinsic absorption due to oxide and hydride can be a problem with these materials unless levels in the glass are kept below ~1 ppm. The optical and general properties of these glasses are reviewed by Savage (1985) and some of this data is given below and in tables 3 and 4. They are much less physically and thermally robust than oxide glasses but they are chemically durable and are able to be used as optical components.

4.1 Bulk Sulphides

Major glass formation occurs amongst the sulphide glasses as follows; As-Tl-S (Flaschen et al 1960a), As-I-S (Flaschen et al 1960b), Ge-As-S (Savage and Neilsen 1965a), As-Te-S (Kolomiets 1964), Ge-P-S, Si-Sb-S (Hilton et al 1964) and Ge-Sb-I-S (Turjanitsa et al 1972). The Ge-As-S glass system offers a wide compositional range allowing glasses with differing optical, thermal and mechanical properties to be prepared and thus can be used to illustrate the performance of sulphide glasses. Binary glasses can be made containing 10% to 40% of As or 15% to > 30% of Ge and ternary glasses can be made containing as little as 30% of S. Materials containing more than 60% of S are visually transparent, the position of the cut on wavelength depending upon the As/Ge/S ratio. They also exhibit a complex IR spectrum between 7 and 15 μm due to combination tones of the S_8 fundamental absorptions (Bernstein and Powling 1950). Glasses containing less than 60% S exhibit a transmittance similar to that seen in figure 3. The dashed curve in this figure clearly shows the problem of extrinsic absorption due to water, oxide and carbon impurities. Some physical property data is given in tables 3 and 4 and further data is reported by Andreichin et al (1976). Glass transition temperatures range from 139° to 394°C, the thermal expansion coefficients range from 11 to >26 x 10^{-6}/°C, knoop hardness ranges from 200 to 280 Kg/mm^2 and Young's Modulus ranges from 20 x 10^9 Nm^{-2} to 41 x 10^9 Nm^{-2}. Other data on sulphide glasses has been reviewed by Savage and Nielsen (1965b), Hilton 1966 and Hilton 1970.

4.2 Bulk Selenides

Major glass formation occurs amongst the selenide glasses as follows; As-Tl-Se, As-S-Se (Flaschen et al 1960a), As-Sb-Se, As-Tl-Se (Kolomiets 1964), Ge-As-Se (Kolomiets 1964, Savage and Nielsen 1964), Ge-P-Se, Si-Sb-Se (Hilton et al 1964) and Ge-Sb-Se (Hilton et al 1966). Most effort has been applied to the Ge-As-Se and Ge-Sb-Se glass systems. The Ge-As-Se glass system offers a wide compositional range and is used to illustrate the performance of selenide glasses. Data on the physical properties of Ge-As-Se glasses is contained in tables 3 and 4 and is reported by Hilton et al (1966), Webber and Savage (1976) and Savage et al (1977). Glasses can be made from pure Se, with Se and up to 30% Ge, with Se and up to 60% As and with as little as 30% Se in the ternary system. Overall the T_g varies from 37° to 401°C, the thermal expansion coefficient from 31.5 to 10.9 x 10^{-6} /°C, the knoop hardness from 100 to 200 Kg/mm^2, the Young's Modulus from 14 x 10^9 Nm^{-2} to 27 x 10^9 Nm^{-2}, the fracture toughness from 5.5 Nmm$^{-3/2}$ to

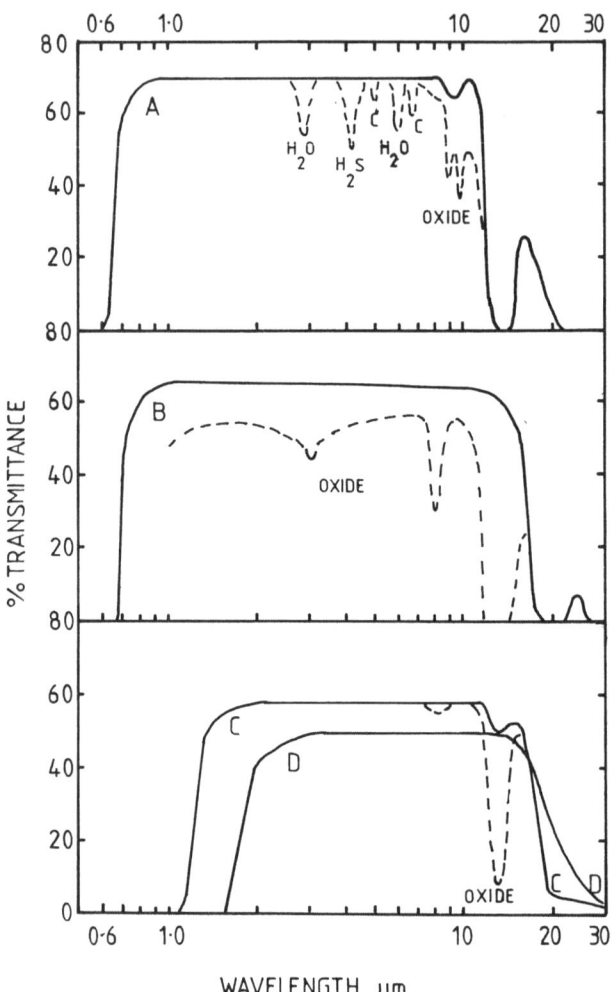

FIGURE 3.
A comparison of the transmittance ranges of sulphide, selenide, selenide-
telluride and telluride glasses also showing the effects of extrinsic
absorption.

A Ge 30 As 20 S 50   atomic % 1.86 mm thick
B Ge 34 As 8 Se 58   atomic % 1.80 mm thick
C Ge 30 As 13 Se 27 Te 30   atomic % 2.3 mm thick
D Ge 10 As 50 Te 40   atomic % 1.62 mm thick.

TABLE 4   THERMAL AND MECHANICAL PROPERTIES OF CHALCOGENIDE GLASSES

Glass Composition Atomic %	Tg °C	Thermal Expansion Coefficient $\times 10^{-6}/°C$	Density Kg/m^{-3} $\times 10^3$	Hardness K=Knoop V=Vickers Kg/mm^2	Thermal Conductivity Cal/cm Sec °K	Rupture Modulus MPa	Young's Modulus GPa	K_{Ic} N mm$^{-3/2}$	Viscosity Fulcher Equation $10^5 - 10^{13}$ P
As40 S60 type B	-	26.1	3.15	109 (K)	-	-	-	-	-
Ge15 As25 S60	-	19.4	3.05	159 (K)	-	-	-	-	-
Ge25 As15 S60	425	12.8	3.00	200 (K)	-	-	-	-	-
Ge30 As15 S55	400	9.6	3.17	216 (K)	-	-	-	-	-
Ge40 As15 S45	-	7.7	3.53	276 (K)	-	-	-	-	-
As40 Se60	178	21.0	4.62	-	-	-	-	-	$\log_{10} \eta = -4.44 + 2764/(T°C - 22.25)$
Ge20 Se80	154	24.8	4.37	147 (V)	-	-	-	-	-
Ge10 As20 Se70	159	24.8	4.47	154 (V)	-	-	16.5	6.7 ± 0.4	-
Ge10 As30 Se60	210	19.0	4.51	176 (V)	-	-	18.0	7.1 ± 0.6	-
Ge10 As40 Se50	222	20.9	4.49	173 (V)	-	-	15.9	7.4 ± 0.8	-
Ge20 As10 Se70	209	20.5	4.41	186 (V)	-	-	16.1	-	-
Ge30 As10 Se60	345	13.7	4.36	236 (V)	-	-	18.6	7.7 ± 0.4	-
Ge30 As15 Se55	351	12.8	4.42	245 (V)	-	-	21.3	-	-
Ge30 As20 Se50	361	11.7	4.47	266 (V)	-	-	22.1	-	-
Ge33 As12 Se55	-	13.0	4.40	170 (K)	0.60	17.2	21.8	-	$\log_{10} \eta = -4.97 + 2824/(T°C - 122.41)$
Ge28 Sb12 Se60	277	15.8	4.67	150 (K)	0.72	17.3	-	-	$\log_{10} \eta = -4.71 + 4070/(T°C - 116.13)$
Ge30 As13 Se57	342	13.0	4.40	237 (V)	-	-	-	-	$\log_{10} \eta = -5.91 + 4627/(T°C - 67.49)$
Ge30 As13 Se47 Te10	308	13.2	4.56	234 (V)	-	-	-	-	$\log_{10} \eta = -9.74 + 6466/(T°C - 5.06)$
Ge30 As13 Se37 Te20	285	12.9	4.77	228 (V)	-	-	-	-	$\log_{10} \eta = -8.19 + 4868/(T°C - 35.52)$
Ge30 As13 Se27 Te30	262	12.8	4.91	226 (V)	-	-	-	-	-
Si25 As25 Te50	-	13.0	4.76	167 (K)	-	-	-	-	-
Ge10 As20 Te70	-	18.0	-	111 (K)	-	-	-	-	-
Si15 Ge10 As25 Te50	-	10.0	-	179 (K)	-	-	-	-	-

9.4 Nmm$^{-3/2}$ and the four point bending strength from 1×10^7 Nm^{-2} to 2.5×10^7 Nm^{-2}. The most robust glasses in terms of their thermal and mechanical properties are located in the region of the phase diagram near Ge 30% As 10 to 20% Se 50 to 60%. The transmittance of these glasses is similar to that shown in figure 3 where the effect of oxide extrinsic absorption is also very apparent.

4.3 Bulk Tellurides

Major glass formation occurs amongst the tellurides as follows; Si-As-Te (Hilton and Brau 1963), As-I-Te (Peck and Dewald 1964), Ge-As-Te, Ge-P-Te (Savage and Nielsen 1966) (Hilton et al 1966), Ge-Te, As-Te (Savage 1972a), Si-Ge-As-Te (Savage 1972b) and quaternaries based on Si-As-Te (Anthonis et al 1973/74). The largest region of glass formation occurs in the Si-As-Te system, in the binaries from 50% to 55% As and 10% to 15% Si together with ~30% to 80% Te in the ternary region yielding T_g values between 112°C and 414°C. However, silicon is difficult to incorporate into the glasses using the sealed tube preparative route resulting in attack of the SiO$_2$ ampoules and thus in extrinsic oxide absorption problems in the glass products. This difficulty would be overcome by a vapour growth approach. At the time that this work was done interest was centred on melt synthesised material for bulk optics so that given the difficulties of refractory materials and small glass forming regions in other telluride systems little more work was done. The glass forming regions in the Ge-As-Te system yielded relatively unstable glasses in a rather narrow glass forming region. However, for comparison purposes the transmittance of one of these glasses is shown in figure 3 and some property data on telluride glasses is given in tables 3 and 4.

In order to extend the phonon cut off to longer wavelengths and avoid melting and thermal stability problems, selenide-telluride glasses were considered for bulk applications (Savage et al 1980) and are also advantageous for fibre applications. Data for a number of glasses is given in tables 3 and 4 and the transmittance of an optimised material for bulk applications is shown in figure 3.

4.4 Chalcogenide Fibres

A number of workers have predicted minimum losses for sulphide glass fibres, 10^{-1} to 10^{-2} dB/km at 5.5 μm for Ge-P-S glass (Shibata et al 1980), 5×10^{-2} dB/km at 4 to 5 μm for As$_2$S$_3$ glass. A fibre loss of 170 dB/km at 5.25 μm was reported for As$_2$S$_3$ material by Miyashita and Terunuma (1982) and 64 dB/km at 2.4 μm by Kanamori et al (1983). Saito et al (1985) have reported a fibre bundle of 200-1000 As-S fibre cores 90 μm in diameter and showing a minimum loss of 0.6 dB/m. Thus sulphide glasses appear to offer some potential as low loss materials in the mid infrared. An indication of some of the losses achieved in early 1984 is given in table 5 (Devyatykh and Dianov 1984). As$_2$S$_3$ fibres 100-500 μm in diameter with a loss at CO$_2$ laser wavelengths of 1 dB/m were reported by Bornstein et al (1982). Brehm et al (1982) reported Ge 30% As 15% Se 55% glass plastic clad fibres 200 μm in diameter, 100 m long, with a breaking strength >1 daN and a loss in the 4 to 11 μm band of ~10 dB/m. More recently Bornstein et al (1984) working with As$_2$Se$_3$ and Ge 28% Sb 12% Se 60% glasses obtained fibre losses better than 3 dB/m at CO$_2$ wavelengths and transmitted power density of 0.8 Kw/cm^2 without fibre damage. Pitt et al (1985) have reported work on Ge-As-Se and As-Se-Te glasses for fibre applications. Long lengths of fibre up to 500 μm in diameter were drawn using a pressurised crucible technique. The

TABLE 5

Core	Cladding	Minimum Loss Wavelength μ m	Minimum Loss dB/m
As_2S_3	–	3.5	0.7
As_2Se_3	–	3.5-4.5	0.58
Ge5% As38% Se57%	–	3.5-6.0	0.60
As_2S_3	As33 S 67	3.5	0.9
As_2Se_3	As_2S_3	3.5	0.9
As_2Se_3	Teflon ∅-42	4.5	0.9
Ge5% As38% Se57%	Teflon ∅-42	3.5	0.53
Ge5% As38% Se57%	Teflon ∅-42	5.0	1.0

Minimum loss in fibres, Devyatykh and Dianov (1984).

tensile strength of poxy-acrylate coated fibres of composition Ge 5% As 38% Se 57% was found to be 285 MPa while the loss at 9.3 μm was 5 dB/m and at 10.6 μm was 10 dB/m. However, a composition of As 40% Se 40% Te 20% was identified as the most promising in terms of reducing the reported losses by an order of magnitude.

5.0 FUTURE DIRECTIONS

High purity CVD techniques have been used in the vapour synthesis of silicate glass for fibre applications enabling attenuations near the theoretical limit to be achieved. Borosilicate type fibre has been made by a melt process but the attenuation has not been reduced to the level obtained from the vapour phase process due to the difficulty of achieving adequate purity. In addition it is much easier to control a particular refractive index profile by the vapour phase than by the solid phase fabrication technique. The fluoride glasses can be melted at low temperatures <900°C easing impurity pick up problems. Nevertheless each constituent fluoride needs to be purified to a very high standard and then melted together with others to form a glass which is then drawn to a fibre. With much care and attention during all of these processes it is likely to be possible to achieve ~1 dB/km but it is questionable whether 10^{-2} dB/km or less can be routinely possible on a commercial basis. If this does not prove possible then alternative compositions more suited to CVD fabrication techniques, such as perhaps chlorides or fluoro-chlorides, will need to be investigated to allow the full promise of ultra low loss to be achieved in halide glasses.

If vapour phase synthesis were to prove impossible and solid phase synthesis were incapable of yielding an attenuation level of $\leqslant 10^{-2}$ dB/km then alternative glasses would need to be considered. In this event the chalcogenide glasses represented by Ge-P-S or As-S may prove of interest due to the low predicted loss of 10^{-1} to 10^{-2} dB/km and the fact that the vapour synthesis would not be a problem except from safety considerations. Also Dumbaugh (1984a, b, c) has reported stable glasses in the $Bi_2O_3-Ga_2O_3-CdO$ and $Bi_2O_3-Ga_2O_3-PbO$ systems with IR transmittance range to 7 or 8 μm. This transmittance potential is similar to that of many fluoride glasses and the physical properties of the new oxide glasses are not unattractive. The refractive index is ~2.4, the knoop hardness is ~225, the thermal expansion coefficient is 8 to 11.2 x 10^{-6}/°C and the chemical durability properties are acceptable. Thus if the vapour synthesis of these glasses proves easier than that of the halide glasses then they may well be considered for ultra low loss applications if the scatter losses are shown to be acceptably low.

REFERENCES

Andreichin R, Nikiforova M, Skordeva E, Yurakova L, Grigorovici R, Manaila R, Papescu M and Vancu A, 1976, J Non Cryst Solids 20, 101-122.
Angell C A and Ziegler D C, 1982, Appl Opt 21, 2096-2098.
Anthonis H E, Kreidl N J and Ratzenback W H, 1973/74, J Non Cryst Solids 13, 13.
Baldwin C M, Almeida R M and Mackenzie J D, 1981, J Non Cryst Solids 43, 309-344.
Bendow B and Drexhage M G, 1981, Proc SPIE 266, 16.

Bendow B and Drexhage M G, 1982, Opt Eng 21, 118–121.

Bernstein H J and Powling J J, 1950, Chem Phys 18, 1018.

Bornstein A, Croitoru N and Marom E, 1982, Proc SPIE 320.

Bornstein A, Croitoru N and Marom E, 1984, Proc SPIE 484, 99–104.

Brehm C, Cornebois M, Le Sargent C and Parant J P, 1982, J Non Cryst Solids 47, 251–254.

Cooper E I and Angell C A, 1983, J Non Cryst Solids 56, 75–80.

Devyatykh G G and Dianov E M, 1984, Proc SPIE 484, 105–107.

Drexhage M G, Moynihan C T and Saleh Boulos M, 1980, Mater Res Bull 15, 213–218.

Drexhage M G, EL-Bayoumi O H and Lipson H, 1983, J Non Cryst Solids, 56, 51–56.

Drexhage M G, 1985, in Treatise in Materials Science. Glass 4 Ed Doremus.

Dumbaugh Jr W H, 1984a, USP 4 483 931, 1984b USP 4 456 692, 1984c, Proc SPIE 505 97–101.

Flaschen S S, Pearson A D and Northover W R, 1960a, J Amer Ceram Soc 43, 274.

Flaschen S S, Pearson A D and Northover W R, 1960b, J Appl Phys 31, 219.

France P W, Williams J, Carter J F and Beales J K, 1983, Paper II 2nd Int Symp on Halide Glasses, Troy N Y.

Fonteneau G, Slim H and Lucas J, 1982, J Non Cryst Solids 50, 61–69.

Gorre L E and Pastor R C, 1983, Mat Res Bull 18, 1391–1398.

Hicks W and Kiritsy, P1961, Fibre Optics Handbook, (Mosaic Fabrications Inc).

Hilton A R and Brau M, 1963, Infrared Physics 3, 69–76.

Hilton A R, Jones C E and Brau M, 1964, Infrared Physics 4, 213.

Hilton A R, 1966, Applied Optics 5, 1877.

Hilton A R, Jones C E and Brau M, 1966, Phys Chem Glasses 7, 105.

Hilton A R, 1970, J Non Cryst Solids 2, 28.

Hirschowitz B I, Cartiss L E, Peters C W and Pallard H M, 1958, Gastroenterology 35, 50.

Hu H and Mackenzie J D, 1982, J Non Cryst Solids 51, 269–272.

Hu H, Fuding M A and Mackenzie J D, 1983, J Non Cryst Solids 55, 169–172.

Jones M W and Kao K C, 1969, J Phys E: Sci Instrum 2 331–335.

Kanamori T, Terunuma Y and Miyashita T, 1983, Integrated Optics and Optical Fibre Communications, Tokyo.

Kao K C and Hockham G A, 1966, Proc IEEE 113, 1151–1158.

Kao K C and Davis T W, 1968, J Phys E: Sci Instrum 1, 1063–1072.

Kao K C, Davis T W and Worthington R, 1970, The Radio and Electronic Engineer 39, 105–111.

Kapany N S, 1959a, J Opt Soc Am 49, 779.

Kapany N S, 1959b, Nature 184, 881.

Kapany N S and Mergerian D, 1960, Infrared Imaging Systems 5, 139.

Kapany N S and Simms R J, 1965, J Opt Soc Am 55, 963.

Kapany N S, 1967, Fibre Optics Principles and Applications (New York: Academic).

Kapron F P, Keck D B and Maurer R D, 1970, Appl Phys Lett 10, 423–425.

Klocek P, 1982, Lasers and Applications, Oct 43–6.

Kolomiets B T, 1964, Phys Stat Sol 7, 359.

Lucas J, Chanthanasinh M, Poulain M, Brun P and Weber M J, 1978, J Non Cryst Solids 27, 273–283.

Lucas J, 1982, Proc SPIE 320, 22.

Lucas J, Tregoat D, El Houari A and Fonteneau G, 1984, Proc SPIE, 484, 36–40.

Mackenzie J D, 1983, 2nd Int Symp on Halide Glasses, Rensselaer, N Y.

382

Matecki M, Poulain M and Poulain M, 1983, J Non Cryst Solids 56, 81-86.
Mechalsky J J, 1983, Paper 32, 2nd Int Symp Halide Glasses, Troy N Y.
Miyashita T and Terunuma Y, 1982, J Appl Phys 21, L75-L76.
Parker J M, Clare A G and Sedden A B, 1985, Int Symp on Halide Glasses,
    Rennes France.
Peck W F and Dewald J F, 1964, J Electrochem Soc 1, 561.
Pitt N J, Sapsford G S and Clapp T V, 1986, Proc SPIE 618.
Poulain M, Chanthanasinh M and Lucas J, 1977, J Mater Res Bull 12, 151-156.
Poulain M and Lucas J, 1978, Verres et Refract 32, 505.
Rawson H, 1967, Inorganic Glass Forming Systems (London Academic).
Rosiewicz A and Gannon J R, 1981, Electron Lett 17, 184-185.
Saito M, Takizawa M, Sakuragi S and Tanei F, 1985, Applied Optics 24,
    2304-2308.
Savage J A and Nielsen S, 1964, Phys Chem Glasses 5, 82.
Savage J A and Nielsen S, 1965a, VII International Congress on Glass,
    Brussels.
Savage J A and Nielsen S, 1965b, Infrared Physics 5, 195.
Savage J A and Nielsen S, 1966, Phys Chem Glasses 7, 56.
Savage J A, 1972a, J Mat Sci 11, 121.
Savage J A, 1972b, J Mat Sci 7, 64.
Savage J A, Webber P J and Pitt A M, 1977, Applied Optics 16, 2938.
Savage J A, Webber P J and Pitt A M, 1980, Infrared Physics 20, 313.
Savage J A, 1982, 1st Int Symp on Halide and Other Non Oxide Glasses,
    Cambridge UK.
Savage J A, 1985, Glass Current Issues, Ed A F Wright and J Dupuy, Series
    E, Applied Sciences No 92, Martinus Nijhoff, Netherlands 281-306.
Shibata S, Terunuma Y and Manabe T, 1980, Japan J Appl Phys 19, L603-L605.
Shultz I, 1957, Naturwissenschaften 44, 536.
Sigel Jr G H and Tran D C, 1984, Proc SPIE 484, 2-6.
Simmons C J, Azali S A and Simmons J H, 1983, Paper 47 2nd Int Symp on
    Halide Glasses, Troy, N Y.
Sun K H, 1949, USP 2, 466, 507.
Sun K H and Huggins M L, 1950, USP 2, 511, 224.
Sun K H, 1979, Glass Technol 20, 36.
Tick P, 1985, Lasers and Applications, Oct 80.
Tran D C, Sigel Jr G H, Levin K H and Cinther R J, 1982, 18, 1046-1048.
Tran D C, Sigel Jr G H and Bendow B, 1984, J Lightwave Technology LT-2,
    566-586.
Turjanitsa I D, Mihalinets I M, Kaperljas B M and Kapinets I F, 1972, J Non
    Cryst Solids 11, 173.
Van Vitert L G and Wemple S H, 1978, Appl Phys Lett 33, 57-59.
Webber P J and Savage J A, 1976, J Non Cryst Solids 20, 271.
Weber M J, Layne C B, Saroyan R A and Milam D, 1976, Opt Commun 18, 171.
Weinberg M C, Neilsen G F and Smith G L, 1983, J Non-Cryst Solids 56,
    45-50.
Yamane M, Kawazae H, Inoue S and Maeda K, 1985, Mat Res Bull 20, 905-911.

DISCUSSION

C: J. Lucas. When one speculates on the potential of ultra-
-transparency in these chalcogenide glasses, one has to
consider the electronic absorption, in addition to the multiphonon and Rayleigh terms. Also in the field of chalcogenide glasses, the best results that have been claimed for
fibers are a few tens of dB/Km in the 3 μm region by
Furukawa and 0.3 dB/m in the 6 μm region by Laboratoires de
Marcoussis (C.G.E.) for Ge-As-Se-Te fibers.

Q: A. Bruce. Yesterday, J. Lucas reported some very interesting
new glasses which contain both chalcogenide and halide com
ponents. Could you comment on the potential of such glasses
and the general miscibility of halides and chalcogenides?
Do you think that such compositions could be utilized as a
means of removing oxide and hydride defects from chalcogenides?

A: J. Savage. These new glasses appear very promising. Their
optical and physical properties will probably be a compromise between those of halides and chalcogenides. Bromide
and iodide ions are quite soluble in chalcogenide melts.
Those compositions seem to have the possibility for a
built-in RAP process of chalcogenide glasses.

INTERACTIONS WITH OTHER FIELDS

DONALD R. ULRICH

Air Force Office of Scientific Research
Washington, D.C. 20332

1. INTRODUCTION
   The objective of this paper is to discuss the interaction
of other fields with halide glasses and the impact of these
interactions. The fields to be discussed are several fields
of chemistry, sol-gel science and polymer fiber optics.

2. CHEMICAL DRIVERS
   Major advances during the next decade will require an emphasis which relates chemical processes to glass formation,
purification and physical behavior. Five chemical drivers
have been identified which will impact this shift to new
directions in halide glass research. These are: polymer network theory and gelation, application of theoretical chemistry,
polymeric glasses--analogs, chemical kinetics, and fluorine and
halogen organochemistry.

3. IMPACT OF CHEMICAL DRIVERS
   The impact will be most evident on new materials fabricated
with low temperature solution processing, vapor deposition
processes, or perhaps even through unique application of super-
critical fluid approaches. Very promising approaches are
available through sol-gel processing. However, the fundamental
chemical problems which will enable sol-gel processing to become a viable approach are poorly understood for gel-derived
oxide glasses. Only within the past year have studies been
undertaken on fluoride glasses derived from gels.

4. SOL-GEL DERIVED FLUORIDE GLASSES
   Under USAF sponsorship Professor Uhlmann of M.I.T. is investigating the sythesis of fluoride glasses using sol-gel
techniques. Some consideration is being given to a modification of the colloid route to glass fabrication which was introduced by Scherer (1) for oxide glasses. As will be discussed
later the colloid approach requires interpretation in terms
of continuous polymer network formation. However, most of the
work is directed to the following approach to forming fluoride
glasses.
   Oxide gels containing metal cations in the atomic ratios
desired in subsequently-produced fluoride glasses are prepared
using wet chemical techniques. The gels are subsequently dried
to form bodies with a large volume fraction of inter-connected
porosity (e.g., in the range of 50%). The porous structure is
then exposed to a source of fluoride ions while heated to

temperatures where the kinetics of reaction are high. Densification of the resulting fluoride material will then be effected
by the familiar process of viscous sintering.

The rationale for this approach is based on the high specific
surface areas (and hence ready accessibility of the skeletal
material to reacting liquids or vapors) which can be produced
in dried gels. It is also based on the fact that the oxides of
interest can be converted to fluorides at temperatures in the
range of 300-400° C (and, indeed, the fluoride glass group at
Rennes in France routinely prepares melted glasses by convert-
ing oxides to fluorides using ammonium bifluoride in the melting process).

The attractive features offered by such a sol-gel approach
to fluoride glass processing range from local chemical homo-
geniety and the capability of low temperature processing to the
reduction of crystallization-associated scattering and expansion of the compositional range of glass formation. The
achievement of these objectives is, however, far from straightforward. The achievement of chemical reaction to convert
oxides to fluorides while maintaining interconnected structures
with fine-scale porosity may prove problematical, as may the
step of densifying without crystallizing.

The synthesis route being explored involves extensive use of
double alkoxides. Use of double alkoxides provides a means of
introducing alkaline earth cations such as barium into the gel
structure without encountering the phenomenon of phase separation (as is commonly found when these cations are introduced
as salts). Double alkoxides also provide a convenient route
for indroducing trivalent cations such as lanthanum into the
structure. In nearly all cases, the double alkoxides of concern represent unconventional materials. Further, since nearly all fluoride glasses of immediate interest contain four or
more cations, synthesis of the oxide gel precursors involves
the use of at least two different double alkoxides. Since
the kinetics of hydrolysis and condensation are in general
different for different double alkoxides, care must be taken
to produce chemically homogeneous gels. In the approach presently being pursued, this involves preliminary partial
hydrolysis of the slower-hydrolyzing double alkoxide prior to
adding the faster-hydrolyzing species plus additional water.
The alternative approach is to use double alkoxides with
different alkoxy groups to yield materials with essentially
equal rates of hydrolysis. Both approaches have been successfully employed with single alkoxides of different cations to
produce chemically homogeneous materials. Their extension to
double alkoxides seems reasonable and promising.

Presently is not possible to predict the effect of the
fluoridation treatment on the pore size distribution of the
starting gels. It is expected, however, that the extent of
pore closure in this stage of the process will depend upon the
chemistry of the gel as well as upon the conditions of fluoridation. It is also expected that post-fluoridation heat treatment will be required to produce fully dense glasses for subsequent drawing into fibers. Each of these processes, as well
as the competition between viscous sintering (the basic process

of pore closure) and crystallization, must be explored.

5. POLYMER NETWORK THEORY AND GELATION

The recent work of the late Paul Flory on the application of polymer network theory to gel-derived oxide glasses will have a major impact on the sol-gel approach to fluoride glasses.

The most extensively investigated sol gel material has been SiO_2 prepared from TEOS. It has been shown that pronounced difference in structure result depending whether processing is carried out under acid or base conditions. It has also been demonstrated that the water-to-TEOS ratio is an important parameter. It has been sometimes suggested that under high pH and higher water content colloidal particles are developed, while under conditions of low pH and low water content linear polymers form.

It must be recognized, however, that these views represent rather considerable approximations to reality and are undoubtedly incorrect in detail (2,3). While fine colloidal size particles are indeed formed at high pH and high water content, these are by no means the dense particulates familiar to colloidal chemists. Indeed, the best evidence suggests that these should be thought of as dense, lacey particulates. Further, under conditions of low pH while more polymeric gels are formed, considering the funcionality of TEOS and its hydrolysis products (functionality=4), it is unlikely that long linear chains will be formed. Insight obtained from the study of many organic systems with functionalities greater than 2 including fluorides have indicated that branching takes place early in polymerization and structures with appreciable three-dimensional crosslinking are obtained (4). As such reactions continue, gelation is observed. On this basis, the polymeric structures produced with TEOS under low pH conditions should probably be expected to display appreciable branching and crosslinking.

One can obtain gels under both high pH and low pH conditions, but the character of these gels differs appreciably--most notably in the pore size distribution of the dried gels. Such differences have important consequences for subsequent processing of the dried gels to produce glass.

The detailed characterisics of the chemistry and structure remain to be properly elucidated even in the case of TEOS. In case of other alkoxide or other organometallic precursors the available state of knowledge is much poorer and it can properly be said that almost nothing is known in satisfactory detail about the chemistry and sequence of structural development in such systems, especially halides.

6. FLUORINE CHEMISTRY AND HALOGEN ORGANOCHEMISTRY

The area of fluoride glasses will be stimulated by application of fluoroorganometallic chemistry, or perhaps halogen organochemistry. New experimental challenges exist for the synthesis of halogen-containing compounds which may offer a quantum step in purity and oxygen content control.

Fluorine chemistry is a resource which remains untapped. A focus of study which is easily overlooked, but which in the writer's opinion is a most critical parameter, is reactivity.

This is the determination of the reaction which will be most conducive to the synthesis of fluoro compounds which will be useful as reagents and copolymers and be capable of conversion to halide glasses. The work of Lagow of the University of Texas, Austin, on direct fluorination has lead to new concepts and experimental methods for controlling the elemental fluorine with organic and inorganic compounds and polymers. Work in directions such as this will point to synthetic directions for the formation of fluoride gels.

The consideration of other halogen chemistries will lead to nonfluoride halide gels such as bromides, chlorides, iodides, and mixtures of halides. A better understanding of the molecular dynamics of fluorooganometallics may lead to expanded and improved vapor deposited halides.

As an example of Lagow's recent work, SF_4 is a fluorine precursor which reacts with glass very well. Recent experiments on the treatment of open glass surfaces with this intermediate indicates that 4 to 6% of fluorine can be substituted into silica glass optical fibers. Other very active intermediates are Si_2F_6 and SiF_3 alkyl analogs. The application of fluorine intermediate chemistry to sol gel or vapor deposition approaches is expected to enhance fluoride glass preparation and purity. No work has been pursued with fluoride intermediates of interest to halide glass chemistry such as aluminun fluoride, barium fluoride or lanthanum fluoride(5).

7. POLYMERIC GLASSES--ANALOGS

A better understanding has been developed of the differences and similarities of polymeric and inorganic glasses. Uhlmann of M.I.T. and Flory of Stanford provided a sound basis for understanding the role of polymer glasses in future inorganic glass research(6).

Uhlmann has proposed developing models of the type used successfully to describe thermosetting organic polymers to predict structural developments in glass produced by wet chemical techniques. These models would be extended to describe the sequence of polymerization, for example, in alkoxide based systems. Such modeling should provide an effective transfer between polymer and glass science(7).

8. CHEMICAL KINETICS

Research to understand the kinetics of polymerization, hydrolysis, and crystallization will contribute to major advances in the preparation of glasses by wet chemical techniques. For electrooptics and nonlinear optics, paths of ease are offered for the clean elimination of impurities. The issues of stepwise hydrolysis, the nature of copolymers formed, and the determination of reactivity ratios will expand the science of glass prepared from metal-organics. A long-term outcome of understanding the kinetics of the competition between viscous sintering and crystallization will be the low temperature processing of dried gels and amorphous powders to full density before crystallization.

9. APPLICATION OF THEORETICAL CHEMISTRY

Semiquantitative methods for computational chemistry have

recently been shown to handle systems of chemical interest in the halide glass area. Studies of reactions of fluorine atoms and fluoride ions with Dewar's MNDO method have shown that the state-of-the-art should soon be adequate to begin applying theoretical chemical calculations to large fluorine-containing molecules of interest to glass chemists (8,9).

10. POLYMER OPTICAL FIBERS

The Japanese are vigorously pursuing development in polymer optical fibers compared to the NATO countries. Their work concentrates on the visible and near-infrared. In particular the claim is made that perdeuterated polymethylmethacrylate glass is starting to close in on common silicates(10). Replacing hydrogen in the core polymer with a combination of fluorine and deuterium brings about a reduction in CH and other molecular vibrational absorption in the IR region and its overtones in the near-IR to visible region(11). Fluorination of aromatic hydrogens of styrene very effectively suppresses the water vapor absorption of the polymer, especially in the near-IR region. There is, however, evidence that the inherent scattering in polymeric glasses due to turbidity will prevent losses being reduced to the order of magnitude of silcates. The application for the aforementioned fibers is for short distance optical signal medium for certain computer-to-terminal data links such as in office automation systems.

In other recently reported Japanese work, large core, high NA PMMA's and polycarbonates cladded with poly-4-methylpentene have been reported for use as heat resistant fiber sensors in automobile engines (to 120°C). The loss is very high, being reported to be 0,8 db/m at 770 nm (12).

11 SUMMARY

Major advances in halide glasses during the next decade will require interactions with other fields. This will require an emphasis which relates chemical processes to glass formation, purification and physical and optical behavior. Several chemical drivers were identified which will impact the gestation of new and improved materials and fabrication processes. Polymer optical fibers, while not appearing to be competitive from a infrared communications point of view, will be competitive in low cost, visible to near-IR applications. Although not specifically discussed, interactions with other fields also include application in optical processing and sensors as well as interfaces with the new and rapidly developing field of nonlinear optical polymers and organics.

REFERENCES

1. Scherer G: private communication.
2. Flory PJ: Network Theory and Gelation. Hench L and Ulrich D (ed): Science of Ceramic Chemical Processing. New York: Wiley 1986.
3. Uhlmann DR: private communication.

4. Flory PJ: Principles of Polymer Chemistry. Ithaca, NY: Cornell University Press, 1953.
5. Lagow R: private communication.
6. Uhlmann DR and Flory PJ: Polymer Glasses and Oxides. Mackenzie JD and Varner J (ed): Frontiers of Glass Science. New York: North Holland, 1980.
7. Uhlmann DR, Zelinski BJJ and Wnek G: The Ceramist as Chemist--Opportunities for New Materials. Brinker CJ, Clark, DE and Ulrich DR (ed): Better Ceramics Through Chemistry. New York: Elsevier, 1984.
8. Davis LP and Burrgraf LW: Applications of MNDO to Silicon Chemistry. Hench L and Ulrich D (ed): Science of Ceramic Chemical Processing. New York: Wiley, 1986.
9. Gordon MS: private communication.
10. Carr SW and Marks T: private communication.
11. Kaino T: Plastic Optical Fibers for Near-IR Transmission. Japan-U.S. Polymer Symposium Preprints, Kyoto, Japan, October, 1985.
12. Kriedl J: private communication.

DISCUSSION

C: P. Klein. It may be risky to extend the sol-gel technology to fluoride glasses for IR fiberoptics. The thermosetting nature of the gels, as you have described, may work against us. In fact thermosets do not melt and, in order to prevent scattering, one needs true melts.

A: D. Ulrich. I agree, but the information which can be derived from the analogy may prove itself useful.

Q: A.Bruce. A number of advantages are achieved by using the gel route for oxide glasses. These include lower fabrication temperature, compositional homogeneity and retention of volatile components. However, in view of the already low melting temperatures of fluoride systems and their high fluidity, these advantages appear to be redundant in the present case. Is the only obvious advantage of a gel method, at this stage, to improve fluoride glass purity?

A: D. Ulrich. At this stage, we do not yet know the best route for fluoride glass fabrication. Although the gel method may not ultimately be successful, there is probably some valuable information to be obtained which will help in the overall understanding of fluorides.

GLASS FORMATION IN HALIDE SYSTEMS

G. de Leede
Eindhoven University of Technology
Eindhoven, The Netherlands

Halide glasses appear to be of potential interest for making ultra-low-loss optical fibres. Fundamental understanding of the mode of formation of the glassy state can be very helpful in developing new materials and compositions.

In this presentation, it is shown that the electronegativities (x) of the constituent elements in halide compounds can be used to predict glass formation. The scale to be used here is the one suggested by Pauling. Within this scale, electronegativities of the elements vary between 3.98 (F) and 0.79 (Cs), while 2.20 was assigned to H by definition. Values for the elements of which the fluorides are known as good glass progenitors are:

Be: 1.57   Zr: 1.55   Hf: 1.57   Al: 1.61   (x)

All other cations of fluoride glass progenitors possess electronegativities ranging from 1.20 (rare earths) to 1.80 (Ga, In).

From these results we calculate cation-anion electronegativity differences (Δx) of 2.2 to 2.8. These differences are -as expected- much larger than for oxide glass formers, where $1.15 < \Delta x < 1.85$ was found. On a scale that represents mixed polar/covalent bonding, using Δx as a parameter, we now have two 'areas of glass formation':

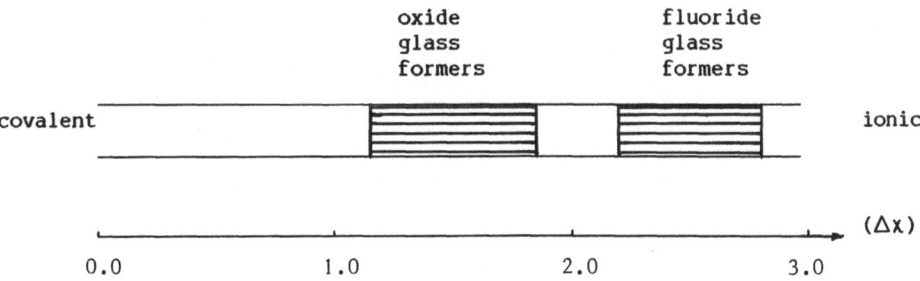

If we calculate Δx for other halide glass formers we find values comparable to those of the oxide glass formers:

$ZnCl_2$: 1.51 $BiCl_3$: 1.14 $ThCl_4$: 1.76 $ZnBr_2$: 1.31   (Δx)

In this respect it seems questionable to classify these systems as ionic.

The magnitude of the ratio of single-bond energies to melting temperature (Rawson, 1956) can be used as an <u>additional criterion</u> to explain most of the exceptions to the electronegativity criterion. So when these criteria are used simultaneously, predictions will be accurate.

It is well-known that strong covalent bonds in compounds of mixed bonding enhance glass-forming ability. From this fact and the above results a <u>new criterion</u> is proposed here to replace the electronegativity criterion: according to the '<u>coordinated polymeric model</u>' (Sanderson, 1967) atomization energies of nonmolecular solids can be calculated from electronegativity data. The total atomization energy and bond energy are thought to consist of a <u>covalent and an ionic part</u>. From the available data it is easy to calculate the covalent part of the bond energy: $E_{b,cov}$. The new criterion is defined as the ratio of covalent part of the single-bond energy to melting temperature: $E_{b,cov}/T_m$. Some values of this parameter are listed in table I. Values for total bond energy/T_m are listed as well, because both ratios seem to be important factors in glass formation.

compound	$E_{b,cov}/T_m$	E_b/T_m	glass progenitor
GeO_2	0.028	0.039	yes
SiO_2	0.025	0.053	yes
Al_2O_3	0.020	0.035	yes
SnO_2	0.019	0.033	yes
ZnO	0.009	0.016	no
CaO	0.006	0.015	no
Na_2O	0.005	0.017	no
BeF_2	0.027	0.109	yes
ZnF_2	0.018	0.037	yes
MnF_2	0.012	0.043	yes
MgF_2	0.010	0.037	yes
BaF_2	0.004	0.028	no
$ZnCl_2$	0.049	0.086	yes
$AgCl$	0.018	0.030	yes
$CdCl_2$	0.016	0.035	yes
$NaCl$	0.005	0.024	no
KCl	0.004	0.025	no
$ZnBr_2$	0.039	0.062	yes
KBr	0.004	0.023	no
AgI	0.017	0.033	yes
CdI_2	0.013	0.021	yes
KI	0.005	0.021	no

Table 1. Ratios of total bond energy (E_b) to melting temperature and covalent part of it.

Roughly, for good glass formers $E_{b,cov}/T_m > 0.02$, for poor glass progenitors $0.01 < E_{b,cov}/T_m < 0.02$, while for $E_{b,cov}/T_m < 0.01$ no glass formation is found. Further tests have to be made to evaluate the accuracy of this criterion.

PREPARATION OF FLUORONITRIDE GLASSES

G. H. FRISCHAT and D. AHLF
PROFESSUR FÜR Glas, INSTITUT FÜR NICHTMETALLISCHE WERKSTOFFE,
TECHNISCHE UNIVERSITÄT CLAUSTHAL, F. R. OF GERMANY

Despite   their unique optical properties most heavy metal
fluoride glasses are substances with low mechanical strength,
poor chemical durability and also low thermal stability. One
way of improving these properties is to substitute the cations
in the glasses by other ones, e. g. by adding AlF_3, YF_3 or HfF_4
to the base composition ZrF_4-BaF_2-NaF. However, although this
way has been tried very often, success is still restricted.
The other way is to change the anionic constitution of the
glasses. Substituting oxygen by nitrogen in the case of oxide
glasses resulted in strongly improved physical and chemical
properties [1 - 3]. This tightening effect can be attributed
to the higher degree of network linkage of N^{3-} ions compared
with O^{2-} ions. In the case of fluorophosphate glasses [4] and
also for halide glasses [5] some preliminary results on the
effect of nitrogen incorporation have also been reported.
Although the introduced N content was $\leq$ 0.3 wt.-%, properties
like T_g or microhardness showed an increasing tendency.

In the case of silicate-based oxide glasses 10 wt.-% of nitrogen or even more could be incorporated in the glass network.
This work tries to find out whether this is also possible in
the case of fluoride glasses. If a N^{3-} ion is substituted for
$3F^-$ ions in the network (see the schematic in the Figure) a
strong tightening effect and hence a strong influence on
physical and chemical properties
could be expected.

Schematic structural array of
N in a fluoride glass.

The base glass composition of this work is (in mole-%) 39 ZrF_4-
36 BaF_2-13 YF_3-12 AlF_3. Incorporation of nitrogen can be done
by substituting AlN for AlF_3, ZrN for ZrF_4, and Ba_3N_2 for BaF_2,
respectively. The addition of these substances is done stepwise. Different ways of melting were tried, e. g. simultaneous

melting of all raw materials, separate melting of the fluorides and later addition of the nitrides, addition of $NH_4F \cdot HF$,
etc. Melting temperatures were varied between 850 and 1050 °C,
melting times between 30 and 270 min.
The resulting products were inspected both with an optical
microscope and a scanning electron microscope. The N content
was determined by means of the Kjeldahl method. The results
obtained so far are not fully satisfactory. Thus many samples
crystallized at least in part. Moreover, the homogeneity of
the glassy parts obviously was also poor in some cases. There
was also a gap between the added and the analyzed N contents.
To overcome these difficulties a twin-roller quenching device
with a maximum cooling rate of up to $10^5 Ks^{-1}$ will be used in
the near future.

The authors appreciate the financial assistance given by the
Deutsche Forschungsgemeinschaft (DFG), Bonn - Bad Godesberg.

REFERENCES

1. Frischat, GH   , Schrimpf, C.: Preparation of nitrogen-
   containing $Na_2O-CaO-SiO_2$ glasses. J. Amer. Ceram. Soc. 63
   (1980), 714 - 715.

2. Schrimpf, C , Frischat, GH   : Property-composition relations of N_2-containing $Na_2O-CaO-SiO_2$ glasses. J. Non-
   Cryst. Solids 56 (1983), 153 - 160.

3. Frischat, GH, Sebastian, K: Leach resistance of nitrogen-
   containing $Na_2O-CaO-SiO_2$ glasses. J. Amer. Ceram. Soc. 68
   (1985), C305 - C307.

4. Fletcher, JP, Risbud, SH: Formation and characterization of
   nitrogen containing fluorophosphate glasses. Mat.Sci. Forum
   5 (1985), 167 - 174.

5. Vaughn, WL, Risbud, SH: New fluoronitride glasses in zirco-
   nium-metal-F-N-systems. J. Mat. Sci. Letters 3 (1984),
   162 - 164.

PL MEASUREMENTS ON FLUOROZIRCONATE GLASSES

M.BRAGLIA, G.COCITO, L.COGNOLATO, M.FERRARIS, G.GREGO, E.MODONE, G.PARISI

CSELT - Centro Studi E Laboratori Telecomunicazioni S.p.A.
Via G. Reiss Romoli, 274 - 10148 TORINO (Italy)

Several fluoro-zirconate glasses, ZBLA type, have been prepared in CSELT.
Fluorides, as raw materials and the technique of melting in Pt crucible
were initially adopted.
Some improvements have been introduced in order to overcome the volatili-
zation of batch components, causing percentage composition random varia-
tion, and to mantain raw material purity.
EPR measurements have been carried out to detect 3d-metals contamination.
Fig. 1 shows four EPR spectra revealing Cu^{2+} and Fe^{3+} ion traces. Instru-
mental sensitivity for Fe^{3+} is about 1 ppm.
The most recent sample (IV) reveals only Fe^{3+} trace.
Similar evolution appears on DSC, XRD, FT-IR and SEM analysis.
Photoluminescence measurements confirm the presence of Fe^{3+} ions in ZBLA
glass.
In fact, using a He-Cd laser ultraviolet emission line as exciting medium
($\lambda = 325$ nm), a PL emission around 525 nm can be detected (Fig. 2).
This band is four-peaked and it seems related to $^4T_1-^6A_1$ transition of
Fe^{3+} ions (Fig. 3). The peak multiplicity can be explained supposing a
transition on $^6A_1$ four-degenerate level.
Same results have been obtained from PL spectra of ZBLA glasses doped
with 200 - 5000 ppm Fe^{3+} ions.
PL signal at $\lambda = 525$ nm disappears in the most recent samples.
Other measurements are in progress with the aim of determining the limit
of the method.

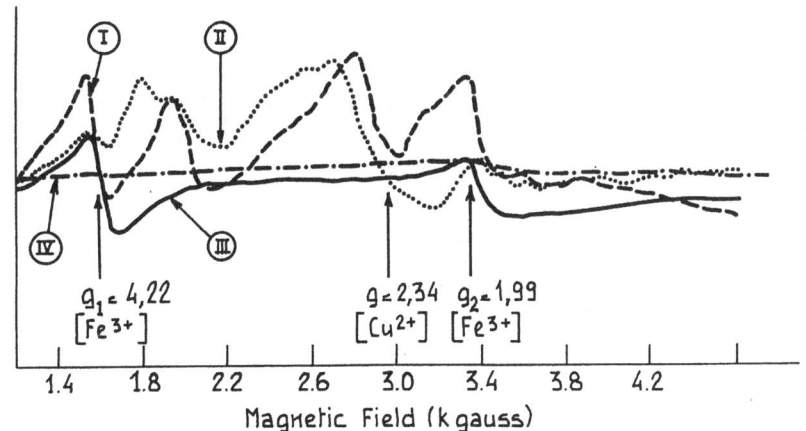

FIGURE 1. EPR spectra.

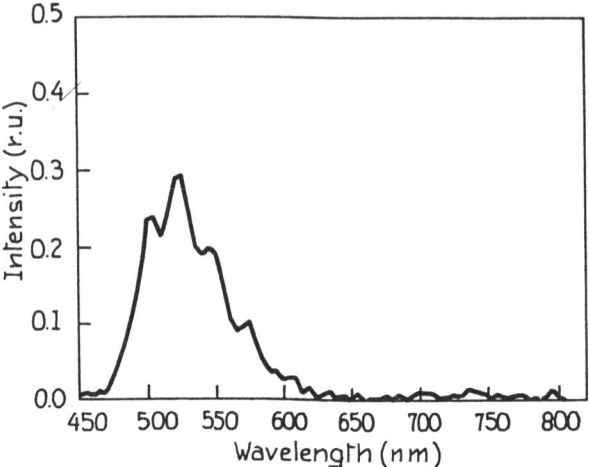

FIGURE 2. PL spectrum.

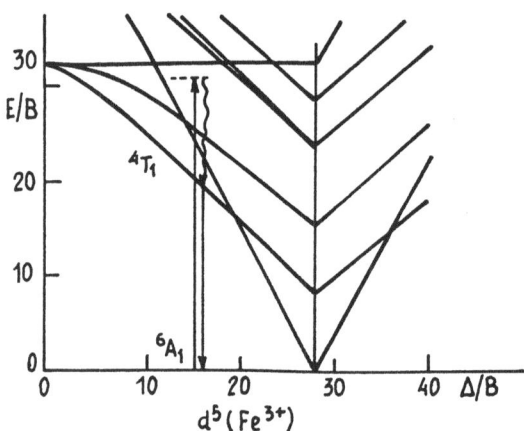

FIGURE 3. PL transition according to Tanabe Sugano diagrams.

CONCLUDING REMARKS

RUI M. ALMEIDA

Centro de Física Molecular - Complexo I, Instituto Superior
Técnico, Av. Rovisco Pais, 1000 Lisboa, Portugal

The main objectives of the NATO Advanced Research Workshop
on Halide Glasses for Infrared Fiberoptics were:

1) To review the state-of-the-art in the field of halide
   glasses in bulk and fiber form, with a particular emphasis on the preparation and properties of fluorozirconate glasses and fibers.

2) To identify new directions of research for vitreous
   halides of potential interest during the next five years.

3) To critically assess the potential of halide glasses for
   infrared fiberoptics.

4) To indicate possible cooperative research efforts between different NATO countries.

5) To put all the above in the form of a book as quickly as
   possible after the workshop.

Let us now briefly examine what we managed to accomplish
during the 5-day meeting:

1) The state-of-the-art was fairly completely reviewed, with
   particular emphasis on fluorozirconates. A little more
   emphasis would even have been placed on the fiber aspects
   if it weren't for the absence of two or three experts in
   the field who were unable to attend.

2) As far as new directions of research during the next
   five years, although most speakers did not address the
   issue directly, some guidelines have clearly emerged,
   namely:

   a) One has to search for new, improved glass composi-
      tions, although present propects for inclusion of
      chlorides, bromides and iodides do not yet appear
      good.

   b) There is urgent need of new phase diagram data for
      fluorides and other halides as well. Maybe the pro-
      gram currently sponsored through the American Ceramic
      Society will provide some relief for this need.

   c) Melting conditions and purity still have to be dras-
      tically improved.

d) One should develop single-mode fibers for long haul telecommunications, which brings about the need for vapor deposition methods in the preparation of fluoride glasses.

3) A critical assessment was made of the potential of halide glasses for infrared fiberoptics:

a) One learned from Japan that short haul fiber systems are now beginning to operate for tasks such as remote temperature sensing, gas sensing, remote spectroscopy and laser power delivery.

b) With respect to long haul telecommunications, requiring ultra-low loss fibers, attenuation is now down to $\sim$ 1 dB/Km at 2.55 µm for short lengths of fiber (U.S.A. and Japan), corresponding to the attaining of the 1 dB/Km landmark. According to some predictions (U.K.), the practical limit for ZrF_4-based glass fibers may lie at $\sim$ 0.035 dB/Km, about five times better than silica. Although this in itself may not be as good as predicted, if it proves to be true, there is still a chance that one may eventually reach the long wavelength side of the OH peak in the future. In addition, the fact that it is already possible to get down to such low levels of attenuation undoubtedly represents a tremendous capability for improvement during a period of only five years of fiber research for these materials.

4) The identification of possible cooperative efforts was indeed done, although mostly on an individual basis and informally. However, giving the unavailability of a minimally complete list of those, no attempt will be made to identify them at this point.

5) The book is now released, after a fairly reasonable time span.

In conclusion, this NATO Advanced Research Workshop was a very useful meeting and NATO should be praised for its highly successful ARW programme.

PARTICIPANTS

- Dr. F. Birkedahl
  R&D Optical Fibers
  NKT Electronik
  Brondbyvestervej 95
  DK - 2605 Brondby
  DENMARK

- Prof. J. Lucas
  Laboratoire de Chimie Minérale D
  Université de Rennes
  Avenue du Géneral Leclerc
  35031 Rennes Cedex
  FRANCE

- Prof. C. Jacoboni
  E.R.A. 609 U.A. 449
  Faculté des Sciences
  Université du Maine
  72017 Le Mans
  FRANCE

- Dr. H. Poignant
  C.N.E.T.
  Route de Tregastel
  B.P. 40
  22301 Lannion
  FRANCE

- Dr. D. Tregoat
  C.G.E.
  Laboratoires de Marcoussis
  Division Matériaux
  Route de Nozay
  91460 Marcoussis
  FRANCE

- Prof. G.H. Frischat
  Technical University of Clausthal
  Zehntnerstrasse 2A
  D-3392 Clausthal-Zellerfeld
  FEDERAL REPUBLIC OF GERMANY

- Dr. H.W. Schneider
  Siemens AG, Dept. ZFE FKE 31
  Otto-Hahn-Ring 6
  D-8000 Muenchen 83
  FEDERAL REPUBLIC OF GERMANY

- Dr. D. Pruss
  Department of Basic Development, Optics
  Dragerwerk AG
  Postfach 1339
  2400 Lubeck 1
  FEDERAL REPUBLIC OF GERMANY

- Prof. N.G. Alexandropoulos
  Department of Physics
  Fifth Laboratory of Physics (x-rays)
  University of Ioannina
  453 32 Ioannina
  GREECE

- Dr. G. Cocito
  C.S.E.L.T.
  Via G. Reiss Romoli, 274
  10148 Torino
  ITALY

- G. de Leede
  Vakgroep F.T.
  Technische Hogeschool Eindhoven
  Den Dolech 2
  5612 AZ Eindhoven
  The NETHERLANDS

- Prof. L.O. Svaasand
  Division of Physical Electronics
  University of Trondheim
  The Norwegian Institute of Technology
  N-7034 Trondheim
  NORWAY

- Prof. R.M. Almeida
  Centro de Física Molecular - Complexo I
  Instituto Superior Técnico
  Av. Rovisco Pais
  1000 Lisboa
  PORTUGAL

- M.C. Gonçalves
  Centro de Física Molecular - Complexo I
  Instituto Superior Técnico
  Av. Rovisco Pais
  1000 Lisboa
  <u>PORTUGAL</u>

- J.L. Grilo
  Centro de Física Molecular - Complexo I
  Instituto Superior Técnico
  Av. Rovisco Pais
  1000 Lisboa
  <u>PORTUGAL</u>

- Prof. Y. Skarlatos
  Physics Department
  Bogazici University
  Bebek - Istanbul
  <u>TURKEY</u>

- Dr. P.W. France
  Britsh Telecom Research Labs
  Martlesham Heath
  Ipswich IPS 7RE
  <u>UNITED KINGDOM</u>

- Prof. J.M. Parker
  Department of Ceramics, Glasses and Polymers
  Sheffield University
  Elmfield
  Northumberland Road
  Sheffield S10 2TZ
  <u>UNITED KINGDOM</u>

- Dr. J.A. Savage
  Royal Signals and Radar Establishment
  St. Andrews Road
  Malvern, Worcs WR14 4NN
  <u>UNITED KINGDOM</u>

- Prof. A.C. Wright
  Department of Physics
  Reading University
  Reading RG6 2AF
  <u>UNITED KINGDOM</u>

- Dr. D.R. Ulrich
  Air Force Office of Scientific Research/NC
  Bolling Air Force Base, Building 410
  Washington, D.C. 20332-6448
  U.S.A.

- Prof. J.D. Mackenzie
  Materials Science and Engineering Department
  6531 Boelter Hall
  U.C.L.A.
  Los Angeles, CA 90024
  U.S.A.

- Dr. A.J. Bruce
  ATT Bell Laboratories
  600 Moutain Ave.
  Murray Hill, NJ 07974
  U.S.A.

- Dr. M.G. Drexhage
  U.S. Air Force
  Rome Air Development Center / ESMO
  Hanscom AFB, MA 01731
  U.S.A.

- Dr. P.H. Klein
  Code 6822
  Naval Research Laboratory
  Washington, D.C. 20375-5000
  U.S.A.

- Prof. C.T. Moynihan
  Department of Materials Engineering
  Rensselaer Polytechnic Institute
  Troy, NY 12180-3590
  U.S.A.

- Prof. C.G. Pantano
  123 Steiddle Building
  Pennsylvania State University
  University Park, PA 16865
  U.S.A.

- Dr. M. Robinson
  Hughes Research Laboratories
  3011 Malibu Canyon Road
  Malibu, CA 90265
  U.S.A.

- Dr. A. Sarhangi
  Corning Glass Works
  Sullivan Park - DV - 2-5
  Corning, NY 14831
  U.S.A.

- Dr. C. F. Rapp
  Owens Corning Fiberglass Technical Center
  Grandville, OHIO 43023
  U.S.A.

- Dr. R.C. Folweiler
  GTE Laboratories, Inc.
  Optical Fiber Technologies
  40 Sylvan Road
  Waltham, MA 02254
  U.S.A.

- Prof. R. Aranda
  Departamento de Física Fundamental
  Faculdade de Ciências
  Universidade de Zaragoza
  Zaragoza 6
  SPAIN

- Prof. R. Reisfeld
  Department of Inorganic and Analytical Chemistry
  The Hebrew University of Jerusalem
  Givat Ram 91904 Jerusalem
  ISRAEL

- Dr. S. Yoshida
  Research and Development Division
  The Furukawa Electric Company, Ltd.
  9-15, Futaba 2-chome
  Shinagawa-ku, Tokyo
  142 JAPAN

- Dr. T. Izumitani
  Hoya Corporation
  572, Miyazawa-cho
  Akishima-city
  Tokyo 196
  JAPAN

INDEX

410